Advanced Electrode Materials Dedicated for Electroanalysis

Advanced Electrode Materials Dedicated for Electroanalysis

Sławomira Skrzypek
Mariola Brycht
Barbara Burnat

Basel • Beijing • Wuhan • Barcelona • Belgrade • Novi Sad • Cluj • Manchester

Sławomira Skrzypek
Department of Inorganic
and Analytical Chemistry
University of Lodz
Lodz
Poland

Mariola Brycht
Department of Inorganic
and Analytical Chemistry
University of Lodz
Lodz
Poland

Barbara Burnat
Department of Inorganic
and Analytical Chemistry
University of Lodz
Lodz
Poland

Editorial Office
MDPI AG
Grosspeteranlage 5
4052 Basel, Switzerland

This is a reprint of articles from the Special Issue published online in the open access journal *Materials* (ISSN 1996-1944) (available at: www.mdpi.com/journal/materials/special_issues/ advanced_electrode_electroanalysis).

For citation purposes, cite each article independently as indicated on the article page online and using the guide below:

Lastname, A.A.; Lastname, B.B. Article Title. *Journal Name* **Year**, *Volume Number*, Page Range.

ISBN 978-3-7258-2114-3 (Hbk)
ISBN 978-3-7258-2113-6 (PDF)
https://doi.org/10.3390/books978-3-7258-2113-6

© 2024 by the authors. Articles in this book are Open Access and distributed under the Creative Commons Attribution (CC BY) license. The book as a whole is distributed by MDPI under the terms and conditions of the Creative Commons Attribution-NonCommercial-NoDerivs (CC BY-NC-ND) license (https://creativecommons.org/licenses/by-nc-nd/4.0/).

Contents

About the Editors vii

Preface ix

Mariola Brycht, Barbara Burnat and Sławomira Skrzypek
Advanced Electrode Materials Dedicated for Electroanalysis
Reprinted from: Materials 2024, 17, 3762, doi:10.3390/ma17153762 1

Izabela Bargiel, Joanna Smajdor, Anna Górska, Beata Paczosa-Bator and Robert Piech
A Novel Voltametric Measurements of Beta Blocker Drug Propranolol on Glassy Carbon Electrode Modified with Carbon Black Nanoparticles
Reprinted from: Materials 2021, 14, 7582, doi:10.3390/ma14247582 7

Joanna Wasąg and Malgorzata Grabarczyk
Copper Film Modified Glassy Carbon Electrode and Copper Film with Carbon Nanotubes Modified Screen-Printed Electrode for the Cd(II) Determination
Reprinted from: Materials 2021, 14, 5148, doi:10.3390/ma14185148 20

Natalia Festinger, Aneta Kisielewska, Barbara Burnat, Katarzyna Ranoszek-Soliwoda, Jarosław Grobelny and Kamila Koszelska et al.
The Influence of Graphene Oxide Composition on Properties of Surface-Modified Metal Electrodes
Reprinted from: Materials 2022, 15, 7684, doi:10.3390/ma15217684 39

Marcio Cristiano Monteiro, João Paulo Winiarski, Edson Roberto Santana, Bruno Szpoganicz and Iolanda Cruz Vieira
Ratiometric Electrochemical Sensor for Butralin Determination Using a Quinazoline-Engineered Prussian Blue Analogue
Reprinted from: Materials 2023, 16, 1024, doi:10.3390/ma16031024 52

Mariola Brycht, Andrzej Leniart, Sławomira Skrzypek and Barbara Burnat
Incorporation of Bismuth(III) Oxide Nanoparticles into Carbon Ceramic Composite: Electrode Material with Improved Electroanalytical Performance in 4-Chloro-3-Methylphenol Determination
Reprinted from: Materials 2024, 17, 665, doi:10.3390/ma17030665 64

Karla Caroline de Freitas Araújo, Emily Cintia Tossi de Araújo Costa, Danyelle Medeiros de Araújo, Elisama V. Santos, Carlos A. Martínez-Huitle and Pollyana Souza Castro
Probing the Use of Homemade Carbon Fiber Microsensor for Quantifying Caffeine in Soft Beverages
Reprinted from: Materials 2023, 16, 1928, doi:10.3390/ma16051928 79

Paweł Krzyczmonik, Marta Klisowska, Andrzej Leniart, Katarzyna Ranoszek-Soliwoda, Jakub Surmacki and Karolina Beton-Mysur et al.
The Composite Material of (PEDOT-Polystyrene Sulfonate)/Chitosan-AuNPS-Glutaraldehyde/as the Base to a Sensor with Laccase for the Determination of Polyphenols
Reprinted from: Materials 2023, 16, 5113, doi:10.3390/ma16145113 93

Anna Cirocka, Dorota Zarzeczańska and Anna Wcisło
Good Choice of Electrode Material as the Key to Creating Electrochemical Sensors—Characteristics of Carbon Materials and Transparent Conductive Oxides (TCO)
Reprinted from: Materials 2021, 14, 4743, doi:10.3390/ma14164743 111

Anna Domaros, Dorota Zarzeczańska, Tadeusz Ossowski and Anna Wcisło
Controlled Silanization of Transparent Conductive Oxides as a Precursor of Molecular Recognition Systems
Reprinted from: *Materials* **2022**, *16*, 309, doi:10.3390/ma16010309 . . . . **126**

Andrzej Leniart, Barbara Burnat, Mariola Brycht, Maryia-Mazhena Dzemidovich and Sławomira Skrzypek
Fabrication and Characterization of an Electrochemical Platform for Formaldehyde Oxidation, Based on Glassy Carbon Modified with Multi-Walled Carbon Nanotubes and Electrochemically Generated Palladium Nanoparticles
Reprinted from: *Materials* **2024**, *17*, 841, doi:10.3390/ma17040841 . . . . **141**

Paweł Kościelniak, Marek Dębosz, Marcin Wieczorek, Jan Migdalski, Monika Szufla and Dariusz Matoga et al.
The Use of an Acylhydrazone-Based Metal-Organic Framework in Solid-Contact Potassium-Selective Electrode for Water Analysis
Reprinted from: *Materials* **2022**, *15*, 579, doi:10.3390/ma15020579 . . . . **154**

About the Editors

Sławomira Skrzypek

Prof. Skrzypek earned her Master's degree in 1986 and PhD in 1997 and completed her habilitation in 2012. She was awarded the title of full professor in 2019. Since 2014, she has led the Department of Inorganic and Analytical Chemistry, Faculty of Chemistry, University of Lodz. She served as Dean of the Faculty of Chemistry, University of Lodz, from 2016 to 2024 and was a member of the Audit Committee of the Polish Chemical Society (PTChem) from 2018 to 2024. Since 2021, she has chaired the Electrochemistry Section of PTChem, and from 2017 to 2024, she served as the treasurer of the Analytical Chemistry Committee of the Polish Academy of Sciences. Her research focuses on the electrochemical study and determination of pesticides and other biologically active compounds in solid electrodes, surface modifications of electrode materials, and the determination of compounds with guanidine groups as catalysts for the reduction of hydrogen.

Mariola Brycht

Dr. Brycht is appointed as an assistant professor in the Department of Inorganic and Analytical Chemistry, the Faculty of Chemistry, University of Lodz, where she earned her master's (2011) and doctoral (2015) degrees and completed her habilitation (2024). During 2018–2019, she worked as a post-doctoral fellow at Charles University in the UNESCO Laboratory of Environmental Electrochemistry (Prague, Czechia). She is a member of the International Society of Electrochemistry, the European Chemical Society, and the Polish Chemical Society. Her scientific interests include the development and characterization of novel carbon-based electrode materials for sensing biologically active hazardous organic compounds (e.g., different classes of drugs and pesticides), and the miniaturization of electrochemical detection devices.

Barbara Burnat

Dr. Burnat earned her master's degree in 2003 and her PhD in physical chemistry with a specialization in electrochemistry in 2008, both from the University of Lodz. She is appointed as an assistant professor in the Department of Inorganic and Analytical Chemistry, the Faculty of Chemistry, University of Lodz. She is a member of the International Society of Electrochemistry, the European Chemical Society, the Polish Chemical Society, and the Polish Society of Biomaterials. Her research interests include metallic biomaterials, corrosion, surface modification, sol–gel processes, nanomaterials, electrodeposition, microscopic surface analysis, and electroanalysis.

Preface

This reprint compiles research articles originally published in the Special Issue of *Materials* entitled "Advanced Electrode Materials Dedicated for Electroanalysis". The subject of this collection spans the development and application of advanced electrode materials, offering insights into their roles in enhancing the performance of electrochemical sensors.

This reprint covers a wide range of topics, providing deep insights into surface- and bulk-modified electrodes, microelectrodes, biosensors, molecular recognition systems, electrodes for electrocatalytic systems, and solid-contact ion-selective electrodes. By compiling these studies, we aim to provide an overview of the latest advancements and trends in the design, fabrication, and characterization of novel electrode materials, as well as their applications in electrochemical sensing.

The primary motivation for assembling this volume is to create a consolidated resource reflecting the significant progress being made in the area of advanced electrode materials. As the demand for more sensitive, selective, and reliable electrochemical sensors grows, the need for innovative materials and approaches becomes critical. This reprint serves as a valuable reference for researchers and practitioners, facilitating knowledge exchange and fostering further innovation.

This reprint is intended for a diverse audience, including researchers, scientists, and engineers working in the fields of materials science, chemistry, and electrochemistry. It is also a useful resource for graduate and postgraduate students specializing in these disciplines. By providing a detailed exploration of advanced electrode materials and their applications, this compilation offers valuable insights for both academic and industrial professionals.

We extend our deepest gratitude to the contributing authors, whose expertise and dedication made this collection possible. We also thank the peer reviewers for their critical evaluations and constructive feedback, which have greatly enhanced the quality of the articles. Special acknowledgment is due to the editorial team of Materials for their unwavering support and meticulous efforts in bringing this compilation to fruition. We also appreciate the support from various institutions and funding agencies that have facilitated the research presented in these articles.

We hope that this reprint will serve as an indispensable resource for the scientific community, sparking further research and innovation in the field of advanced electrode materials. We believe that the insights and discoveries documented within these pages will significantly contribute to the advancement of electrochemical sensors and electroanalysis.

Sławomira Skrzypek, Mariola Brycht, and Barbara Burnat
Editors

Editorial

Advanced Electrode Materials Dedicated for Electroanalysis

Mariola Brycht *[ ], Barbara Burnat *[ ] and Sławomira Skrzypek [ ]

University of Lodz, Faculty of Chemistry, Department of Inorganic and Analytical Chemistry, Tamka 12, 91-403 Lodz, Poland; slawomira.skrzypek@chemia.uni.lodz.pl
* Correspondence: mariola.brycht@chemia.uni.lodz.pl (M.B.); barbara.burnat@chemia.uni.lodz.pl (B.B.)

1. Introduction

The development of advanced electrode materials has significantly enhanced the capabilities of electrochemical devices, enabling their application in diverse fields such as environmental monitoring, medical diagnostics, food safety, and industrial processes. Ideal electrode materials are expected to exhibit high electrical conductivity and rapid electron transfer across a broad range of redox systems and maintain structural and electrochemical stability over a wide potential range. Additionally, simplicity and low production costs are highly desirable attributes. Meeting these criteria ensures that the electrode material can serve as an effective modification platform for subsequent electroanalytical applications. Numerous types of electrode materials have been developed and successfully applied in electroanalysis, finding applications in environmental, healthcare, and pharmaceutical analyses as electrochemical detectors, micro-/nano-electrochemical devices, and chemical and biochemical sensors. Given their broad applicability, there is a constant demand for innovative advanced electrode materials that offer high selectivity, sensitivity, operational simplicity, low production cost, and miniaturization potential.

This Special Issue, entitled "Advanced Electrode Materials Dedicated for Electroanalysis", brings together high-quality feature papers that provide insights into and highlight the latest progress and innovative developments in electrode materials. The topics covered include the fabrication and processing of various advanced electrode materials as well as their characterization and potential electrochemical applications.

2. Advanced Electrode Materials Discussed in Articles Published in This Special Issue

The collection of eleven articles featured in this Special Issue reflects the significant progress and innovation in the development and application of electrode materials for electrochemical analysis. These papers can be divided into seven categories according to their contents: surface-modified electrodes, bulk-modified electrodes, microelectrodes, biosensors, molecular recognition systems, electrodes for electrocatalytic systems, and solid-contact ion selective electrodes.

2.1. Surface-Modified Electrodes

Surface modifications of electrodes can significantly enhance their electrochemical performance, making them highly effective for various analytical applications. These modifications typically involve the application of different materials to the electrode surface to improve sensitivity, selectivity, and overall performance. Common techniques for surface modification include drop-casting, electrodeposition, and layer-by-layer assembly. Among these, drop-casting is the most commonly used technique.

The article by Bargiel et al. [1] presents the development of an electrochemical sensor based on a glassy carbon electrode (GCE) modified with a hybrid material composed of carbon black nanoparticles and Nafion. This drop-casting modification, combined with a preconcentration step, significantly increases the active surface area and improves the limit of detection (LOD) and sensitivity for the beta blocker drug propranolol. The high

accuracy and potential for routine laboratory use were validated with real sample analyses, including pharmaceutical tablets and human urine. Praised for its simplicity, low cost, and excellent analytical performance, the method outperforms many existing techniques, such as spectrophotometry and high-performance liquid chromatography (HPLC), in terms of LOD and sensitivity. These findings suggest that the developed voltammetric method is a valuable tool for propranolol determination in both pharmaceutical and biological samples.

The work by Wasag and Grabarczyk [2] presents the development of two electrochemical sensors for the anodic stripping voltammetric determination of ultra trace concentrations of Cd(II) in environmental water samples. These sensors include a GCE and carbon nanotubes modified screen-printed electrode (CN/SPE), both with an electrochemically deposited copper film (CuF). The modified CuF/CN/SPE, along with an accumulation step, demonstrated a slightly lower LOD and better performance in terms of sensitivity and field applicability. The method was validated with certified reference materials and real water samples, demonstrating the electrodes' suitability for environmental monitoring. Both sensors showed satisfactory reproducibility and selectivity. These results suggest that both developed electrodes are effective tools for the sensitive and accurate determination of Cd(II) in various environmental contexts and offer a non-toxic alternative to traditional mercury electrodes.

The study by Festinger et al. [3] on graphene oxides as electrode surface modifiers reveals the complexity and variability of these materials. In this work, noble metal (gold and platinum) disk electrodes were modified by drop-casting suspensions of two types of graphene oxides (GO I and GO II) at different concentrations. The resulting surface-modified electrodes were characterized for their spectral, structural, and electrochemical properties. Despite having similar topographies, elemental analysis and electrochemical studies revealed significant differences in the oxygen content and performance between the two types of graphene oxide tested. Surprisingly, the authors found that GO I, acquired as graphene oxide, is more reduced than GO II, which was purchased as reduced graphene oxide. This study underscores the need to standardize graphene oxide-based materials to ensure repeatability and reliability in electrochemical applications.

2.2. Bulk-Modified Electrodes

Bulk modifications involve incorporating functional materials into the bulk of the electrode to improve its overall properties. These modifications can significantly enhance the electrochemical performance by increasing the effective surface area and improving electron transfer capabilities. Compared to surface modifications, bulk modifications often provide more robust and stable enhancements, as the active materials are distributed throughout the electrode rather than just on the surface. This can lead to improved durability and longer lifespan of the electrode.

Monteiro et al. [4] developed a ratiometric electrochemical sensor based on a carbon paste electrode (CPE) bulk-modified with a quinazoline-engineered Prussian blue analogue (qnz-PBA) for the determination of the herbicide butralin. This sensor utilizes the stable signal of qnz-PBA as an internal reference to minimize deviations across multiple assays, thus enhancing the precision and accuracy of the measurements. The performance of the CPE modified with qnz-PBA was successfully validated using lettuce and potato samples, confirming its accuracy and applicability in real-world scenarios. This ratiometric sensor offers simplicity, cost-effectiveness, and excellent analytical performance. The authors stated that the developed ratiometric sensor is an effective tool for the sensitive and precise determination of butralin in both agricultural and environmental samples and is a reliable and accurate alternative to chromatographic methods used for butralin determination.

The work by Brycht et al. [5] presents the development of a carbon ceramic electrode (CCE) bulk-modified with bismuth(III) oxide nanoparticles (Bi_2O_3NPs) for the determination of 4-chloro-3-methylphenol (PCMC), a priority environmental pollutant. The CCE modified with Bi_2O_3NPs (Bi-CCE), characterized using microscopic and electrochemical techniques, demonstrated a more compact and less porous surface compared to the unmodified CCE and a higher effective surface area, indicating an increased number of electroactive sites. Additionally, the incorporation of Bi_2O_3NPs significantly enhances the electrochemical properties of the CCE, providing an extended linear detection range, improved sensitivity and a lower LOD compared to the unmodified CCE. The presence of Bi_2O_3NPs improves electron transfer and reduces background current. The sensor's performance was validated using river water samples, demonstrating excellent recovery rates and selectivity. The Bi-CCE exhibited high reproducibility and stability over three months. These results suggest that the Bi-CCE is an effective tool for the sensitive and accurate determination of PCMC and other phenolic compounds, with potential applications in environmental monitoring and pollutant detection.

2.3. Microelectrodes

Microelectrodes offer unique advantages due to their small size, including enhanced mass transfer rates and reduced ohmic losses. These properties make them highly suitable for sensitive and precise analytical applications. Their small dimensions lead to a higher current density and faster response times compared to macroelectrodes, making them ideal for real-time monitoring and detection in confined environments.

Araújo et al. [6] present the development of a homemade carbon fiber microelectrode (CF-μE) for quantifying caffeine in soft beverages. The fabricated microelectrode, characterized for its electrochemical properties, demonstrated significantly enhanced sensitivity and a lower LOD for caffeine compared to other analytical methods. The CF-μE exhibited a high mass transfer rate and a sigmoidal voltammetric profile, confirming its microelectrode characteristics. The sensor's stability and reproducibility were confirmed over multiple tests. The method was validated with real soft beverage samples, showing satisfactory results consistent with the literature values and HPLC validation. The homemade CF-μE offers a cost-effective, portable, and reliable alternative for caffeine determination in the beverage industry, highlighting its potential for broader application in quality control and environmental monitoring.

2.4. Biosensors

Biosensors incorporate biological elements, such as enzymes, to provide high specificity for target analytes. These sensors are particularly useful for detecting biologically relevant compounds in complex matrices. The integration of biological recognition elements with electrochemical transducers allows for the selective and sensitive detection of various analytes, making biosensors highly valuable in medical diagnostics, environmental monitoring, and food safety.

Krzyczmonik et al. [7] developed an electrochemical enzyme-based biosensor for the determination of polyphenols. The biosensor, constructed on a GCE, utilized a composite material consisting of poly(3,4-ethylenedioxy-thiophene), poly(4-lithium styrenesulfonic acid), chitosan, gold nanoparticles (AuNPs), and glutaraldehyde, which was further modified by immobilizing laccase using glutaraldehyde as a cross-linker. The composite material demonstrated boosted electrical conductivity, enhanced stability of the chitosan layer while maintaining high biocompatibility, and improved surface morphology, confirmed through wide range of complementary analytical techniques. The addition of AuNPs increased the effective surface area and facilitated the oxidation of polyphenols. The biosensor exhibited high catalytic activity and excellent performance towards the oxidation and detection of polyphenols such as catechol, gallic acid, and caffeic acid. The practical applicability of the developed biosensor was validated using white wine samples. This innovative biosensor offers a cost-effective, highly sensitive, and stable alternative for polyphenol determination in various environmental and biological samples.

2.5. Molecular Recognition Systems

The development of molecular recognition systems has become a primary objective in modern electrochemistry. These systems, which play a crucial role in various analytical applications, are created by depositing molecules with specific properties onto the surfaces of electrode materials. Typically, semiconductors (such as silicon) or dielectrics (such as glass or ceramics) are used as the base materials, which acquire valuable properties by forming conductive or semiconductive structures on their surfaces. Most commonly, a chemically defined layer of inorganic oxides or carbon materials with distinct electrical properties is applied to the surface of the base material.

The work by Cirocka et al. [8] presents the search for optimal electrode materials to serve as platforms for future sensors. A wide group of electrode materials was tested, including fluorine-doped tin oxide (FTO), silicon modified with carbon nanowalls, and silicon and glass modified with nanocrystalline boron-doped diamond layers with varying B/C ratios. The electrochemical properties and wettability of these electrode materials were evaluated and compared to commercially available carbon-based electrodes, such as boron-doped diamond electrode and GCE. Among the tested electrode materials, FTO was identified as the optimal electrode material due to its excellent electrochemical properties, high chemical stability, and valuable optoelectronic characteristics, making it a prime candidate for further research and development in various analytical and industrial applications.

The study by Domaros et al. [9] on the modification of transparent conductive oxide electrodes with alkoxysilanes demonstrates the potential for creating selective molecular recognition systems. In this work, FTO electrodes were silanized using 3-aminopropyltrimethoxysilane, trimethoxy(propyl)silane, and trimethoxy(octyl)silane under various reaction conditions (time and temperature). The modification process included single and double alkoxysilane modifications, as well as two-step mixed alkoxysilane modifications. The obtained electrodes were characterized in terms of electrochemical properties and wettability. The research highlights how different alkoxysilane structures and modification conditions can control surface properties and charge transfer processes, paving the way for tailored analytical tools with enhanced selectivity and sensitivity.

2.6. Electrodes for Electrocatalytic Systems

Electrodes play a crucial role in electrocatalytic systems, serving as the interface where electrochemical reactions occur. Their importance lies in their ability to facilitate electron transfer, which is essential for efficiently catalyzing reactions. The development of novel electrode materials aims to optimize these systems for better performance, efficiency, and durability in a wide range of applications, including fuel cells, electrolyzers, batteries, and electrochemical sensors. A specific emphasis is placed on the electrocatalytic oxidation of small organic compounds, such as methanol, ethanol, isopropanol, formaldehyde, and formic acid, on various modified electrodes. Many electrocatalytic systems employ noble metals such as platinum and palladium as key components, due to their unique catalytic properties.

Leniart et al. [10] report on the development of an advanced electrochemical platform based on a GCE modified with multi-walled carbon nanotubes (MWCNTs) and palladium nanoparticles (PdNPs). The MWCNTs were applied to the GCE surface using the drop-casting method, while PdNPs were produced electrochemically via a potentiostatic method using various programmed charges from an ammonium tetrachloropalladate(II) solution. This charge-controlled electrodeposition method enabled precise control over the amount and size of the deposited PdNPs. Detailed characterization and electrochemical assessment of GCE/MWCNTs/PdNPs revealed that the size and dispersion of PdNPs significantly influence the catalytic activity towards formaldehyde oxidation. Additionally, the long-term stability of the modified electrodes highlights their potential for practical applications.

2.7. Solid-Contact Ion Selective Electrodes

Solid-contact ion-selective electrodes (SC-ISEs) represent a significant advancement in the field of ion detection, offering improved performance, stability, and versatility across multiple applications. SC-ISEs are a type of ion-selective electrode that utilize a solid-state material as the transducer element between the ion-selective membrane and the electron-conducting substrate. Unlike traditional ISEs that rely on liquid-filled internal solutions, SC-ISEs incorporate solid-contact layers, which can be conducting polymers, carbon-based materials, etc. However, the search for the ideal transducer material is still ongoing. The perfect transducer material should exhibit a reversible transition from ionic to electronic conduction, high exchange current density, stable chemical composition, and possibly high hydrophobicity to minimize the formation of water at the transducer-membrane interface. Recently, the use of metal–organic frameworks (MOFs), a sub-class of highly ordered and porous materials with two- or three-dimensional structures, seems to be promising material as an ion-to-electron transducer.

The article of Kościelniak et al. [11] describes preliminary studies on the development of a solid-contact ion-selective electrode for detecting potassium in environmental water. The authors implemented two versions of a stable cadmium acylhydrazone-based MOF (JUK-13 and JUK-13_H_2O), differing in guest molecules, as the ion-to-electron transducers. Both MOFs significantly improved the potentiometric response and stability of the electrode. The K-JUK-13_H_2O-ISE demonstrated a good Nernstian slope and excellent long-term potential stability, making it a reliable tool for environmental water analysis. Its successful application in determining potassium in certified reference materials highlights its precision and accuracy.

3. Conclusions

The collection of articles featured in this Special Issue reflects the significant progress and innovation in the development and application of electrode materials for electrochemical analysis. The studies presented here offer deep insights into surface-modified electrodes, bulk-modified electrodes, microelectrodes, biosensors, molecular recognition systems, electrodes for electrocatalytic systems, and solid-contact ion selective electrodes, highlighting the interdisciplinary nature of this field. Collectively, the articles published within this Special Issue emphasize the importance of electrode material selection, surface modification, and comprehensive characterization in advancing electroanalysis. We hope that the findings and discussions presented here will inspire further research and innovation in developing advanced electrode materials for electrochemical applications. As the Guest Editors, we would like to extend our gratitude to all the authors, reviewers, and the editorial team for their contributions to this Special Issue. We believe that the knowledge shared within these pages will significantly impact the future of electrochemical analysis and its applications.

Conflicts of Interest: The authors declare no conflicts of interest.

References

1. Bargiel, I.; Smajdor, J.; Górska, A.; Paczosa-Bator, B.; Piech, R. A Novel Voltametric Measurements of Beta Blocker Drug Propranolol on Glassy Carbon Electrode Modified with Carbon Black Nanoparticles. *Materials* **2021**, *14*, 7582. [CrossRef] [PubMed]
2. Wasąg, J.; Grabarczyk, M. Copper Film Modified Glassy Carbon Electrode and Copper Film with Carbon Nanotubes Modified Screen-Printed Electrode for the Cd(II) Determination. *Materials* **2021**, *14*, 5148. [CrossRef] [PubMed]
3. Festinger, N.; Kisielewska, A.; Burnat, B.; Ranoszek-Soliwoda, K.; Grobelny, J.; Koszelska, K.; Guziejewski, D.; Smarzewska, S. The Influence of Graphene Oxide Composition on Properties of Surface-Modified Metal Electrodes. *Materials* **2022**, *15*, 7684. [CrossRef] [PubMed]
4. Monteiro, M.C.; Winiarski, J.P.; Santana, E.R.; Szpoganicz, B.; Vieira, I.C. Ratiometric Electrochemical Sensor for Butralin Determination Using a Quinazoline-Engineered Prussian Blue Analogue. *Materials* **2023**, *16*, 1024. [CrossRef] [PubMed]
5. Brycht, M.; Leniart, A.; Skrzypek, S.; Burnat, B. Incorporation of Bismuth(III) Oxide Nanoparticles into Carbon Ceramic Composite: Electrode Material with Improved Electroanalytical Performance in 4-Chloro-3-Methylphenol Determination. *Materials* **2024**, *17*, 665. [CrossRef] [PubMed]
6. de Freitas Araújo, K.C.; de Araújo Costa, E.C.T.; de Araújo, D.M.; Santos, E.V.; Martínez-Huitle, C.A.; Castro, P.S. Probing the Use of Homemade Carbon Fiber Microsensor for Quantifying Caffeine in Soft Beverages. *Materials* **2023**, *16*, 1928. [CrossRef] [PubMed]

7. Krzyczmonik, P.; Klisowska, M.; Leniart, A.; Ranoszek-Soliwoda, K.; Surmacki, J.; Beton-Mysur, K.; Brożek-Płuska, B. The Composite Material of (PEDOT-Polystyrene Sulfonate)/Chitosan-AuNPS-Glutaraldehyde/as the Base to a Sensor with Laccase for the Determination of Polyphenols. *Materials* **2023**, *16*, 5113. [CrossRef] [PubMed]
8. Cirocka, A.; Zarzeczańska, D.; Wcisło, A. Good Choice of Electrode Material as the Key to Creating Electrochemical Sensors—Characteristics of Carbon Materials and Transparent Conductive Oxides (TCO). *Materials* **2021**, *14*, 4743. [CrossRef] [PubMed]
9. Domaros, A.; Zarzeczańska, D.; Ossowski, T.; Wcisło, A. Controlled Silanization of Transparent Conductive Oxides as a Precursor of Molecular Recognition Systems. *Materials* **2023**, *16*, 309. [CrossRef] [PubMed]
10. Leniart, A.; Burnat, B.; Brycht, M.; Dzemidovich, M.-M.; Skrzypek, S. Fabrication and Characterization of an Electrochemical Platform for Formaldehyde Oxidation, Based on Glassy Carbon Modified with Multi-Walled Carbon Nanotubes and Electrochemically Generated Palladium Nanoparticles. *Materials* **2024**, *17*, 841. [CrossRef] [PubMed]
11. Kościelniak, P.; Dębosz, M.; Wieczorek, M.; Migdalski, J.; Szufla, M.; Matoga, D.; Kochana, J. The Use of an Acylhydrazone-Based Metal-Organic Framework in Solid-Contact Potassium-Selective Electrode for Water Analysis. *Materials* **2022**, *15*, 579. [CrossRef] [PubMed]

Disclaimer/Publisher's Note: The statements, opinions and data contained in all publications are solely those of the individual author(s) and contributor(s) and not of MDPI and/or the editor(s). MDPI and/or the editor(s) disclaim responsibility for any injury to people or property resulting from any ideas, methods, instructions or products referred to in the content.

Article

A Novel Voltametric Measurements of Beta Blocker Drug Propranolol on Glassy Carbon Electrode Modified with Carbon Black Nanoparticles

Izabela Bargiel, Joanna Smajdor, Anna Górska, Beata Paczosa-Bator and Robert Piech *

Department of Analytical Chemistry and Biochemistry, Faculty of Materials Science and Ceramics, AGH University of Science and Technology, Al. Mickiewicza, 30-059 Kraków, Poland; izabela.bargiel1991@gmail.com (I.B.); smajdorj@agh.edu.pl (J.S.); gorska.anna7@gmail.com (A.G.); paczosa@agh.edu.pl (B.P.-B.)
* Correspondence: rpiech@agh.edu.pl

Citation: Bargiel, I.; Smajdor, J.; Górska, A.; Paczosa-Bator, B.; Piech, R. A Novel Voltametric Measurements of Beta Blocker Drug Propranolol on Glassy Carbon Electrode Modified with Carbon Black Nanoparticles. *Materials* **2021**, *14*, 7582. https://doi.org/10.3390/ma14247582

Academic Editor: Alina Pruna

Received: 27 October 2021
Accepted: 3 December 2021
Published: 9 December 2021

Publisher's Note: MDPI stays neutral with regard to jurisdictional claims in published maps and institutional affiliations.

Copyright: © 2021 by the authors. Licensee MDPI, Basel, Switzerland. This article is an open access article distributed under the terms and conditions of the Creative Commons Attribution (CC BY) license (https://creativecommons.org/licenses/by/4.0/).

Abstract: A new voltametric method for highly sensitive propranolol (PROP) determination was developed. A glassy carbon electrode modified with a hybrid material made of carbon black (CB) and Nafion was used as the working electrode. The preconcentration potential and time were optimized (550 mV and 15 s), as well as the supporting electrolyte (0.1 mol L^{-1} H$_2$SO$_4$). For 15 s preconcentration time, linearity was achieved in the range 0.5–3.5 µmol L^{-1} and for 120 s in 0.02–0.14 µmol L^{-1}. Based on the conducted calibration (120 s preconcentration time) limit of detection (LOD) was calculated and was equal to 7 nmol L^{-1}. To verify the usefulness of the developed method, propranolol determination was carried out in real samples (tablets and freeze-dried urine). Recoveries were calculated and were in the range 92–102%, suggesting that the method might be considered as accurate. The repeatability of the signal expressed as relative standard deviation (RSD) was equal to 1.5% (n = 9, PROP concentration 2.5 µmol L^{-1}). The obtained results proved that the developed method for propranolol determination might be successfully applied in routine laboratory practice.

Keywords: propranolol; carbon black; Nafion; voltammetry; modified electrode

1. Introduction

Propranolol (PROP) belongs to the group of non-selective β-blockers. Its mechanism of action bases on the inhibition of β1 and β2 receptors. Pharmacological inhibition of these receptors inhibits its stimulation. It limits the influence of epinephrine and norepinephrine on tissues that possess β-receptors (e.g., in the heart, vessels, and bronchi). In practice, β-blockers reduce heart rate and contraction force and lead to reduction of blood pressure. Propranolol might be characterized by a wide range of clinical applications, e.g., treatment of hypertension, primary and secondary prevention of myocardial infarction, prevention of migraine, reduction of anxiety, control of arrhythmias [1–3].

From the chemical point of view, propranolol is an organic compound described as 1-[(1-methylethyl) amino]-3-(1-naphthalenyloxy). In the literature various analytical propranolol determination methods have been reported, among them were spectrophotometry [4,5], spectrofluorimetry [6,7], high performance liquid chromatography (HPLC) [8–10], and capillary electrophoresis [11,12]. Another method that is commonly used for propranolol determination is voltammetry. In comparison with the above mentioned methods, voltammetry might be characterized by very low detection limit, high sensitivity, low interferences impact, relatively low cost of analysis and no need to use toxic chemicals. The most important part of each voltammetric system is the working electrode (WE). For propranolol determination, different types of solid electrodes were used, e.g., glassy carbon electrode (GCE) [13,14], graphite electrode (GE) [15,16], carbon paste electrode (CPE) [17,18], boron doped diamond electrode (BDDE) [19], and screen printed electrode

(SPE) [20,21]. A recent trend in electrochemical methods has focused on the modification of solid electrodes (GCE, CPE, GE, and SPE) in order to improve their performance. Surface modifiers should exhibit certain properties, like, for example, good electrical conductivity, high specific surface area, and easy electron transfer. Therefore, different types of materials might be used for this purpose: carbon nanomaterials [22,23], metal nanoparticles [24,25], conducting polymers [26,27], etc.

However, use of the functionalized carbon nanomaterials such as carbon nanotubes or graphene for voltammetric measurements is associated with the risk of obtaining heterogeneous layers with different nanotubes orientations, which may cause the problem of low repeatability of obtained signals. Carbon nanotubes per se may also differ from each other considering the differences in its activation process or various numbers of active centers or function groups on its surface, that also can affect working conditions. Nowadays, electrode modifiers consisted of noble metals nanoparticles getting more attention, but its manufacturing process is also quite demanding, requiring the usage of strong acid under the conditions that can generate toxic products. Other disadvantage of such solution is the quite high price of such modifiers.

The aim of this work was developing of a new, highly sensitive and simple method for propranolol determination. For this purpose, a glassy carbon electrode modified with carbon black and Nafion was used. Developed modifier is an example of hybrid material (combination of carbon nanomaterial and polymer) that combines advantages of both components. Undoubted advantage of carbon black combined with Nafion as a modifier layer is obtaining wide working surface due to its physical parameters. The consequence of enlarging the electrode surface is clearly visible when comparing the detection limits of voltammetric sensors based on other modification materials. The use of Nafion as a dispersion component results not only in expanding the working surface, but also assures shorter electrode preparation time of about 15 min for the measurement process due to its quick drying process. This significantly improves the measurement process compared to other popular solvents used in electrode preparation, which have to be left to dry completely for a few hours. The simplicity of the proposed sensor and low cost of manufacturing are also crucial factors of choosing such a design solution.

2. Experimental

2.1. Measuring Apparatus

For all voltametric measurements, a multipurpose Electrochemical Analyzer M161 and the electrode stand M164 (MTM-ANKO, Krakow, Poland) with the EAGRAPH software (1.0, Krakow, Poland) were used. The standard three-electrode voltammetric quartz cell with volume of 20 mL was composed of a glassy carbon electrode modified with carbon black as the working electrode (CBGC), a double junction silver chloride reference electrode Ag/AgCl/KCl (3 mol L^{-1}), and a platinum rod as an auxiliary electrode. Homogenization of the supporting electrolyte was ensured using magnetic Teflon-coated bar (stirring speed of about 500 rpm). pH-meter (N-512 elpo, Polymetron, Wroclaw, Poland) was used to measure solutions pH value. All experiments were carried out at room temperature.

2.2. Chemicals

Standard stock solution of propranolol (Sigma Aldrich, Darmstadt, Germany) was obtained by dissolving an appropriate weight of standard in proportion of water and ethanol (1:1) and stored in fridge (10 mL, 0.01 mol L^{-1}). Sulfuric acid (96%) was purchased from Merck (Darmstadt, Germany), methanol (99%) and ethanol (96%) was purchased from POCH (Gliwice, Poland). The Triton X-100 was purchased from Windsor Laboratories Ltd. (Kingston, Jamaica). Interferents: citric acid, lactose monohydrate, starch, magnesium stearate, talc, cellulose, titanium dioxide, glucose, caffeine, ascorbic acid, uric acid, acetaminophen were purchased from Merck (Darmstadt, Germany). Carbon black nanoparticles with the surface area of 100 $m^2 g^{-1}$ and average particle size of 30 nm were obtained from 3D-nano (Kraków, Poland). The ion-exchange polymer Nafion (5% solution in a mixture of lower aliphatic alcohols) was obtained from Sigma-Aldrich (Darmstadt, Germany). Freeze-dried human urine was purchased from Medichem (Hobokem, NJ, USA). All reagents were of analytical grade and used without further purification. All solutions were prepared with double-distilled water.

2.3. Pharmaceutical Sample Preparation

Pharmaceutical samples such as propranolol WZF (Polfa Warszawa, Warszawa, Poland) and propranolol Accord (Accord Healthcare, London, UK) were investigated to measure the propranolol content. Pharmaceuticals were obtained from a local pharmacy. For measurements, samples were prepared by crushing three tablets in a mortar and quantitatively transferring to the volumetric flask (10 mL) and dissolving in water and ethanol (1:1). After complete dissolving and homogenization solution was ready for analysis. Solutions with lower concentrations were prepared daily.

The amount of propranolol in the samples was measured by the standard addition method and validated with the recovery parameter.

2.4. Urine Sample Preparation

Urine sample was prepared by dissolution of freeze-dried human urine with 5 mL of double distilled water and shaking on the ultrasonic washer (Emag, Leipzig, Germany) until complete dissolution of the powder. Then 900 µL of urine and 100 µL of methanol were transferred to the Eppendorf flask (1.5 mL) and stirred on the table centrifuge (Eppendorf, Hamburg, Germany) at 2000 rpm for 1 min. Measurements were performed in the supporting electrolyte consisting of 0.1 mol L^{-1} H_2SO_4 with the addition of 100 µL of previously prepared urine sample using the standard addition method and validated with recovery parameter.

2.5. Standard Procedure of Measurements

The differential pulse voltammetry (DPV) technique was applied for highly sensitive quantitative measurements of propranolol. The electrode was coated with 10 µL of homogenized carbon black solution layer daily (carbon black suspended in Nafion, 1 mg mL^{-1}). After preparation, glassy carbon electrode modified with carbon black nanoparticles and Nafion was used for propranolol (PROP) determination in the supporting electrolyte consisting of 0.1 mol L^{-1} H_2SO_4 (pH 1.8, total volume of 10 mL). Voltammograms were registered in the potential range from 500 mV to 1275 mV, with preconcentration potential E_{acc} of 550 mV (preconcentration time t_{acc} = 15 s). Other instrumental parameters of DPV technique are as follow: potential step E_s = 4 mV, pulse amplitude dE = 50 mV, time step potential 20 s (10 ms waiting time t_w and 10 ms sampling time t_s).

3. Results and Discussion

3.1. Voltammetric Characterization of Glassy Carbon Electrode Modified with Carbon Black

The parameters of carbon black surface on the glassy carbon electrode were investigated in 1 mmol L^{-1} potassium ferricyanide $(Fe(CN)_6^{-3}/Fe(CN)_6^{-4}$, solution in 1 mol L^{-1} KCl (both from POCH, Gliwice, Poland) using cyclic voltammetry. The range of the scan rate values was from 10 to 250 mV s^{-1}. The glassy carbon electrode modified with carbon black nanoparticles in Nafion working surface where the PROP oxidation process takes place was calculated using the dependence between the ferricyanide peak current and the square root of the scan rate. In order to compare the performance of the electrodes, this parameter was calculated both for modified and unmodified electrode. For the electrode modified by Nafion and carbon black, the size of active surface was of about 0.1058 cm^2, whereas for the unmodified electrode surface the size was significantly lower of about 0.0152 cm^2, which indicates that the size of modified electrode working surface is approximately seven times higher than unmodified.

The propranolol behavior was investigated on the CBGC electrode using a cyclic voltammetry technique. Measurements were performed in the supporting electrolyte with addition of 10 µmol L^{-1} PROP. The effect of the scan rate changing in the range from 10 to 250 mV s^{-1} on propranolol oxidation process is presented in Figure 1. The absence of the reduction peak in the cathodic scan implies that propranolol oxidation process on CBGC electrode is irreversible. In order to explain the mechanism of PROP oxidation, the dependences of its peak current versus the scan rate and the square root of the scan rate were plotted. The linear correlation was obtained from the peak potential on the square root of the scan rate plot, that suggests that the propranolol oxidation process takes place by diffusion. The propranolol oxidation process on the glassy carbon electrode modified by carbon black and Nafion is connected with the reaction on the Nafion layer. Considering the propranolol pKa value of 9.42 and the supporting electrolyte with 0.1 M H_2SO_4, the reaction group of propranolol exists in cationic form (Scheme 1). The positively charged PROP exchanges protons with the sulphonic group of Nafion, which improves its efficiency of accumulation on the electrode surface.

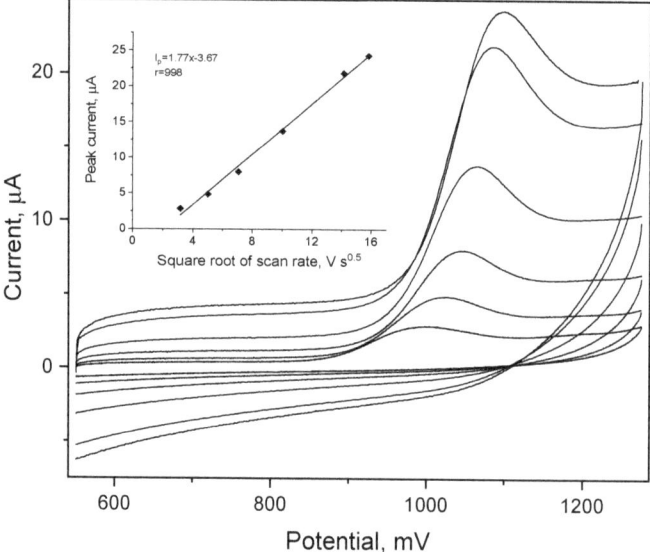

Figure 1. Cyclic voltammograms of 10 µmol L^{-1} propranolol in 0.1 mol L^{-1} H_2SO_4 (pH 1.8) measured on the glassy carbon electrode modified with carbon black nanoparticles. Scan rate values: 10, 25, 50, 100, 200, and 250 mV s^{-1}.

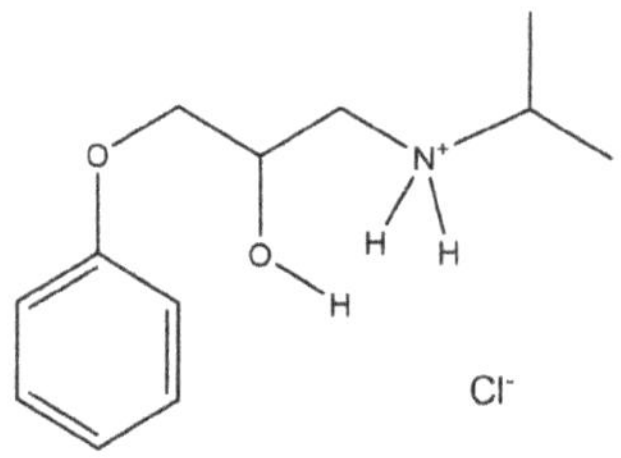

Scheme 1. Cationic form of propranolol formed during the electrochemical reaction.

The propranolol preconcentration on the glassy carbon electrode modified by the carbon black takes part in the way of adsorption, but the transport from the Nafion modifier layer to the electrode surface is a diffusion-controlled process. The electrode surface modified by carbon black nanoparticles due to its physical properties is characterized by bigger active surface, which allows to accumulate more analyte than on the glassy carbon electrode surface.

Moreover, the plot of the peak current vs. logarithm of the scan rate was developed, with obtained linear regression equation of:

$$E_k = 0.0133 \ln v + 0.0267 \; [V], \; r = 0.999 \tag{1}$$

Considering the obtained values of the regression equation and assuming that the oxidation process of propranolol is irreversible, it is possible to calculate the number of electrons exchanged during the electrode reaction using the Laviron equation [28]:

$$E_k = E^0 + \left(\frac{RT}{\alpha nF}\right) \ln\left(\frac{RTk^0}{\alpha nF}\right) + \left(\frac{RT}{\alpha}nF\right) \ln v \tag{2}$$

where α is the transport coefficient, k^0 is the electrochemical rate constant, n is the number of exchanged electrons, v is the scan rate value, E^0 is the formal potential, T is temperature value, F is the Faraday constant and R is the gas constant. Assuming α value as 0.5 and the value of the αn coefficient equal to 0.97, the number of the electrons that participate in the electrochemical propranolol oxidation could be calculated as 2.

The number of electrons exchanged during the propranolol oxidation reaction can also be determinate using following equation:

$$\alpha n = \frac{0.048}{\left|E_p - E_{p1/2}\right|} \tag{3}$$

The αn value calculated from the equation was equal to 0.96. Assuming α as 0.5, the number of electrons exchanged during the oxidation reaction could be calculated as 2, which confirms previous calculations result.

To clarify the oxidation mechanism, investigation of the amount of proton that participates in the oxidation process was performed using different pH values of the supporting electrolyte in the range from 1.8 to 7.1 (Figure 2). The propranolol peak was shifting toward more positive potentials along with decreasing pH values. The dependence of the peak potential value versus the supporting electrolyte pH is linear, according to the following equation.

$$E_p = 0.062 \; pH + 1.15 \; V \tag{4}$$

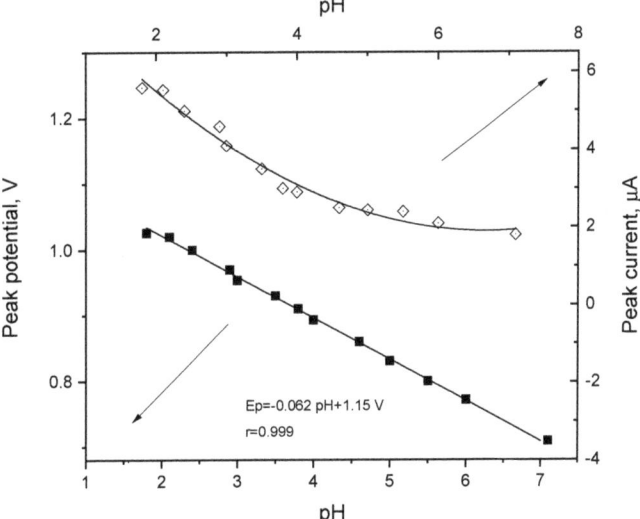

Figure 2. Propranolol peak and potential dependence on supporting electrolyte pH.

The value of the slope equals to 0.062 V pH^{-1}, which implies that the amount of exchanged electrons and protons is equal during the propranolol oxidation. The proposed mechanism of possible propranolol oxidation on CBGC electrode is presented on Scheme 2.

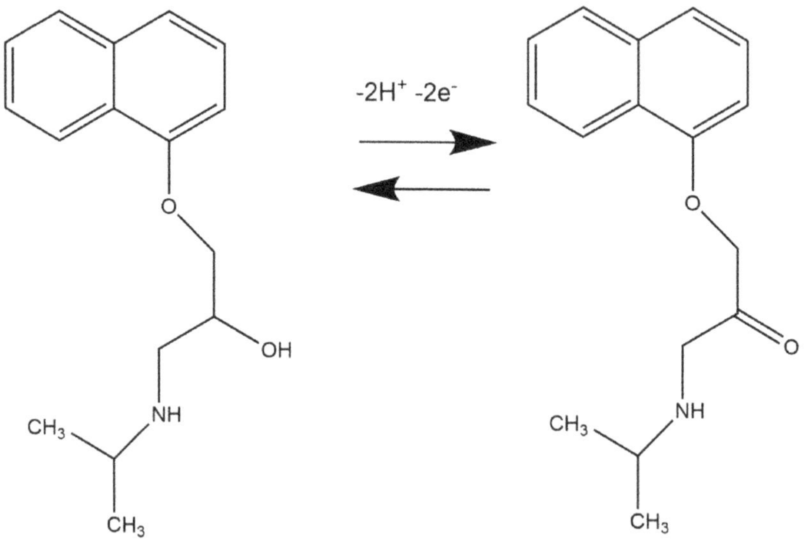

Scheme 2. Possible propranolol oxidation mechanism on glassy carbon electrode modified with carbon black nanoparticles.

A comparison in Linear Sweep Voltammetry (LSV) propranolol measurements between modified and unmodified glassy carbon electrode was performed (Figure not included). A linear correlation between propranolol peak current and square root of scan rate was observed, which indicates that its oxidation process on bare glassy carbon electrode is also diffusion controlled. The plot of the peak current vs logarithm of the scan rate was developed, with the obtained linear regression equation.

Considering the obtained values of the regression equation and assuming that the oxidation process of propranolol is irreversible, it was possible to calculate the number of electrons exchanged during the electrode reaction. Assuming α value as 0.5 and the value of the αn coefficient equal to 0.64, the number of the electrons that participates in the electrochemical propranolol oxidation could be calculated as 1.

3.2. Influence of Modifier Layer Volume on Propranolol Peak

In order to examine the influence of modifier volume applied on the glassy carbon electrode surface on the propranolol signal, the appropriate experiment was conducted. The ion-exchange properties of the Nafion depends strongly on the film thickness. Too thick film may decrease the diffusion of the propranolol to the electrode surface, where the exact electrochemical reaction occurs, therefore the obtained signal decreases in its size. Thus, optimizing the amount of modifier layer on the electrode surface is necessary. For this purpose, each GC electrode was modified with a different volume of CB-Nafion dispersion: 0, 2, 5, 7.5, 10, 15, 20 µL (Figure 3). As it might be observed, modification of GCE significantly improved propranolol signal. During the measurements, a few parameters were considered, such as the capacitive current value, the relation of the peak current to the background current, and also the peak shape and its good distinction from the background current. Considering all these parameters, the most favorable characteristic of the propranolol peak with the highest peak current (5.78 µA) was obtained for the layer of 10 µL, therefore this amount of carbon black modification layer was chosen as optimal for further studies. In comparison, peak current register on bare GCE was equal to 0.29 µA, which means that by modification, propranolol signal was improved almost 20 times.

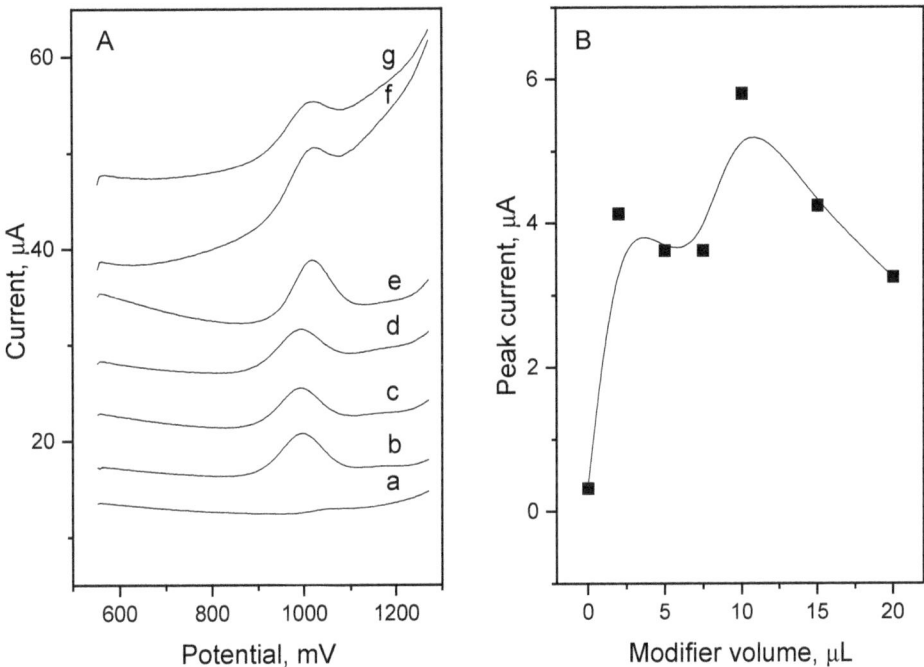

Figure 3. Influence of modifier layer volume on propranolol peak; (**A**) voltammograms obtained for modifier amount of (a) 0 µL, (b) 2 µL, (c) 5 µL, (d) 7.5 µL, (e) 10 µL, (f) 15 µL, and (g) 20 µL and (**B**) the value of corresponding peak current.

3.3. Influence of Preconcentration Time and Potential on Propranolol Peak

To provide high sensitivity of the performed measurements, the influence of preconcentration potential and time on the propranolol peak values was investigated. The preconcentration potential was investigated in the range from −100 to 750 mV (data not included). Examined values did not significantly affect the potential and current of propranolol peak, therefore the 550 mV was chosen as a propranolol preconcentration potential in furthers studies.

The plot of relationship between propranolol peak current value and preconcentration time value obtained on CBGC electrode is presented in Figure 4. For all investigated PROP concentration values, increasing the preconcentration time results in an increased value of peak current. The maximum obtained peak current was for the propranolol concentration of 10 µmol L^{-1} and it was equal to 25.49 µA (t_{acc} 240 s), while for 2.0 µmol L^{-1} it was equal to 19.69 (t_{acc} 240 s). For the analytical performance studies time of 15 s was picked to accumulate the analyte on the CBGC electrode surface, in order to ensure the good quality of the signal obtained simultaneously with a short duration of a single analysis.

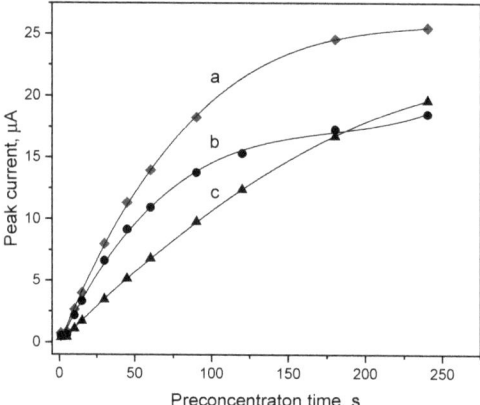

Figure 4. Dependence of the propranolol peak current on preconcentration time in the range from 0 to 240 s for (a) 10 μmol L^{-1} (b) 5.4 μmol L^{-1}, and (c) 2 μmol L^{-1} propranolol concentration in 0.1 mol L^{-1} H$_2$SO$_4$ (pH 1.8).

3.4. Influence of Supporting Electrolyte on Propranolol Peak

To maintain optimal conditions of propranolol determination, its peak properties (concentration of 10 μmol L^{-1}) were examined in miscellaneous base electrolytes, such as: 0.1 mol L^{-1} ammonia buffer (pH 8.2, Chempur, Piekary Śląskie, Poland), 0.1 mol L^{-1} acetate buffer (pH 3.8, Chempur, Piekary Śląskie, Poland), 0.1 mol L^{-1} H$_2$SO$_4$ (pH 1.8), 0.1 mol L^{-1} KH$_2$PO$_4$ (Merck, Darmstadt, Germany), 0.1 mol L^{-1} KCl, and 0.1 mol L^{-1} HCl (Figure not included). Signals obtained in a supporting electrolyte consisted of sulfuric acid characterized by the optimal properties of obtaining the propranolol peak, considering the relationship of its peak current value and background current value. The signal obtained in this environment was also characterized by good peak shape and high repeatability. Furthermore, to ensure the best possible measurement conditions, the influence of sulfuric acid concentration in a range of 0.025 to 0.5 mol L^{-1} on the propranolol peak was examined (Figure not included). The change in this parameter was shown to not significantly affect the maximum current, therefore the supporting electrolyte of 0.1 mol L^{-1} H$_2$SO$_4$ was selected for later studies.

3.5. Influence of Potential Interferents on Propranolol Peak

Study of interferences is an important part of developing a new analytical method. It allows to determine the influence of potential interferents (that might be found in sample's matrix) on analyte signal. In the experiment, the influence of the following metals on the propranolol signal was investigated: Mg(II), Ca(II), Na(I), K(I) (50 μmol L^{-1} added), Cu(II), Pb(II), Cd(II), Zn(II), Mo(IV), Mn(II) (5 μmol L^{-1} added). Moreover, organic compounds and potential ingredients of the pharmaceutical formulation and urine were tested, such as: citric acid (50 μmol L^{-1} added), lactose monohydrate, starch, magnesium stearate, talc, cellulose, titanium dioxide, glucose, caffeine, ascorbic acid, uric acid, acetaminophen (20 μmol L^{-1} added), and Triton X-100 (2.5 ppm added). Each measurement was carried out in 0.1 mol L^{-1} H$_2$SO$_4$, the preconcentration potential and time were equal to 550 mV and 15 s, respectively, the propranolol concentration was 10 μmol L^{-1}. Among the tested substances only a few of them had influence on propranolol signal. In the case of Cd(II) and Mo(II), their concentration equal to 5 μmol L^{-1} caused a 15% and 13% decrease in the peak current, respectively. The presence of Mg(II) resulted in a 25% increase in the signal (concentration 50 μmol L^{-1}). Cu(II) ions caused a 22% decrease in the peak current when its concentration was equal to 5 μmol L^{-1}. Remaining interferents had no or negligibly small influence on propranolol signal. In Table 1 changes in propranolol peak current before and after interferents dosing are presented.

Table 1. Propranolol peak current change in presence of interferents.

Interferent\Concentration of Interferent	Peak Current Value, µA					Signal Change, %
	0 µmol L^{-1}	5 µmol L^{-1}	20 µmol L^{-1}	50 µmol L^{-1}	2.5 ppm	
Mg (II)	5.59	-	-	6.99	-	+25
Ca (II)	5.50	-	-	5.87	-	+7
Na (I)	5.52	-	-	5.74	-	+4
K (I)	5.56	-	-	5.90	-	+6
Cu (II)	5.57	4.35	-	-	-	−22
Pb (II)	5.55	5.77	-	-	-	+4
Cd (II)	5.59	4.75	-	-	-	−15
Zn (II)	5.63	5.35	-	-	-	−6
Mo (IV)	5.49	4.77	-	-	-	−13
Mn (II)	5.09	5.01	-	-	-	−2
Citric acid	5.58	-	-	5.40	-	−3
Lactose monohydrate	5.56	-	5.40	-	-	−3
Starch	5.49	-	5.43	-	-	−1
Magnesium stearate	5.56	-	5.56	-	-	0
Talc	5.58	-	5.47	-	-	−2
Cellulose	5.54	-	5.43	-	-	−2
Titanium dioxide	5.62	-	5.63	-	-	0
Glucose	5.63	5.29	-	-	-	−6
Caffeine	5.61	-	5.22	-	-	−7
Ascorbic acid	5.60	-	5.43	-	-	−3
Uric acid	5.57	-	5.46	-	-	−2
Acetaminophen	5.58	-	5.36	-	-	−4
Triton X-100	5.79	-	-	-	5.18	−11

3.6. Calibration and Real Samples Studies

Propranolol DP voltammograms of 0.02 to 3.5 µmol L^{-1} with a preconcentration time in the range from 15 to 120 s was registered and presented in Figure 5.

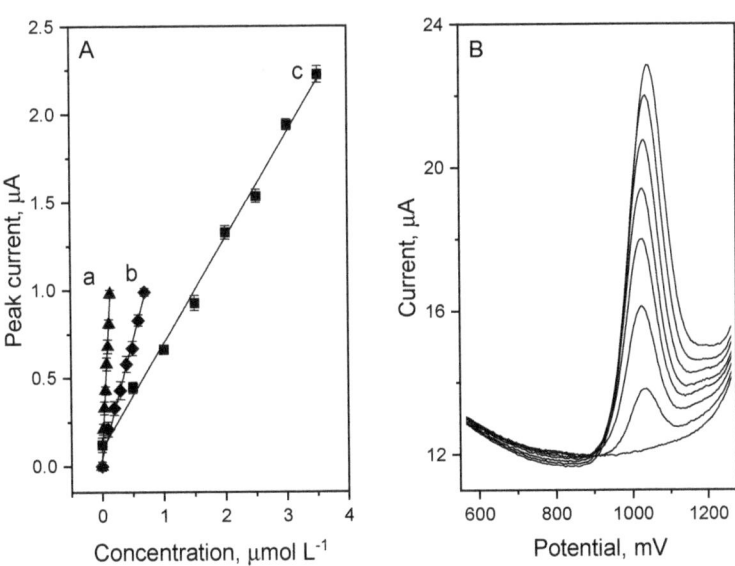

Figure 5. DPV calibration curves of propranolol registered for preconcentration times (a) 2 min, (b) 1 min, and (c) 15 s in 0.1 mol L^{-1} H$_2$SO$_4$ (pH 1.8) (**A**) and corresponding voltammograms for preconcentration time of 1 min and concentration range 0 to 0.7 µmol L^{-1} (**B**).

The linear dependence between PROP concentration and peak current value for short preconcentration time of 15 s was in the range from 0.5 up to 3.5 µmol L^{-1}, with the detection limit of 0.12×10^{-6} mol L^{-1} (signal to noise relation = 3) and sensitivity of 0.59 µA µM^{-1} (Ip = 0.598x + 0.098, R = 0.998). In order to achieve lower detection limit, parameter of preconcentration time was elongated to 120 s. Obtained calibration curve with linearity from 0.02 to 0.14 µmol L^{-1} let to accomplish the detection limit of 0.007×10^{-6} mol L^{-1} (signal to noise relation = 3) and sensitivity of 6.58 µA µM^{-1} (Ip = 6.583x + 0.042, R = 0.997). The reproducibility of the presented propranolol determination method was calculated from the obtained voltammograms and specified as RSD with the value of 1.5% for 9 repetition of measurements of 2.5 µmol L^{-1} propranolol concentration. Comparison of propranolol detection limits for its different determination methods reported in the literature is showed in Table 2.

Table 2. Comparison of other propranolol determination methods.

Method	Detection Limit	Source
Spectrophotometry	0.34 µmol L^{-1}	[5]
RP-HPLC [1]	1.04 µmol L^{-1}	[8]
SWV (GC/MWCNTs) [2]	26 nmol L^{-1}	[13]
SWV (AgNP-IL-FG-NF/GCE) [3]	17 nmol L^{-1}	[14]
SWV (EPPG/Graphen/CP) [4]	20 nmol L^{-1}	[15]
CV (TiO$_2$/MWCNT/PGE) [5]	21 nmol L^{-1}	[16]
SWV (BDDE) [6]	0.18 µmol L^{-1}	[19]
DPV (SPE) [7]	13 nmol L^{-1}	[20]
SWV (C:N electrode) [8]	0.75 µmol L^{-1}	[29]
Polarography	5 nmol L^{-1}	[30]
Spectrofluorimetry	11.9 nmol L^{-1}	[31]
Spectrofluorimetry	30.8 nmol L^{-1}	[32]
LC/MS [9]	0.19 nmol L^{-1}	[33]
Chemiluminescence	3.4 µmol L^{-1}	[34]
Chemiluminescence	0.14 µmol L^{-1}	[35]
SWV (MWCNT/SR) [10]	78 nmol L^{-1}	[36]
CV (CPE/CuO) [11]	2.91 µmol L^{-1}	[37]
DPV (GC/CB)	7 nmol L^{-1}	This work

[1] Reversed-phase chromatography; [2] square wave voltammetry with glassy carbo electrode modified with multi-walled carbon nanotubes; [3] square wave voltammetry with glassy carbon electrode modified with functionalized-graphene, ionic liquid and silver nanoparticles; [4] square wave voltammetry with edge plane pyrolytic graphite electrode modified with graphene and conductive polymer; [5] cyclic voltammetry with pencil graphite electrode modified with TiO$_2$ and multiwalled carbon nanotubes; [6] square wave voltammetry with boron doped diamond electrode; [7] differential pulse voltammetry with screen printed electrode; [8] square wave voltammetry with nitrogen-containing tetrahedral amorphous carbon; [9] liquid chromatography coupled with mass spectrometry; [10] square wave voltammetry with multiwalled carbon nanotubes, graphite and silicone rubber electrode; [11] cyclic voltammetry with copper-oxide nanoparticle modified carbon paste electrode.

To assess its validity, the proposed method was applied to the sensitive propranolol determination in authentic pharmaceutical samples containing the studied drug and freeze-dried human urine sample using the standard addition method. Two commonly accessible drugs: Propranolol WZF (10 mg of propranolol per tablet) and Propranolol Accord (10 mg of propranolol per tablet) were investigated. The sample preparation for analysis was as described in point 2.3. The results obtained with the recovery parameter are presented in Table 1. The value of recovery parameter ranged between 92 and 102% suggests the usefulness of the proposed method for highly sensitive propranolol determination in pharmaceutical samples.

In addition, a human freeze-dried urine sample was tested in order to check the suitability of the method for sensitive determination of propranolol in human body fluids. The urine sample was prepared as described in point 2.4 and the determination of propranolol was performed using the standard addition method. The results obtained with the recovery parameter measured for each medication are presented in Table 3. The value of recovery parameter ranged between 97 and 106%, and suggests the usefulness of proposed method for high sensitive propranolol determination in urine samples. The sample of the obtained voltammograms with corresponding calibration plot is presented in Figure 6. The PROP peak obtained in urine using glassy carbon electrode modified by carbon black was well shaped and clearly distinguished from the background. By expanding the preconcentration time value, it is possible to reach the propranolol concentration values that are noticed in the real urine sample collected from the patients (1 μg mL^{-1}). For the preconcentration time of 45 s, obtained detection limit was of about 18.2 μmol L^{-1}.

Table 3. Results of propranolol determination in pharmaceutical samples and human urine.

Sample	PROP Added, mg/Tablet	PROP Found ± mg/Tablet	Recovery, %
Propranolol WZF	0	11.5 ± 0.2	-
	15	13.8 ± 0.3	92
	30	30.9 ± 0.2	100
	45	44.9 ± 0.2	100
Propranolol Accord	0	11.7 ± 1.2	-
	15	15.4 ± 0.3	102
	30	28.9 ± 0.5	96
	45	45.6 ± 0.3	101
Sample	PROP added, μg/mL	PROP found ± μg/mL	Recovery, %
Urine diluted 100×	0	Not detected	-
	6	6.36 ± 0.9	106
	12	11.7 ± 0.8	97
	18	18.5 ± 0.5	103
	24	23.8 ± 0.4	99

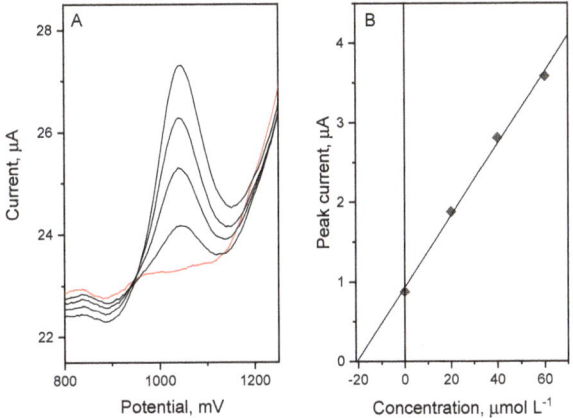

Figure 6. Voltammograms of propranolol determination in urine sample (urine curve marked as red) (A) with corresponding calibration curve (B).

4. Conclusions

In this work, voltametric method of highly sensitive propranolol determination is presented. For the first time of propranolol determination, glassy carbon electrode modified with hybrid nanomaterial based on carbon black and Nafion was used as working electrode. In comparison to previously reported electrode modifiers for highly sensitive propranolol determination, developed sensor might be characterized by ease of its preparation, low-cost, and excellent analytical performance. The conditions for the determination of propranolol were optimized: supporting electrolyte 0.1 mol L^{-1} H$_2$SO$_4$, preconcentration potential, and time equal to 550 mV and 15 s, respectively. Based on conducted calibrations, LOD values were calculated and were equal to 120 nmol L^{-1} for 15 s preconcentration time and 7 nmol L^{-1} for 120 s preconcentration time. Signal repeatability calculated as RSD was equal to 1.5% ($n = 9$, propranolol concentration 2.5 µmol L^{-1}). To verify the usefulness of the developed method, the propranolol concentration was measured in two commercially available pharmaceutical products and in freeze-dried human urine sample. The obtained recovery parameter was in the range 92–106%, which suggests that the method might be assumed to be accurate. Considering the presented results, it might be concluded that the developed voltametric method for propranolol determination using carbon black/Nafion modifier layer could be a useful tool in routine laboratory practice.

Author Contributions: Conceptualization, R.P.; methodology, R.P.; validation, A.G. and B.P.-B.; formal analysis, I.B. and J.S.; investigation, I.B. and J.S.; resources, R.P. and B.P.-B.; writing—original draft preparation, J.S. and A.G.; writing—review and editing, J.S., R.P. and B.P.-B.; visualization, J.S., A.G. and I.B.; supervision, R.P. and B.P.-B.; project administration, R.P.; funding acquisition, R.P. and B.P.-B. All authors have read and agreed to the published version of the manuscript.

Funding: Research project supported by program "Excellence initiative—research university" for the AGH University of Science and Technology.

Institutional Review Board Statement: Not applicable.

Informed Consent Statement: Not applicable.

Data Availability Statement: The data presented in this study are available on request from the corresponding author.

Conflicts of Interest: The authors declare that they have no conflict of interest.

References

1. Bylund, D.B. Propranolol. In *Reference Module in Biomedical Sciences*; Elsevier: Amsterdam, The Netherlands, 2015; pp. 1–8. ISBN 9780128012383.
2. El-Shabrawi, M.; Hassanin, F. Propranolol Safety Profile in Children. *Curr. Drug Saf.* **2011**, *6*, 259–266. [CrossRef]
3. Al-Majed, A.A.; Bakheit, A.H.H.; Abdel Aziz, H.A.; Alajmi, F.M.; AlRabiah, H. Propranolol. *Profiles Drug Subst. Excip. Relat. Methodol.* **2017**, *42*, 287–338. [CrossRef] [PubMed]
4. El-Ries, M.A.; Abou Attia, F.M.; Ibrahim, S.A. AAS and spectrophotometric determination of propranolol HCl and metoprolol tartrate. *J. Pharm. Biomed. Anal.* **2000**, *24*, 179–187. [CrossRef]
5. El-Emam, A.A.; Belal, F.F.; Moustafa, M.A.; El-Ashry, S.M.; El-Sherbiny, D.T.; Hansen, S.H. Spectrophotometric determination of propranolol in formulations via oxidative coupling with 3-methylbenzothiazoline-2-one hydrazone. *Farmaco* **2003**, *58*, 1179–1186. [CrossRef]
6. El-Abasawi, N.M.; Attia, K.A.M.; Abo-serie, A.A.M.; Morshedy, S.; Abdel-Fattah, A. Simultaneous determination of rosuvastatin and propranolol in their binary mixture by synchronous spectrofluorimetry. *Spectrochim. Acta Part A Mol. Biomol. Spectrosc.* **2018**, *198*, 322–330. [CrossRef]
7. Madrakian, T.; Afkhami, A.; Mohammadnejad, M. Simultaneous spectrofluorimetric determination of levodopa and propranolol in urine using feed-forward neural networks assisted by principal component analysis. *Talanta* **2009**, *78*, 1051–1055. [CrossRef]
8. Imam, S.S.; Ahad, A.; Aqil, M.; Sultana, Y.; Ali, A. A validated RP-HPLC method for simultaneous determination of propranolol and valsartan in bulk drug and gel formulation. *J. Pharm. Bioallied Sci.* **2013**, *5*, 61–65. [CrossRef]
9. Salman, S.A.B.; Sulaiman, S.A.; Ismail, Z.; Gan, S.H. Quantitative determination of propranolol by ultraviolet HPLC in human plasma. *Toxicol. Mech. Methods* **2010**, *20*, 137–142. [CrossRef]
10. Kim, H.K.; Hong, J.H.; Park, M.S.; Kang, J.S.; Lee, M.H. Determination of propranolol concentration in small volume of rat plasma by HPLC with fluorometric detection. *Biomed. Chromatogr.* **2001**, *15*, 539–545. [CrossRef] [PubMed]
11. Micke, G.A.; Costa, A.C.O.; Heller, M.; Barcellos, M.; Piovezan, M.; Caon, T.; de Oliveira, M.A.L. Development of a fast capillary electrophoresis method for the determination of propranolol-Total analysis time reduction strategies. *J. Chromatogr. A* **2009**, *1216*, 7957–7961. [CrossRef]
12. Zhou, X.; Li, X.; Zeng, Z. Solid-phase microextraction coupled with capillary electrophoresis for the determination of propranolol enantiomers in urine using a sol-gel derived calix[4]arene fiber. *J. Chromatogr. A* **2006**, *1104*, 359–365. [CrossRef]
13. Oliveira, G.G.; Azzi, D.C.; Vicentini, F.C.; Sartori, E.R.; Fatibello-Filho, O. Voltammetric determination of verapamil and propranolol using a glassy carbon electrode modified with functionalized multiwalled carbon nanotubes within a poly (allylamine hydrochloride) film. *J. Electroanal. Chem.* **2013**, *708*, 73–79. [CrossRef]
14. Santos, A.M.; Wong, A.; Fatibello-Filho, O. Simultaneous determination of salbutamol and propranolol in biological fluid samples using an electrochemical sensor based on functionalized-graphene, ionic liquid and silver nanoparticles. *J. Electroanal. Chem.* **2018**, *824*, 1–8. [CrossRef]
15. Gupta, P.; Yadav, S.K.; Agrawal, B.; Goyal, R.N. A novel graphene and conductive polymer modified pyrolytic graphite sensor for determination of propranolol in biological fluids. *Sens. Actuators B Chem.* **2014**, *204*, 791–798. [CrossRef]

16. Dehnavi, A.; Soleymanpour, A. Titanium Dioxide/Multi-walled Carbon Nanotubes Composite Modified Pencil Graphite Sensor for Sensitive Voltammetric Determination of Propranolol in Real Samples. *Electroanalysis* **2021**, *33*, 355–364. [CrossRef]
17. Broli, N.; Vasjari, M.; Vallja, L.; Duka, S.; Shehu, A.; Cenolli, S. Electrochemical determination of atenolol and propranolol using a carbon paste sensor modified with natural ilmenite. *Open Chem.* **2021**, *19*, 875–883. [CrossRef]
18. Gaichore, R.R.; Srivastava, A.K. Electrocatalytic determination of propranolol hydrochloride at carbon paste electrode based on multiwalled carbon-nanotubes and γ-cyclodextrin. *J. Incl. Phenom. Macrocycl. Chem.* **2014**, *78*, 195–206. [CrossRef]
19. Sartori, E.R.; Medeiros, R.A.; Rocha-Filho, R.C.; Fatibello-Filho, O. Square-wave voltammetric determination of propranolol and atenolol in pharmaceuticals using a boron-doped diamond electrode. *Talanta* **2010**, *81*, 1418–1424. [CrossRef]
20. Khairy, M.; Khorshed, A.A. Simultaneous voltammetric determination of two binary mixtures containing propranolol in pharmaceutical tablets and urine samples. *Microchem. J.* **2020**, *159*, 105484. [CrossRef]
21. Khorshed, A.A.; Khairy, M.; Banks, C.E. Electrochemical determination of antihypertensive drugs by employing costless and portable unmodified screen-printed electrodes. *Talanta* **2019**, *198*, 447–456. [CrossRef]
22. Yin, H.; Ma, Q.; Zhou, Y.; Ai, S.; Zhu, L. Electrochemical behavior and voltammetric determination of 4-aminophenol based on graphene-chitosan composite film modified glassy carbon electrode. *Electrochim. Acta* **2010**, *55*, 7102–7108. [CrossRef]
23. Xu, Q.; Wang, S.-F. Electrocatalytic Oxidation and Direct Determination of L-Tyrosine by Square Wave Voltammetry at Multi-Wall Carbon Nanotubes Modified Glassy Carbon Electrodes. *Microchim. Acta* **2005**, *151*, 47–52. [CrossRef]
24. Maringa, A.; Mugadza, T.; Antunes, E.; Nyokong, T. Characterization and electrocatalytic behaviour of glassy carbon electrode modified with nickel nanoparticles towards amitrole detection. *J. Electroanal. Chem.* **2013**, *700*, 86–92. [CrossRef]
25. Abollino, O.; Giacomino, A.; Malandrino, M.; Piscionieri, G.; Mentasti, E. Determination of mercury by anodic stripping voltammetry with a gold nanoparticle-modified glassy carbon electrode. *Electroanalysis* **2008**, *20*, 75–83. [CrossRef]
26. Smajdor, J.; Paczosa–Bator, B.; Piech, R. Voltammetric Electrode Based on Nafion and Poly(2,3–dihydrothieno–1,4–dioxin)–poly(styrenesulfonate) Film for Fast and High Sensitive Determination of Metamizole. *J. Electrochem. Soc.* **2016**, *163*, B146–B152. [CrossRef]
27. Dong, Y.; Ding, Y.; Zhou, Y.; Chen, J.; Wang, C. Differential pulse anodic stripping voltammetric determination of Pb ion at a montmorillonites/polyaniline nanocomposite modified glassy carbon electrode. *J. Electroanal. Chem.* **2014**, *717–718*, 206–212. [CrossRef]
28. Laviron, E. General expression of the linear potential sweep voltammogram in the case of diffusionless electrochemical systems. *J. Electroanal. Chem.* **1979**, *101*, 19–28. [CrossRef]
29. Lourencao, B.C.; Silva, T.A.; Fatibello-Filho, O.; Swain, G.M. Voltammetric Studies of Propranolol and Hydrochlorothiazide Oxidation in Standard and Synthetic Biological Fluids Using a Nitrogen-Containing Tetrahedral Amorphous Carbon (ta-C:N) Electrode. *Electrochim. Acta* **2014**, *143*, 398–406. [CrossRef]
30. El-Ries, M.A.; Abou-Sekkina, M.M.; Wassel, A.A. Polarographic determination of propranolol in pharmaceutical formulation. *J. Pharm. Biomed. Anal.* **2002**, *30*, 837–842. [CrossRef]
31. Pérez Ruiz, T.; Martínez-Lozano, C.; Tomás, V.; Carpena, J. Simultaneous determination of propranolol and pindolol by synchronous spectrofluorimetry. *Talanta* **1998**, *45*, 969–976. [CrossRef]
32. Tabrizi, A.B. A simple spectrofluorimetric method for determination of piroxicam and propranolol in pharmaceutical preparations. *J. Food Drug Anal.* **2007**, *15*, 242–248. [CrossRef]
33. Partani, P.; Modhave, Y.; Gurule, S.; Khuroo, A.; Monif, T. Simultaneous determination of propranolol and 4-hydroxy propranolol in human plasma by solid phase extraction and liquid chromatography/electrospray tandem mass spectrometry. *J. Pharm. Biomed. Anal.* **2009**, *50*, 966–976. [CrossRef]
34. Townshend, A.; Murillo Pulgarín, J.A.; Alañón Pardo, M.T. Flow injection-chemiluminescence determination of propranolol in pharmaceutical preparations. *Anal. Chim. Acta* **2003**, *488*, 81–88. [CrossRef]
35. Tsogas, G.Z.; Stergiou, D.V.; Vlessidis, A.G.; Evmiridis, N.P. Development of a sensitive flow injection-chemiluminescence detection method for the indirect determination of propranolol. *Anal. Chim. Acta* **2005**, *541*, 149–155. [CrossRef]
36. Dos Santos, S.X.; Cavalheiro, É.T.G.; Brett, C.M.A. Analytical potentialities of carbon nanotube/silicone rubber composite electrodes: Determination of propranolol. *Electroanalysis* **2010**, *22*, 2776–2783. [CrossRef]
37. Shadjou, N.; Hasanzadeh, M.; Saghatforoush, L.; Mehdizadeh, R.; Jouyban, A. Electrochemical behavior of atenolol, carvedilol and propranolol on copper-oxide nanoparticles. *Electrochim. Acta* **2011**, *58*, 336–347. [CrossRef]

Article

Copper Film Modified Glassy Carbon Electrode and Copper Film with Carbon Nanotubes Modified Screen-Printed Electrode for the Cd(II) Determination

Joanna Wasąg [1,*] and Malgorzata Grabarczyk [2]

[1] Department of Materials Engineering, Institute of Engineering and Technical Sciences, Faculty of Natural Sciences and Health, The John Paul II Catholic University of Lublin, 20-950 Lublin, Poland
[2] Department of Analytical Chemistry, Institute of Chemical Sciences, Faculty of Chemistry, Maria Curie-Sklodowska University, 20-031 Lublin, Poland; mgrabarc@poczta.umcs.lublin.pl
* Correspondence: joannawasag@kul.pl

Citation: Wasąg, J.; Grabarczyk, M. Copper Film Modified Glassy Carbon Electrode and Copper Film with Carbon Nanotubes Modified Screen-Printed Electrode for the Cd(II) Determination. *Materials* **2021**, *14*, 5148. https://doi.org/10.3390/ma14185148

Academic Editor: Alessandro Dell'Era

Received: 19 July 2021
Accepted: 6 September 2021
Published: 8 September 2021

Publisher's Note: MDPI stays neutral with regard to jurisdictional claims in published maps and institutional affiliations.

Copyright: © 2021 by the authors. Licensee MDPI, Basel, Switzerland. This article is an open access article distributed under the terms and conditions of the Creative Commons Attribution (CC BY) license (https://creativecommons.org/licenses/by/4.0/).

Abstract: A copper film modified glassy carbon electrode (CuF/GCE) and a novel copper film with carbon nanotubes modified screen-printed electrode (CuF/CN/SPE) for anodic stripping voltammetric measurement of ultratrace levels of Cd(II) are presented. During the development of the research procedure, several main parameters were investigated and optimized. The optimal electroanalytical performance of the working electrodes was achieved in electrolyte 0.1 M HCl and 2×10^{-4} M Cu(II). The copper film modified glassy carbon electrode exhibited operation in the presence of dissolved oxygen with a calculated limit of detection of 1.7×10^{-10} M and 210 s accumulation time, repeatability with RSD of 4.2% ($n = 5$). In the case of copper film with carbon nanotubes modified screen-printed electrode limit of detection amounted 1.3×10^{-10} M for accumulation time of 210 s and with RSD of 4.5% ($n = 5$). The calibration curve has a linear range in the tested concentration of 5×10^{-10}–5×10^{-7} M (r = 0.999) for CuF/GCE and 3×10^{-10}–3×10^{-7} M (r = 0.999) for CuF/CN/SPE with 210 s accumulation time in both cases. The used electrodes enable trace determination of cadmium in different environmental water samples containing organic matrix. The validation of the proposed procedures was carried out through analysis certified reference materials: TM-25.5, SPS-SW1, and SPS-WW1.

Keywords: copper modified electrode; carbon-based electrode materials; screen-printed electrode; electrochemical detection; stripping voltammetry; cadmium determination

1. Introduction

This work developed a novel voltammetric procedure for determination of cadmium using two types of working electrode: a copper film with carbon nanotubes modified screen-printed electrode (CuF/CN/SPE) and a copper film modified glassy carbon electrode (CuF/GCE). For the first time, copper modified electrodes were used to determine ultratrace amounts of Cd(II) ions. The use of copper as a film on the surface of the working electrode is a very important aspect, as now a lot of emphasis is placed on the development of a new type of film electrodes using non-toxic metals. Copper is non-toxic and allowed us to obtain very low detection limit for cadmium. Importantly, this work was also created to draw the attention of scientists to this type of copper electrodes, which has been practically unused until now, but is proving to be a powerful tool in the trace analysis of many metal ions. At work, it is also important to determine cadmium on two different working electrodes under almost identical measurement conditions, which allows the measurements to be transferred to field conditions and the results obtained with both methods can be compared. The procedure was first developed and optimized for the CuF/GCE working electrode and then successful measurements were carried out using the obtained parameters with the novel modified CuF/CN/SPE electrode. As already mentioned, in the literature, we do not find too many works relating to the use of the copper film electrode [1–4]. Copper film modified electrodes seem to be an excellent proposal for the determination of trace amounts of metals also in real samples. Such an electrode allows low detection limits to be obtained, and the fact that it is created in situ from the test solution significantly shortens the measurement time. The SPE electrode, on the other hand, has several advantages over conventional electrodes, such as the simplicity of use, commercial availability, low price, and the possibility of using it in field research as a portable sensor. The reproducibility and sensitiveness of these electrodes are very good, so they can replace classical solid electrodes in the analysis [5,6]. Their effectiveness in analysis gives a chance for their widespread and more frequent use [7–10]. These electrodes are also easily accessible to everyone. There is a wide variety of screen-printed electrode materials in the commercial industry, depending on the specific needs. They can be easily purchased and used in direct field analyzes.

Film metal modified electrodes have become more and more popular in recent years, especially as a replacement for toxic mercury electrodes. These electrodes can be generated on various substrates, but the most common is glassy carbon [1,2,4,11–14]. In recent years, more attention has been paid to screen-printed electrodes, which can be used either as direct working electrodes or as an attractive substrate for the generation of film metal electrodes. The sensitivity of screen-printed electrodes can be increased by the incorporation of desirable functional parameters or specific nanoparticles in the ink before the printing process. The screen-printed electrode used in this work is modified with carbon nanotubes (CN), which have a large surface, excellent electrical conductivity, and good chemical stability [15]. Carbon nanotubes exhibit better electrochemical performance than other carbon-based electrodes. In the literature, there are examples of the use of electrodes that are modified by carbon nanotubes. To name a few uses, they have been used for electrochemical oxidation of inorganic and organic compounds, including pharmaceuticals [16], and catalytic oxidation of thiols [11]. Various modifications with the use of copper are also known in the literature [17–19]. In our work, a copper coating is applied to the surface of the working electrode, which forms an integral part of the electrode. In other works, an interesting solution is the use of hybrid materials based on copper oxide successfully synthesized by an ultrasound sonochemical method and applied as an electrode material for supercapacitor applications [17]. Another interesting example can be the use of metal organic framework (MOF) derived Co-Al layered double hydroxide by Cr(VI) and Pb(II) ion adsorption [18]. These works give us an insight into the effectiveness of the practical use of copper-based materials as a diverse medium for the determination of many metal ions.

The aim of our research was to use modified copper film electrodes generated on various substrates, such as GCE and CN/SPE, and to develop competitive procedures for the determination of trace amounts of cadmium. Cadmium is a familiar hazardous pollutant in the ecological system. It is an element that is relatively sparse in the earth's crust, but poses a serious threat to human and animal health. As a result of human activities, cadmium has become the main chemical pollutant of the environment, and as it is used in many technology processes in various industries and agriculture, its presence is found in air, water, and soil as well as in plants and animal tissues. In industry, cadmium is used for the production of dyes and plastic stabilizers, artificial and galvanic protective coatings, solders and alloys, and cadmium bars. It is also used for the production of alkaline nickel-cadmium batteries, fireworks, and fluorescent paints [20]. Fertilizers (e.g., superphosphates) that are contaminated with this metal in an amount from 10 to 100 mg/kg are a significant source of cadmium in the environment. Its long-term and widespread use leads to continual cadmium contamination of the soil [21]. Once introduced into the environment, cadmium is not subject to degradation and remains in constant circulation. Its long half-life translates into the accumulation of this element in the organisms of plants, animals, and humans. Environmental exposure factors can lead to the absorption of large amounts of cadmium and the toxic effects of this element on the body. In living organisms, even in small amounts, it causes liver diseases, kidney and cardiovascular dysfunction, toxic effects in Alzheimer's disease, and carcinogenic effects on humans [22]. Therefore, it is crucial to obtain information on the amount of Cd(II) ions in real environmental samples as their toxicological effect depends on their concentration and the form of the compound in which cadmium occurs [23].

In our research, we focused on the determination of Cd(II) in water environmental samples, and we wanted to use working electrodes of a new generation for this purpose, allowing for excellent signal reproducibility and high sensitivity of determinations. In the research, anodic stripping voltammetry (ASV) was used, which allows the above-mentioned advantages of film modified solid electrodes to be exploited. Stripping voltammetric analysis methods are widely used in trace analysis of various metals and successfully used to monitor environmental samples [12,13,24–29]. Additionally, these techniques have often been used to designate cadmium as heavy metal. Abbasi et al. [30] summarized the literature on cadmium determination using the striping voltammetry technique up to 2011. In their work, Rojas-Romo et al. [31] summarized the electroanalytical methods applied for Pb(II) and Cd(II) determination using different types of working electrodes and anodic stripping voltammetry. The vast majority of these papers describe the determination of cadmium ions simultaneously with other elements, most often lead. Here, we determine cadmium without accompanying ions. Table 1 compares the proposed procedure with the publications concerning the determination of Cd(II) ions in the works from recent years using the ASV technique.

Table 1. Comparison of the proposed procedure with the previously reported voltammetric methods using ASV for the determination of Cd(II). The works are ranked according to the decreasing limit of detection.

Electrode	Accumulation Time	LOD	Sample	References
polyPCA/GE	125 s	0.142 µM	freshwater and real water	[32]
SWCNTs/Biomass/GCE	120 s	0.103 µM	real water	[33]
GQDs/NF/GCE	150 s	0.126 µM	bivalve mollusks	[34]
BOC/GCE	500 s	0.035 µM	tap water	[35]
Hg(Ag)FE	30 s	0.013 µM	real water	[36]
MFE/GCE	240 s	0.006 µM	the constituent parts of the illegal cigarettes	[37]
IL/GO/GCE	300 s	0.003 µM	tap water	[38]
MWCNT/GCE	not specified	0.002 µM	real water	[39]
BiFE/GCE	60 s	0.008×10^{-1} µM	real water	[31]
GO@Fe3O4@2-CBT/GCE	180 s	0.027×10^{-2} µM	real water	[40]
CuF/GCE	210 s	0.017×10^{-2} µM	real water	[this work]
CuF/CN/SPE	210 s	0.013×10^{-2} µM	real water	[this work]

polyPCA/GE—graphite electrodes modified with poly(*p*-coumaric acid), SWCNTs/Biomass/GCE—glassy carbon electrode modified by a mixture of single walled carbon nanotubes and biomass, GQDs/NF/GCE—glassy carbon electrode modified with graphene quantum dots and Nafion, BOC/GCE—glassy carbon electrode modified bismuth oxycarbide, Hg(Ag)FE—renewable mercury film silver-based electrode, MFE/GCE—glassy carbon electrode modified mercury film, IL/GO/GCE—glassy carbon electrode modified graphene oxide and ionic liquid, MWCNT/GCE—glassy carbon electrode multi-walled carbon nanotube electrode, BiFE/GCE—glassy carbon electrode modified bismuth film, and GO@Fe3O4@2-CBT/GCE—glassy carbon electrode modified with magnetic graphene oxide modified with benzothiazole-2-carboxaldehyde.

As we can see, our procedure has the lowest detection limit compared to other ASV procedures for the determination of Cd(II) ions published in recent years. We achieved this due to the use of new generation copper modified electrodes, CuF/CN/SPE and CuF/GCE, in cadmium analysis for the first time. We obtained detection limits even lower than with the use of mercury electrodes, which, as is well known, enable determination of one of the lowest detection limits in voltammetric methods. The elimination of mercury electrodes from research is another aspect that supports the development of other electrochemical sensors using non-toxic metals. It is, therefore, a major advantage of the tested method described here.

2. Materials and Methods

2.1. Apparatus

A µAutolab analyzer (EcoChemie, Utrecht, The Netherlands) with GPES software was used to perform voltammetric studies. The three-electrode system used for measurement consisted of a glassy carbon working electrode and a modified carbon nanotubes screen-printed working electrode (GCE, 1 mm diameter, and CN/SPE, 4 mm diameter), an Ag/AgCl (saturated NaCl) reference electrode (AutoLab), and platinum wire as an auxiliary electrode (AutoLab). The surfaces of the working electrodes were modified before each measurement in situ with copper. The studies were conducted in a volumetric cell (10 mL volume). The glassy carbon electrode (AutoLab) was polished daily on 2000 grit sandpaper, and afterwards it was polished using 0.3 µm alumina slurry on a Buehler polishing pad and immersed for 30 s in an ultrasonic bath. The modified carbon nanotubes screen-printed electrode was used without any special preparation in the form in which it was purchased (nLab). FEI Quanta 3D FEG scanning electron microscope (SEM) equipped with an energy dispersive X-ray spectrometer EDX Octane Elect Plus was used to accurately identify surface morphology and to take images of the electrode surfaces.

2.2. Reagents

The supporting electrolyte was prepared by diluting concentrated hydrochloric acid to 0.1 M HCl (Suprapure Merck). Standard cadmium of 1 g/L was purchased from Fluka (Buchs, Switzerland). The working solution of Cd(II) with a lower concentration of 1×10^{-4} M was prepared from standard cadmium in 0.01 M HNO_3 solution. The interference effect was tested using standard stock solutions of 1 g/L of Al(III), As(III), As(V), Ca(II), Cr(III), Cr(VI), Fe(III), Mg(II), Mn(II), Ni(II), Pb(II), W(VI), Zn(II), Ti(IV), Sb(III), Mo(VI), Sn(IV), Se(IV), In(III), and Ga(III) from Fluka. The solution of Triton X-100 (non-ionic surfactant), SDS (anionic surfactant), and CTAB (cationic surfactant) were purchased from Fluka, whereas HF (humic acids) was obtained from Aldrich. FA (fulvic acids) and NOM (natural organic matter) from the Suwannee River were purchased from the International Humic Substances Society. Rhamnolipids (biosurfactant) and Amberlite XAD-7 resin were obtained from Sigma. The resin was prepared by rinsing it four times in distilled water and drying at 50 °C before use. All solutions were made using ultra-purified water supplied by a Milli-Q system (Millipore, London, UK).

In the research, certified reference materials were used such as: TM-25.5 (environmental matrix reference material, Environment and Climate Change, Ottawa, ON, Canada), SPS-SW1 (surface water, Spectrapure Standards As, Oslo, Norway), and SPA-WW1 (waste water, Spectrapure Standards As, Oslo, Norway).

2.3. ASV Procedure of Cadmium Determination

For both used electrodes, CuF/CN/SPE and CuF/GCE, the measurements were performed under optimum conditions using hydrochloric acid at a concentration of 0.1 M containing 2×10^{-4} M Cu(II). The experiments were performed using differential pulse anodic stripping voltammetry (DP-ASV) in the following sequence of potentials: +0.4 V for 10 s and −0.7 V for 60 s for CuF/GCE, and +0.4 V for 10 s and −0.75 V for 60 s for CuF/CN/SPE. The first step was performed to electrochemically clean the working electrode. The potential and time of electrochemical cleaning had been optimized and successfully applied in the previous work using CuF/GCE [1,4], and in this work it also proved to be effective in removing traces of earlier measurements from the surface of the solid electrode. During the second potential (accumulation potential), in situ plated copper on the surface glassy carbon electrode and cadmium on the surface of the produced copper film were deposited simultaneously. After a deposition time of 60 s, the differential pulse stripping voltammogram was recorded, after 5 s equilibration time, while the potential was scanned from −0.7 V to −0.4 V for CuF/GCE and from −0.8 V to −0.5 V for CuF/CN/SPE, with a pulse time of 10 ms and a pulse height of 50 mV. The measurements were conducted on the non-deareated solution with no apparent effect on the cadmium signal. During all steps, the solution was stirred using a magnetic stirring bar. The intensity of the obtained signal was proportional to the concentration of Cd(II) in the sample solution.

2.4. Procedure of Preliminary Mixing with Resin

When conducting studies on real water samples, one should take into account the possibility of a negative impact on the measurements of organic substances and surfactants that may be present in such samples. The organic substances and surfactants can adsorb on the electrode surface, subsequently blocking electroactive sites. In our previous studies [14,24], we have proved that such interferences can be effectively eliminated using Amberlite XAD-7 resin with adsorption properties. During the procedure of preliminary mixing with resin, the interfering substances are adsorbed onto the resin, and consequently the CuF/CN/SPE and CuF/GCE electrodes are not blocked and the Cd(II) ions can be efficiently adsorbed on the modified electrode surface. Due to this, the determination can be carried out directly from a natural sample without negative organic matter interferences. An additional advantage is the fact that, in ASV procedures, the resin can be added directly to the measuring cell. In the case of adsorptive stripping voltammetry procedures (AdSV), mixing with the resin has to be performed in an additional step before the actual measurement [14,24]. This is due the fact that, in the case of AdSV methods, it is necessary to introduce a complexing agent into the vessel and, as it has been proven, the determined metals in the form of complexes are often adsorbed on the resin, which results in lower results. In the case of the ASV method, it is not necessary to introduce a complexing agent and the determined metal is not adsorbed on the resin. In this case, 0.1 g of resin was added directly to the measuring cell and the determinations were performed as described in Section 2.3.

3. Results and Discussion

In the earlier literature [1–4], it was documented that the copper film electrode can be another interesting alternative to mercury electrodes, apart from the lead film electrode [13,14] and the bismuth film electrode [31,41]. As proven in this work, a copper film can be generated on both the GCE and CN/SPE substrate. It enables the analysis to be transferred to field conditions, which provides quick and cheap direct analysis of environmental samples. In order to achieve the best performance and lowest detection limit, an optimization study was performed. The parameters influencing the height of the obtained signal were optimized: the pH and concentration of the supporting electrolyte, the concentration of copper, the deposition potential and time, and the pulse time and pulse height of the stripping voltammetry measurement of the trace concentration of Cd(II) ions. The optimization process was carried out first for the electrode CuF/GCE.

3.1. Effect of Composition and Concentration of Supporting Electrolyte

The type and pH of the basic electrolyte used in anodic stripping voltammetry measurements is of great importance for the sensitivity, stability, and repeatability of analytical signals. Several solutions that can act as the supporting electrolyte were tested, including ammonia buffer, acetate buffer, phosphorus buffer, hydrochloric acid, perchloric acid, and acetic acid. In the previous study that used CuF/GCE as a working electrode, 0.1 M HCl with 0.4 M NaCl [1,4] or 0.01 M HCl [2] was selected as a supporting electrolyte. Additionally, in the case of this work, after preliminary tests and attempts to obtain a signal, hydrochloric acid was selected from among the above-mentioned reagents. In all cases, the measurements were performed for a solution with a standard composition, a fixed concentration of 5×10^{-8} M Cd(II), 2×10^{-4} M Cu(II), and 0.1 M of the tested supporting electrolyte, and with a variable pH range in the case of the buffer solution. It was observed that only in the case of hydrochloric acid the cadmium signal was obtained, so this acid was used in further studies.

In addition to the selection of the electrolyte, its concentration in the tested sample also had to be adjusted. The concentration of hydrochloric acid was examined in the range from

0.05 to 0.4 M. The studied solution contained, as previously, 5×10^{-8} M Cd(II), 2×10^{-4} M Cu(II), and an appropriate amount of HCl. It was noted that the highest, narrowest, and symmetric peak was obtained at a concentration of 0.1 M hydrochloric acid in the solution. At a lower concentration of HCl in the solution, the cadmium peak was lower, while at a higher concentration of HCl in the solution, the peak initially remained the same and then decreased. In the next measurement, the hydrochloric acid concentration of 0.1 M was selected.

3.2. Effect of Copper Concentration

The influence of the concentration of Cu(II) in the measured solution used to create the thin film on the surface of the solid electrode on the cadmium signal is shown in Figure 1. As shown, copper concentration affects the signal obtained by voltammetric technique. The analysis was carried out with the solution containing a fixed concentration of 5×10^{-8} M Cd(II) and 0.1 M HCl with a variable concentration of Cu(II) from 1.6×10^{-6} to 3.2×10^{-4} M. The stripping of cadmium sharply increased in the concentration range between 8×10^{-6} and 4×10^{-5} M; at a higher concentration of Cu(II), the cadmium signal continued to increase, but slightly, to a concentration of 1.6×10^{-4} M, and then remained constant. Taking into account the above considerations, the optimal concentration of copper in the test objects was assumed to be 2×10^{-4} M. Additionally, using the Randles-Sevcik equation [42], the active surface areas of the working electrode surfaces were calculated. Using this Equation (1), the peak current (I_p) is defined as:

$$I_p = 0.4463 \left(\frac{F^3}{RT} \right)^{1/2} An^{3/2} D^{1/2} C_o v^{1/2} \tag{1}$$

where: F—Faraday constant (F = 96 485 C mol^{-1}), T—the absolute temperature (T = 298 K), R—the universal gas constant (R = 8.314 J mol^{-1} K^{-1}), A—the electrode surface area (cm^2), n—the number of electrons involved in the redox reaction (n = 2), D—diffusion coefficient (D = 7.2×10^{-6} cm^2 s^{-1}), and C_o—the concentration of Cu(II) (2×10^{-4} M). For the CuF/GCE working electrode geometric area of the surface was equal to 0.00785 cm^2, while the active surface area of the glassy carbon electrode modified with copper equals to 0.00017 ± 0.00001 cm^2, number of repeated measurements = 3 (n). The smaller active area than the geometric area of the electrode confirms the fact that the active sites on the electrode surface are copper sites. The area between the accumulated copper remains inactive.

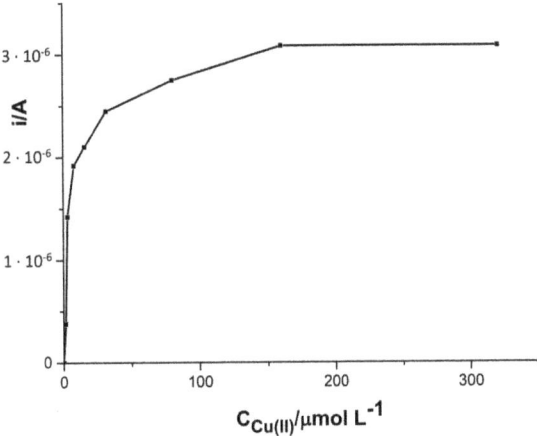

Figure 1. Influence of copper concentration on the Cd(II) signal. Concentration of Cd(II) 5×10^{-8} M. Accumulation potential -0.7 V and accumulation time 60 s.

3.3. Conditions of Accumulation Potential and Time

In order to check the effect of the accumulation potential on the measurements, tests were carried out with the solution containing 5×10^{-8} M Cd(II), 2×10^{-4} M Cu(II), and 0.1 M HCl. During the accumulation potential stage, a copper film is formed and, at the same time, cadmium is accumulated in the form of Cd(0) as a result of the reduction in its Cd(II) ions. In the optimization, the accumulation potential was changed over the range of -0.9 to -0.5 V. The obtained results showed that the cadmium signal was visible for the accumulation potential range from -0.8 to -0.65 V, and the highest peak was obtained at the accumulation potential of -0.7 V. Therefore, for further experiments, the accumulation potential equal to -0.7 V was selected as the most appropriate potential for anodic stripping voltammetry determination of Cd(II) ions.

After adjusting the accumulation potential, the accumulation time was optimized. This parameter has a pronounced effect on sensitivity in stripping techniques. This influence was measured in the accumulation time range 0–260 s. In the tested solution, the concentration was 5×10^{-8} M Cd(II), 2×10^{-4} M Cu(II), and 0.1 M HCl. The influence of accumulation time on the Cd(II) peak current is presented in Figure 2. The accumulation potential was -0.7 V. The value of the voltammetric signal increased almost linearly with the accumulation time prolonged to 210 s. For the longer accumulation time, we can observe a reduction in the cadmium peak and the blurring of its shape. Thus, an accumulation time of 210 s was used as optimal in constructing the calibration curve and calculating the limit of detection, RSD, and the correlation coefficient. However, to shorten the measuring time, an accumulation time of 60 s was used in the measurements during the optimization procedure, interfering testing, and tests with certified reference materials.

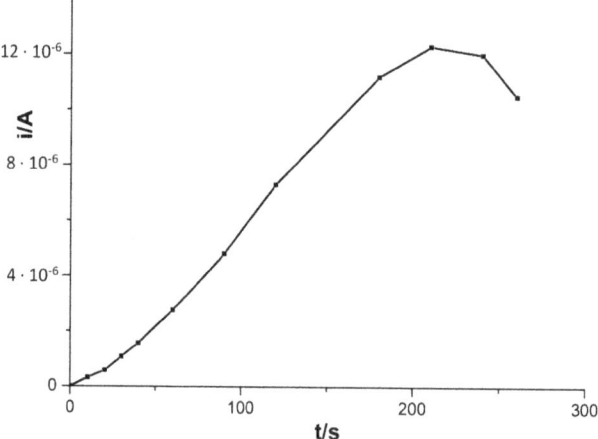

Figure 2. Influence of accumulation time on the Cd(II) signal. Composition of solution 5×10^{-8} M Cd(II), 2×10^{-4} M Cu(II), and 0.1 M HCl. Accumulation potential -0.7 V.

3.4. Pulse Time and Pulse Height

The pulse time and pulse height also have effects on the cadmium peak intensity, so they were also examined. The pulse time was examined from 2 to 20 ms, and it turned out that, with an increase in pulse time above 10 ms, the signal of Cd(II) decreased, and hence for further tests the value of 10 ms was chosen. The variation of the pulse height between 20 and 100 mV showed that with the increase in pulse height to 50 mV, the peak current of cadmium increased linearly. In the higher values, the signal of Cd(II) undergoes gradual blurring. Figure 3 shows the obtained results of cadmium peak current on pulse height (A) and pulse time (B).

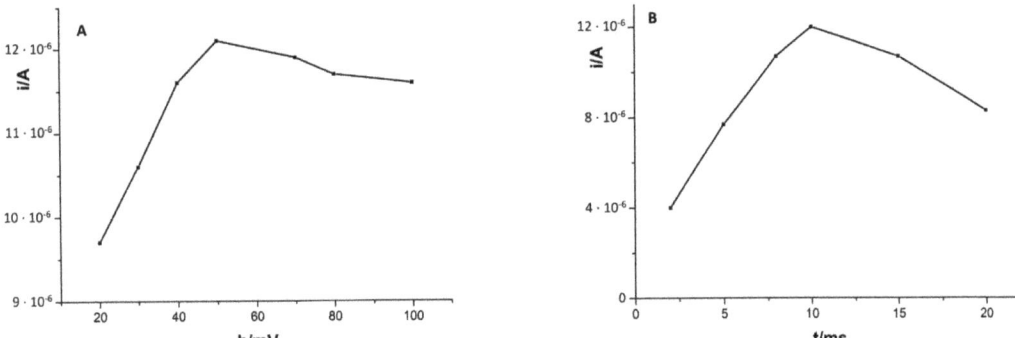

Figure 3. Influence of pulse height (**A**) and pulse time (**B**) on the Cd(II) signal. Composition of solution 5×10^{-8} M Cd(II), 2×10^{-4} M Cu(II), and 0.1 M HCl. Accumulation potential -0.7 V, accumulation time 210 s.

3.5. Analytical Characterization

Based on the previously optimized parameters, such as concentration and type of the supporting electrolyte, copper concentration, accumulation potential and time, and pulse time and height, a series of measurements was carried out to prepare a calibration curve. For this purpose, the solution was prepared: 0.1 M HCl, 2×10^{-4} M Cu(II), to which cadmium additives were added during the measurements with an accumulation time of 210 s and with an accumulation potential of -0.7 V. It was found that the intensity of the peak current derived from cadmium ions increased linearly (correlation coefficient r = 0.999) in the concentration range from 5×10^{-10} to 5×10^{-7} M. The limit of detection calculated from the calibration curve is equal to 1.7×10^{-10} M, with the equation y = 0.191x + 0.918, where y is the peak current (μA) and x is Cd(II) concentration (nM). The sensitivity calculated for comparison with other papers [43] was 1123.529 μA nM^{-1} cm^{-2}. The relative standard deviation (RSD) for all measured concentrations of cadmium from the linear range of the calibration graph was 4.2% (n = 5). Figure 4 presents the linear range of the Cd(II) calibration curve. Figure 5 shows selected voltammograms obtained when creating a calibration curve for low concentrations of cadmium in the sample.

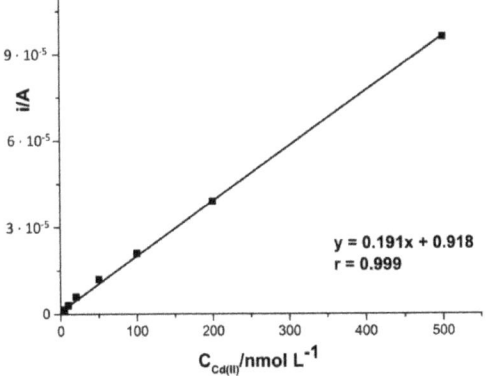

Figure 4. Linear range of the cadmium calibration curve. DPV parameters: accumulation time 210 s, accumulation potential -0.7 V.

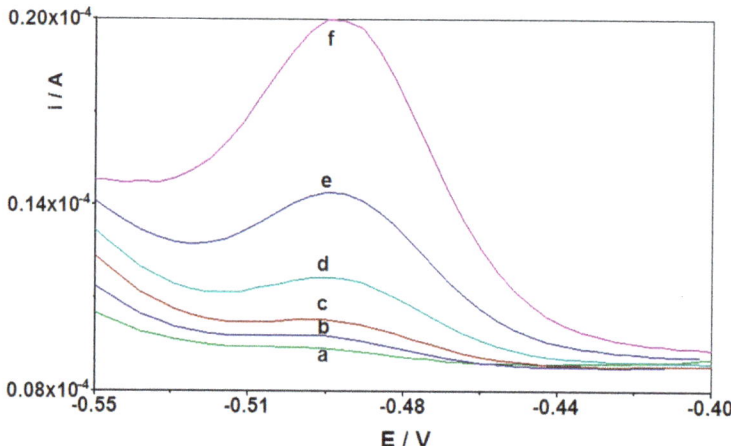

Figure 5. Differential pulse voltammograms obtained in the course of Cd(II) determination at the GCE working electrode recorded for solutions containing: (**a**) background: 1 mL 1 M HCL + 125 µL 1 g/L Cu(II) and distilled water; (**b**) as (**a**) + 1×10^{-9} Cd(II); (**c**) as (**a**) + 2.5×10^{-9} Cd(II); (**d**) as (**a**) + 5×10^{-9} Cd(II); (**e**) as (**a**) + 1×10^{-8} Cd(II); and (**f**) as (**a**) + 2.5×10^{-8} Cd(II). Accumulation potential -0.7 V and accumulation time 210 s.

The reproducibility of the peak current was also determined by successive measurements (n = 5) of the signal of 5×10^{-9} M Cd(II) and was assessed from the experiments performed in five consecutive days as RSD, which was 3.2%.

3.6. Interferences

Before attempting an analysis of real water samples, the influence of potential interference substances and ions on the analytical signal of 5×10^{-8} M Cd(II) was investigated. Two major sources of interference were examined: other metal or metalloid ions and organic substances, surfactants. Interference from other metal or metalloid ions could cause the blocking of the working electrode surface or create intermetallic compounds with other components of the tested solution causing a reduction or complete disappearance of the cadmium signal. The effects of the influence of co-existing metal or metalloid ions were examined using a fixed concentration of Cd(II) with different amounts of foreign ions under standard optimized conditions. The result showed that an up to 200-fold excess of Al(III), As(III), As(V), Ca(II), Cr(III), Cr(VI), Fe(III), Mg(II), Mn(II), Ni(II), W(VI), Zn(II), Ti(IV), Sb(III), Mo(VI), Sn(IV), Se(IV), In(III), and Ga(III) did not have any significant effect on the Cd(II) peak current. The addition of a 100-fold excess of Pb(II) and Sn(IV) caused a $50 \pm 3\%$ decrease in the cadmium signal.

Surfactants and humic substances are other types of interfering substances occurring in natural water samples. They can adsorb on the surface of the electrode, which reduces access to it and may make it difficult to form a metallic film on it [44]. In order to investigate the effect of these substances on the cadmium peak current, experiments with non-ionic surfactant Triton X-100, cationic surfactant CTAB (cetyltrimethylammonium bromide), anionic surfactant SDS (sodium dodecyl sulfate), and biosurfactant Rhamnolipids were carried out. As humic substances, humic acid (HA), fulvic acid (FA), and natural organic matter (NOM) were used in the measurements. In the case of determination of Cd(II) ions, only three of the above-mentioned substances caused a decrease and, consequently, at higher concentrations, the disappearance of the cadmium peak. As observed already, a concentration of 2 ppm CTAB caused a reduction in the cadmium peak by about 80%, while the addition of 2 ppm HA and FA decreased the signal by about 60%. In the case of other organic substances, additions up to 30 ppm (NOM) and 50 ppm (Triton X-100,

SDS, Rhamnolipid) did not significantly affect the cadmium signals, only a deterioration of the peak shape was observed with large amounts of the additives. Taking into account the above considerations and the previously presented lack of negative influence on the measurements of foreign metal ions, the lack of interference from NOM, Triton X-100, SDS, and Rahmnolipid potentially present in natural samples is a great advantage of the described procedure. This makes it possible to use the Cd(II) determination procedure on the CuF/GCE electrode in direct tests from natural samples without the need to prepare them for analysis, and to use the CuF/CN/SPE electrode to conduct research in field conditions. This significantly reduces the costs and time of the performed determinations.

In case of CTAB, FA, and HA, in order to eliminate the negative influence on the signals, the procedure of preliminary mixing the test sample with the resin Amberlite XAD-7, having adsorption properties, was used. This developed method is described in the literature on the subject [14,24,36]. All steps of preliminary mixing with the resin used in this work are described earlier in Section 2.4. Figure 6 presents the results obtained before and after application of preliminary mixing with the resin for the interfering substances CTAB, HA, and FA. In Table 2, we can also see the results obtained when using preliminary mixing with the resin Amberlite XAD-7 and, for comparison, without using this procedure. Thanks to this method, an undisturbed cadmium signal was obtained even at 20 ppm CTAB and FA, and at 10 ppm HA in the tested sample. Thus, we can see a significant improvement and the effectiveness of the resin used in removing interference.

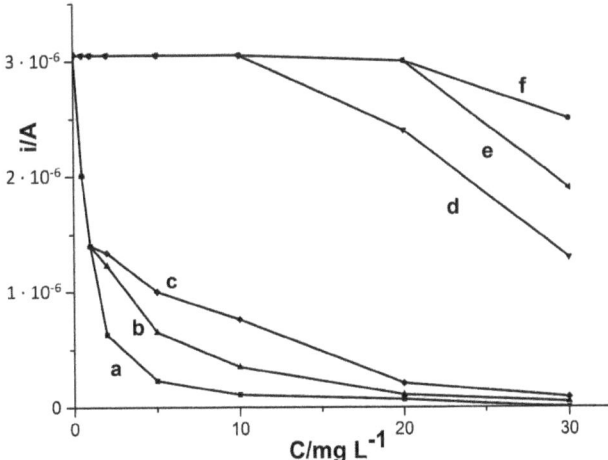

Figure 6. Influence of CTAB (**a,f**), HA (**b,d**), and FA (**c,e**) on the cadmium peak intensity using the procedure without (**a–c**) and with (**d–f**) preliminary mixing with Amberlite XAD-7 resin. Concentration of Cd(II) 5×10^{-8} M, accumulation potential -0.7 V, and accumulation time 60 s.

Table 2. Influence of CTAB, HA, and FA on the Cd(II) voltammetric signal using the procedure with and without preliminary mixing with Amberlite XAD-7 resin. Concentration of Cd(II) was 5×10^{-8} M.

Organic Substance	Maximum Allowable Concentration of Organic Substances That Does Not Interfere with the Cd(II) Signal (ppm)	
	Without Mixing with Resin	With Mixing with Resin
CTAB	1	20
HA	1	10
FA	1	20

3.7. Impact of Temperature

In the next stage of the research, it was checked whether the increase in temperature from 20 to 60 °C had an impact on cadmium signals obtained using new modifications of the CuF/CN/SPE and CuF/GCE electrodes. For this purpose, a series of measurements were carried out for the solution containing a constant cadmium concentration of 5×10^{-8} M, 0.1 M HCl, and 2×10^{-4} M Cu(II) at 20, 30, 40, 50, and 60 °C. For this purpose, an appropriately designed 10 mL voltammetric cell connected to a thermostat was used, which allowed the desired temperature to be maintained. For each temperature, a series of 5 measurements was carried out to check the stability of the obtained signal. The obtained results showed that temperature did not affect the cadmium signal, which means that it did not affect the process of creating new types of electrodes modified with copper: CuF/CN/SPE and CuF/GCE. This is a great advantage of these electrodes that can work in a wide temperature range without adversely affecting the process of surface modification of the working electrodes.

In subsequent studies, it was investigated whether the increase in temperature may improve the elimination of interference from CTAB, HA, and FA, and increase the permissible concentrations of other organic substances, so that, even at higher concentrations in the samples, they would not affect the cadmium peak current. It was also investigated whether the increase in temperature may affect the better performance of the Amberlite XAD-7 resin in the process of removing interferences from organic substances. The measurements were carried out using the conditions and composition of the solution as before: 0.1 M HCl, 2×10^{-4} M Cu(II), 5×10^{-8} M Cd(II), potential -0.7 V, time 60 s, and an appropriate quantity of organic substances and resin. The temperature was varied from 20 to 60 °C during the measurements, performing five repetitions at a given temperature. Based on the obtained results, it was proven that a temperature rise to 50 °C reduces the negative impact of organic substances on the cadmium peak while, when the resin was used, greater recoveries were obtained than at 20 °C. At higher temperatures (60 °C), the cadmium signal slightly decreased. The results for the influence of CTAB, HA, and FA at various temperatures on the voltamperometric cadmium peak current are collected in Table 3. As can be concluded from the obtained data, the use of elevated temperature (up to 50 °C) allows for better sensitivity of the determinations in the presence of interfering substances. The improvement of the signal is not significant but, with a high presence of organic substances in the samples, it is possible to additionally reduce these interferences by manipulating the temperature.

Table 3. Influence of CTAB, HA, and FA on the Cd(II) voltammetric signal at different temperatures. Concentration of Cd(II) 5×10^{-8} M.

Organic Substance	Temperature (°C)	Maximum Allowable Concentration of Organic Substances That Does Not Interfere with the Cd(II) Signal (ppm)	
		Without Mixing with Resin	With Mixing with Resin
CTAB	20	1	20
	30	2	23
	40	3	28
	50	3	30
	60	2	24
HA	20	1	10
	30	1.5	14
	40	2.5	16
	50	2	20
	60	1	12

Table 3. *Cont.*

Organic Substance	Temperature (°C)	Maximum Allowable Concentration of Organic Substances That Does Not Interfere with the Cd(II) Signal (ppm)	
		Without Mixing with Resin	With Mixing with Resin
FA	20	1	20
	30	1.5	25
	40	2	28
	50	2	29
	60	1	25

3.8. Procedure with CuF/CN/SPE Electrode

3.8.1. Morphological, Structural, and Compositional Information of the Electrode Materials

After the optimization of the procedure of cadmium determination using CuF/GCE as a working electrode, additional studies were performed using novel modified screen-printed electrodes CuF/CN/SPE. It turned out that the developed test method can also be effectively applied by using the CuF/CN/SPE electrode without a significant change in the measurement parameters. A novel copper film with carbon nanotubes modified screen-printed electrode was used in the tests. The SPE electrodes are now very popular and are often used in voltamperometric determinations [5–10,15]. They are valued primarily for their reproducibility and sensitivity, effectiveness in analysis, a large active surface, excellent electrical conductivity, and good chemical stability. Additionally, they combine the three-electrode system into one system, which reduces the costs of analysis, and due to their small size they enable analyses to be carried out in the field. Figure 7 presents the voltammograms obtained for the CuF/CN/SPE electrode with the appropriate additives: with cadmium without copper, after adding copper to the solution, and in a solution with only copper without cadmium. As can be seen, without copper, there is no signal from the cadmium present in the solution. Only after adding copper to the tested solution two peaks appear in the voltammogram, one from Cd(II) and one from Cu(II) ions. This is confirmed by the fact that a copper film is formed on the surface of the CuF/CN/SPE electrode, which allows the accumulation of cadmium ions on its surface.

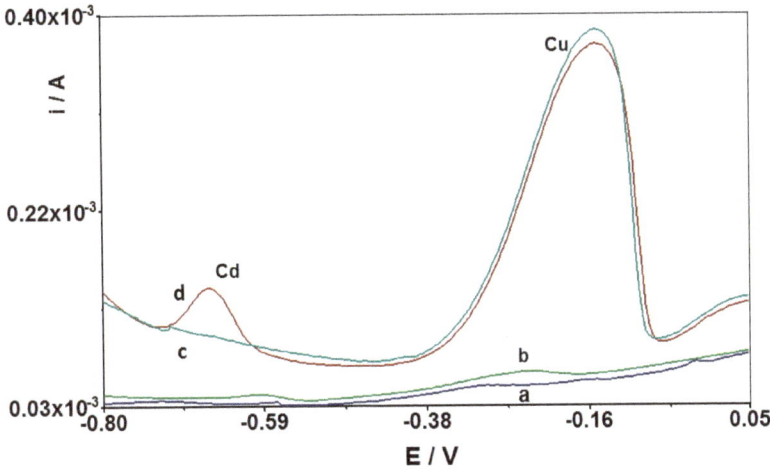

Figure 7. Comparison of differential pulse voltammograms obtained in the course of Cd(II) determination at the CuF/CN/SPE working electrode: (**a**) background, without addition of Cu(II) and Cd(II); (**b**) as (**a**) + 1×10^{-7} M Cd(II); (**c**) as (**a**) + 2×10^{-4} M Cu(II); and (**d**) as (**b**) + 2×10^{-4} M Cu(II). Accumulation potential and time was -0.7 V and 210 s, respectively.

Additionally, Figure 8 presents images of the morphology of CuF/CN/SPE electrode surface unmodified (A) and after copper film modification (B). The images obtained by scanning electron microscope display the effect of covering the working electrode surface with copper, and it was confirmed after comparison of the images of bare and in situ modified electrode surface. After the in situ deposition of copper bright points (clusters of copper) appeared on the electrode surface (Figure 8B). This was confirmed by EDX analysis, which revealed the presence of certain amounts of Cu on the modified electrode surface, and no Cu on the bare electrode surface. The results of the EDX analysis are shown in Figure 9.

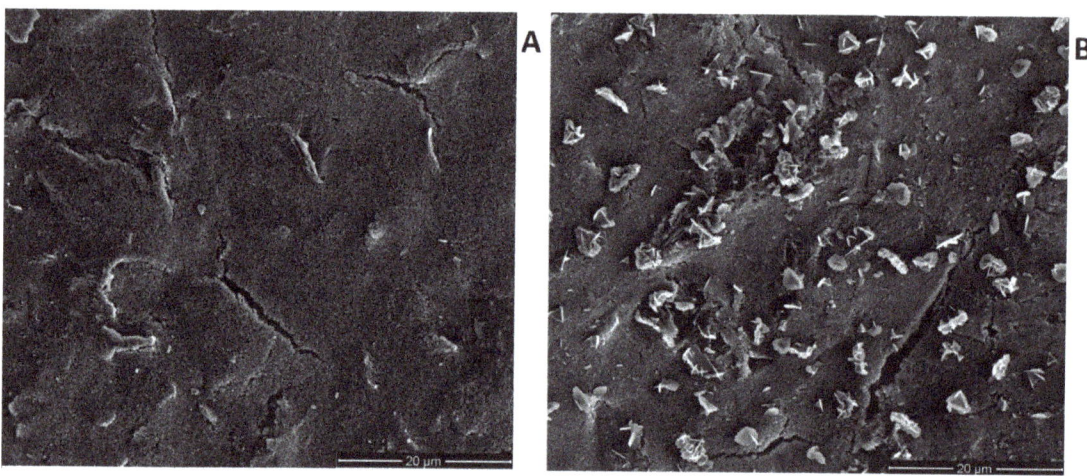

Figure 8. Images obtained through scanning electron microscope for the bare (**A**) and modified with copper film (**B**) CN/SPE working electrode. Copper film modified electrode was prepared in situ from the water solution containing 0.1 M HCl and 2×10^{-4} M Cu(II).

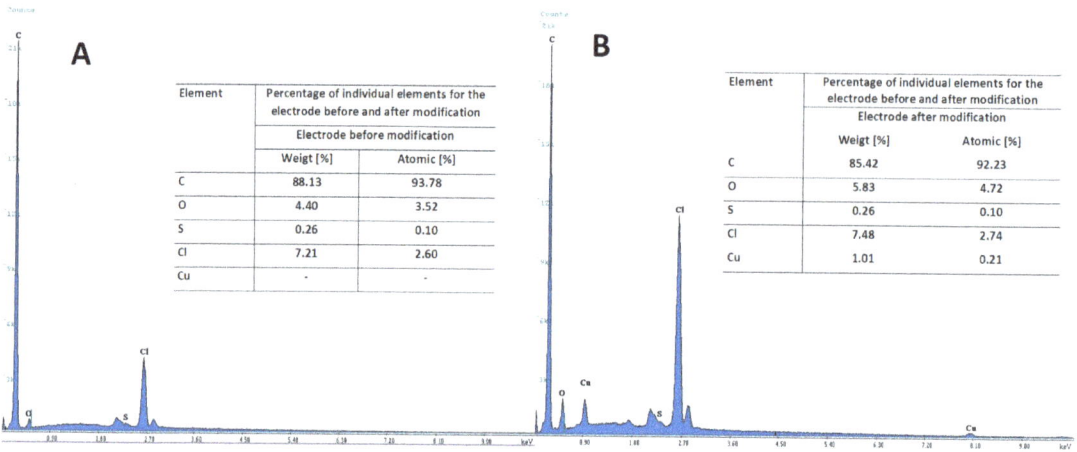

Figure 9. EDX spectra obtained for electrode non modified (**A**) and modified with copper (**B**) from the solution containing 0.1 M HCl and 2×10^{-4} M Cu(II).

As described earlier for CuF/GCE electrode in Section 3.2 using the Randles-Sevcik Equation (1), active surface areas of the CuF/CN/SPE electrode surfaces were calculated [42]. For the CuF/CN/SPE, geometric area of the surface was equal to 0.12560 cm², while the active surface area of the carbon nanotubes screen-printed electrode modified with copper equals to 0.04673 ± 0.00170 cm² (n = 3). The smaller active area than the geometric area of the electrode confirms the results obtained from morphology images and EDX analysis. The active sites of the electrode surface in this case are the copper sites, and the sites outside the copper are in active for cadmium accumulation. This is consistent with the voltammograms presented in Figure 7 confirming that, without copper on the electrode, cadmium does not undergo accumulation.

3.8.2. Analytical Parameters

The parameters influencing the Cd(II) signal height were optimized. The same parameters were tested as in the case of the CuF/GCE electrode: the pH and concentration of the supporting electrolyte, the concentration of copper, the deposition potential and time, and the pulse time and pulse height. The measurements were performed with a fixed concentration of Cd(II) 5×10^{-8} M. The selected composition of the test solution was the same as before: 0.1 M HCl, 2×10^{-4} M Cu(II). After the tests, it was confirmed that the most optimal cadmium signal was obtained for the same parameters as for GCE, and only a slight change can be made to the potential for accumulation of cadmium ions on the surface of the modified CuF/CN/SPE electrode, changing it to -0.75 V. At this potential, a slight improvement in the shape and height of the peak was obtained, but the -0.7 V potential, which generates equally high signals and is equally reproducible, can also be used successfully. The accumulation time remained the same as before and it was 60 s. Cyclic voltammetry (CV) analysis was also performed, and it was proven that the cadmium accumulation process on the working electrode is irreversible. An example voltammogram is shown in Figure 10.

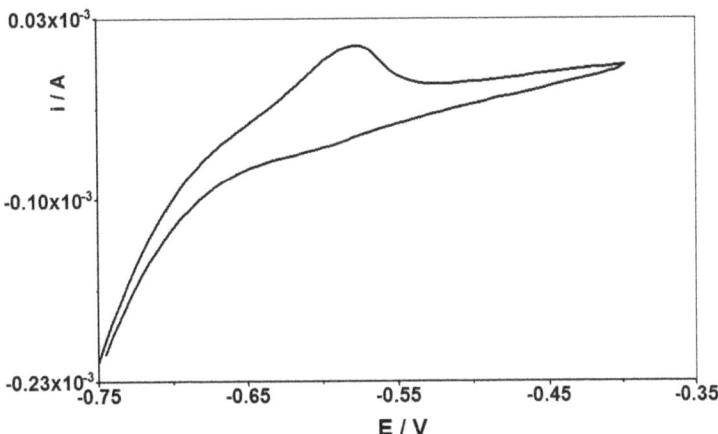

Figure 10. Cyclic voltammograms of 5×10^{-8} M Cd(II) in 0.1 M HCl and 2×10^{-4} M Cu(II) solution at the CuF/CN/SPE electrode, for scan rate of 1 V s^{-1}.

3.8.3. Analytical Characterization

The detection limit obtained in the case of the CuF/CN/SPE electrode was slightly lower from that for the CuF/GCE electrode, amounting 1.3×10^{-10} M, while the linearity range of the calibration curve ranged from 3×10^{-10} to 3×10^{-7} M with an accumulation time of 210 s and accumulation potential of -0.75 V. The equation of the calibration curve was equal to y = 0.333x + 0.396, where y is the peak current (µA) and x is Cd(II) concentration (nM) with correlation coefficient r = 0.999. As for the CuF/GCE electrode, the sensitivity was calculated for the CuF/CN/SPE electrode and was 7.126 µA nM^{-1} cm^{-2}. Figure 11 presents the comparison of the voltammograms obtained for the CuF/GCE and CuF/CN/SPE electrodes. The conducted research shows that the CuF/GCE or

CuF/CN/SPE novel modified electrodes can be applied interchangeably for the determination of cadmium using the voltammetric procedure developed in this study without loss in sensitivity or reproducibility of signals. Both electrodes give similar effects, but CuF/CN/SPE has a lower limit of detection and can be successfully used in field studies of real samples, which is an extremely important aspect in environmental analysis and a great advantage of the described research work. In addition, the CuF/CN/SPE electrode is readily commercially available, making it affordable for any scientist.

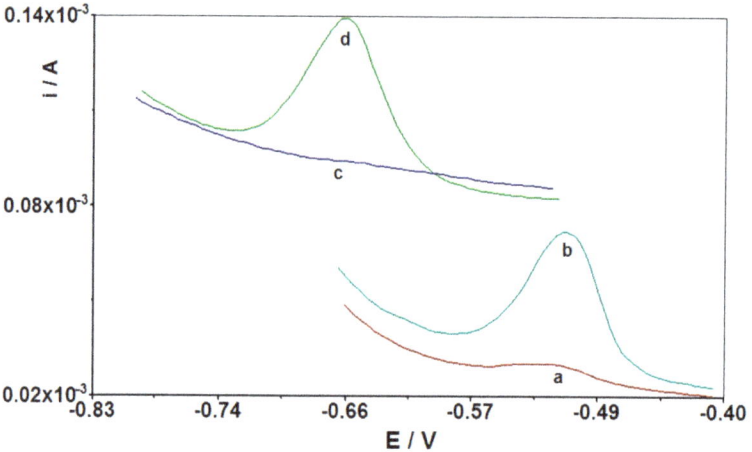

Figure 11. Comparison of differential pulse voltammograms obtained in the course of Cd(II) determination at the electrode CuF/GCE (**a,b**) and CuF/CN/SPE (**b,c**): (**a**) background for CuF/GCE; (**b**) as (**a**) + 1×10^{-7} M Cd(II); (**c**) background for CuF/CN/SPE; and (**d**) as (**c**) + 1×10^{-7} M Cd(II). Accumulation potential and time for CuF/GCE was −0.7 V and 210 s; accumulation potential and time for CuF/CN/SPE was −0.75 V and 210 s.

3.9. Analytical Application

In order to validate the developed procedure, tests were carried out with certified reference materials. The certified references materials TM-25.5 (environmental matrix), SPS-WW1 (waste water), and SPS-SW1 (surface water) were selected. The advantage of these materials is that they contain between 13 and 45 different trace elements, including cadmium. The cadmium concentration in these materials is 24 ng mL^{-1} (TM-25.5), 20 ng mL^{-1} (SPS-WW1), and 0.52 ng mL^{-1} (SPS-SW1). The concentration of the remaining components of the solutions ranged from 0.5 ng mL^{-1} to 2000 ng mL^{-1}; these matrices reflect the composition of environmental samples very well. The measurements were performed using the standard addition method. In the case of SPS-WW1 and SPS-SW1, and an appropriate amount of NaOH was additionally added to neutralize the solution as these materials contain nitric acid. All experiments were performed in five replicates. The recoveries were between 92.25% and 107.69%, whereas the relative standard deviations between 5.8% and 6.5%, which indicates good accuracy of the proposed method. Table 4 presents the results of Cd(II) determination in the certified reference materials.

To confirm the applicability of this procedure to the analysis of environmental samples, the proposed method was applied in the determination of Cd(II) in natural water samples collected from eastern areas of Poland. Tap water and rainwater were also tested. The voltammograms recorded for those samples did not exhibit any cadmium signal, which proves that the concentration of cadmium in the tested samples was below the limit of detection. To confirm the possibility of determining Cd(II) ions in such samples, the analyzed samples were spiked with cadmium. The standard addition method was used to calculate the recovery value. All experiments were carried out in five replicates. The recoveries were between 96.54% and 101.50%, whereas the relative standard deviations between 3.5% and 4.3%, which indicates good accuracy of the developed method. Table 5 presents the results of Cd(II) determination in natural water samples.

Table 4. Analytical results of Cd(II) determination in the certified reference materials without and with addition of Cd(II) ions. The samples were examined using the standard addition method.

Sample	Cd(II) Content in Certified Reference Material (ng mL^{-1})	Cd(II) Found in Certified Reference Material (ng mL^{-1})	Recovery (%)	RSD (n = 5) (%)
TM-25.5	24	22.74	94.75	6.5
SPS-SW1	0.52	0.56	107.69	5.8
SPS-WW1	20	18.45	92.25	5.8

Table 5. Analytical results of Cd(II) determination in natural water samples. The samples were examined using the standard addition method.

Sample	Cd(II) Added (nM)	Cd(II) Found (nM)	Recovery (%)	RSD (n = 5) (%)
Tap water	50	49.23	98.46	3.7
Rain water	50	50.75	101.50	3.5
Bystrzyca river water	50	48.65	97.30	4.2
Lake Zemborzyce	50	48.27	96.54	3.8
San river water	50	50.09	100.18	4.3

4. Conclusions

In this work, the authors present the applicability of the copper film modified glassy carbon electrode and the novel copper film with carbon nanotubes modified screen-printed electrode for anodic stripping voltammetric determination of trace concentrations of cadmium. This is the first work of this type devoted to the determination of cadmium ions on copper modified working electrodes. It was also the first time that the screen-printed electrode was successfully modified with copper. As it turned out, this approach allowed for significant reduction in the detection limits of ultratrace concentrations of cadmium ions. The CuF/GCE electrode was electrochemically deposited onto the glassy carbon solid electrode with simultaneous accumulation of Cd(II) ions. The CuF/CN/SPE electrode combines a solid electrode modified with carbon nanotubes, a platinum auxiliary electrode, and a silver reference electrode in the microcircuit. The advantage of these microelectrodes is better sensitivity and the possibility of using them in field experiments. They are also readily available in many variants depending on the needs of the researcher. Screen-printed electrodes have become an attractive analytical tool also due to the low production costs, appropriate levels of repeatability and their electrochemical properties. The use of the in situ created film electrode reduced the need for medium exchange after deposition of the metal, which significantly shortens the measurement time and reduces the consumption of chemical reagents. This method is simple, cheap, sensitive, selective, and fast, and does not require complicated apparatus. A very low detection limit for cadmium was obtained compared with other voltammetric techniques for the determination of Cd(II) ions (Table 1). In the literature we did not find any papers on the determination of cadmium on the CuF/GCE and with carbon nanotubes modified copper CuF/CN/SPE electrodes. The influence of various interfering substances, such as foreign metal ions and organic substances, was tested to check whether it would be possible to perform tests in aqueous environmental samples. Very satisfactory results were obtained, and in some cases the method using Amberlite XAD-7 resin with adsorptive properties was employed, which significantly improved cadmium recovery from the samples containing surfactants and humic substances. The influence of temperature on the performance of the modified electrodes was investigated, and it was proven that the cadmium signal is stable and the electrodes are perfectly reproducible over a wide temperature range. The method appears to be promising for its adoption in environmental research and field analysis.

Author Contributions: Conceptualization, M.G.; methodology, M.G. and J.W.; software, M.G.; validation, J.W.; formal analysis, M.G. and J.W.; investigation, J.W.; resources, M.G. and J.W.; data curation, J.W.; writing—original draft preparation, J.W.; writing—review and editing, M.G. and J.W.; visualization, J.W.; supervision, M.G.; project administration, J.W. All authors have read and agreed to the published version of the manuscript.

Funding: This research received no external funding.

Institutional Review Board Statement: Not applicable.

Informed Consent Statement: Not applicable.

Data Availability Statement: Data are available in a publicly accessible repository that does not issue DOIs; data are contained within this article. The data received in the project will be stored on a PC, and their backups will be stored on mobile devices (external drive). The official PC is connected to the KUL network, administered by DTI employees. Updates are performed daily. All data will be recorded in laboratory notes (during individual measurements, before entering them into a spreadsheet). The notebooks will be stored in the offices of those responsible for the implementation of the project. Raw data will be stored for a minimum of 10 years from the end of the project implementation.

Conflicts of Interest: The authors declare no conflict of interest.

References

1. Jovanovski, V.; Hrastnik, N.; Hočevar, S. Copper film electrode for anodic stripping voltammetric determination of trace mercury and lead. *Electrochem. Commun.* **2015**, *57*, 1–4. [CrossRef]
2. Kolar, M.; Oražem, T.; Jovanovski, V.; Hočevar, S.B. Copper film electrode for sensitive detection of nitrophenols. *Sens. Actuators B Chem.* **2020**, *330*, 129338. [CrossRef]
3. Gregory, B.W.; Norton, M.L.; Stickney, J.L. Thin-layer electrochemical studies of the underpotential deposition of cadmium and tellurium on polycrystalline Au, Pt and Cu electrodes. *J. Electroanal. Chem. Interfacial Electrochem.* **1990**, *293*, 85–101. [CrossRef]
4. Jovanovski, V.; Hrastnik, N. Insights into the anodic stripping voltammetricbehaviour of copper film electrodes for determination of trace mercury. *Microchem. J.* **2019**, *146*, 895–899. [CrossRef]
5. Oliveira-Roberth, A.; Santos, D.I.V.; Cordeiro, D.D.; Lino, F.M.D.A.; Bara, M.T.F.; Gil, E.D.S.; De Oliveira-Roberth, A. Voltammetric determination of Rutin at Screen-Printed carbon disposable electrodes. *Open Chem.* **2012**, *10*, 1609–1616. [CrossRef]
6. Apetrei, I.M.; Apetrei, C. Study of Different Carbonaceous Materials as Modifiers of Screen-Printed Electrodes for Detection of Catecholamines. *IEEE Sens. J.* **2014**, *15*, 3094–3101. [CrossRef]
7. Sasal, A.; Tyszczuk-Rotko, K. Screen-printed sensor for determination of sildenafil citrate in pharmaceutical preparations and biological samples. *Microchem. J.* **2019**, *149*, 104065. [CrossRef]
8. Porada, R.; Fendrych, K.; Baś, B. Development of novel Mn-zeolite/graphite modified Screen-printed Carbon Electrode for ultrasensitive and selective determination of folic acid. *Measurement* **2021**, *179*, 109450. [CrossRef]
9. Sullivan, C.; Lu, D.; Senecal, A.; Kurup, P. Voltammetric detection of arsenic (III) using gold nanoparticles modified carbon screen printed electrodes: Application for facile and rapid analysis in commercial apple juice. *Food Chem.* **2021**, *352*, 129327. [CrossRef]
10. Tavares, A.P.; de Sá, M.H.; Sales, M.G.F. Innovative screen-printed electrodes on cork composite substrates applied to sulfadiazine electrochemical sensing. *J. Electroanal. Chem.* **2020**, *880*, 114922. [CrossRef]
11. Salimi, A.; Hallaj, R. Catalytic oxidation of thiols at preheated glassy carbon electrode modified with abrasive immobilization of multiwall carbon nanotubes: Applications to amperometric detection of thiocytosine, l-cysteine and glutathione. *Talanta* **2005**, *66*, 967–975. [CrossRef] [PubMed]
12. Gęca, I.; Korolczuk, M. Anodic stripping voltammetry following double deposition and stripping steps: Application of a new approach in the course of lead ion determination. *Talanta* **2017**, *171*, 321–326. [CrossRef]
13. Adamczyk, M.M.; Grabarczyk, M. Simple and Fast Simultaneously Determination of In(III) and Ti(IV) Using Lead Modified Glassy Carbon Electrode by Stripping Voltammetry. *J. Electrochem. Soc.* **2020**, *167*, 126515. [CrossRef]
14. Wasąg, J.; Grabarczyk, M. A Fast and Simple Voltammetric Method Using a Lead Film Electrode for Determination of Ultra-Trace Concentration of Titanium in Environmental Water Samples. *J. Electrochem. Soc.* **2016**, *163*, H1076–H1080. [CrossRef]

15. Yakobson, B.; Smalley, R.E. Some unusual new molecules—Long, hollow fibers with tantalizing electronic and mechanical properties—Have joined diamonds and graphite in the carbon family. *Am. Sci.* **1997**, *85*, 324–337.
16. Agüí, L.; Yáñez-Sedeño, P.; Pingarrón, J.M. Role of carbon nanotubes in electroanalytical chemistry: A review. *Anal. Chim. Acta* **2008**, *622*, 11–47. [CrossRef]
17. Acharya, J.; Raj, B.G.S.; Ko, T.H.; Khil, M.S.; Kim, H.Y.; Kim, B.S. Facile one pot sonochemical synthesis of CoFe2O4/MWCNTs hybrids with well-dispersed MWCNTs for asymmetric hybrid supercapacitor applications. *Int. J. Hydrogen Energy* **2019**, *45*, 3073–3085. [CrossRef]
18. Poudel, M.B.; Awasthi, G.P.; Kim, H.J. Novel insight into the adsorption of Cr(VI) and Pb(II) ions by MOF derived Co-Al layered double hydroxide @hematite nanorods on 3D porous carbon nanofiber network. *Chem. Eng. J.* **2021**, *417*, 129312. [CrossRef]
19. Ojha, G.P.; Muthurasu, A.; Tiwari, A.P.; Pant, B.; Chhetri, K.; Mukhiya, T.; Dahal, B.; Lee, M.; Park, M.; Kim, H.-Y. Vapor solid phase grown hierarchical CuxO NWs integrated MOFs-derived CoS2 electrode for high-performance asymmetric supercapacitors and the oxygen evolution reaction. *Chem. Eng. J.* **2020**, *399*, 125532. [CrossRef]
20. Jakubowski, M.; Muszer, J.; Slota, P. The risk of exposure on cadmium in chosen the branches of industry. *Med. Pracy* **1995**, *46*, 109–122.
21. Satarug, S.; Baker, J.R.; Urbenjapol, S.; Haswell, M.; Reilly, P.E.; Williams, D.J.; Moore, M.R. A global perspective on cadmium pollution and toxicity in non-occupationally exposed population. *Toxicol. Lett.* **2002**, *137*, 65–83. [CrossRef]
22. Ostrowska, P. Cadmium, occurrence, pollution sources and recycling methods. *Miner. Resour. Manag.* **2008**, *24*, 255–260.
23. Duarte, K.; Justino, C.I.L.; Freitas, A.C.; Gomes, A.M.P.; Duarte, A.C.; Rocha-Santos, T.A.P. Disposable sensors for environmental monitoring of lead, cadmium and mercury. *Trends Anal. Chem.* **2014**, *64*, 183–190. [CrossRef]
24. Grabarczyk, M.; Wasąg, J. Ultratrace Determination of Indium in Natural Water by Adsorptive Stripping Voltammetry in the Presence of Cupferron as a Complexing Agent. *J. Electrochem. Soc.* **2016**, *163*, H218–H222. [CrossRef]
25. Nodehi, M.; Baghayeri, M.; Veisi, H. Preparation of GO/Fe3O4@PMDA/AuNPs nanocomposite for simultaneous determination of As3+ and Cu2+ by stripping voltammetry. *Talanta* **2021**, *230*, 122288. [CrossRef] [PubMed]
26. Sanchayanukun, P.; Muncharoen, S. Chitosan coated magnetite nanoparticle as a working electrode for determination of Cr(VI) using square wave adsorptive cathodic stripping voltammetry. *Talanta* **2020**, *217*, 121027. [CrossRef]
27. Baś, B.; Jedlińska, K.; Węgiel, K. New electrochemical sensor with the renewable silver annular band working electrode: Fabrication and application for determination of selenium(IV) by cathodic stripping voltammetry. *Electrochem. Commun.* **2014**, *49*, 79–82. [CrossRef]
28. Zuziak, J.; Jakubowska, M. Voltammetric determination of aluminum-Alizarin S complex by renewable silver amalgam electrode in river and waste waters. *J. Electroanal. Chem.* **2017**, *794*, 49–57. [CrossRef]
29. Grabarczyk, M.; Koper, A. Direct Determination of Cadmium Traces in Natural Water by Adsorptive Stripping Voltammetry in the Presence of Cupferron as a Chelating Agent. *Electroanalysis* **2011**, *24*, 33–36. [CrossRef]
30. Abbasi, S.; Bahiraei, A.; Abbasai, F. A highly sensitive method for simultaneous determination of ultra trace levels of copper and cadmium in food and water samples with luminol as a chelating agent by adsorptive stripping voltammetry. *Food Chem.* **2011**, *129*, 1274–1280. [CrossRef]
31. Rojas-Romo, C.; Aliaga, M.E.; Arancibia, V.; Gomez, M. Determination of Pb(II) and Cd(II) via anodic stripping voltammetry using an in-situ bismuth film electrode. Increasing the sensitivity of the method by the presence of Alizarin Red S. *Microchem. J.* **2020**, *159*, 105373. [CrossRef]
32. Lima, T.M.; Soares, P.I.; Nascimento, L.A.D.; Franco, D.L.; Pereira, A.C.; Ferreira, L.F. A novel electrochemical sensor for simultaneous determination of cadmium and lead using graphite electrodes modified with poly(p-coumaric acid). *Microchem. J.* **2021**, *168*, 106406. [CrossRef]
33. Dali, M.; Zinoubi, K.; Chrouda, A.; Abderrahmane, S.; Cherrad, S.; Jaffrezic-Renault, N. A biosensor based on fungal soil biomass for electrochemical detection of lead (II) and cadmium (II) by differential pulse anodic stripping voltammetry. *J. Electroanal. Chem.* **2018**, *813*, 9–19. [CrossRef]
34. Pizarro, J.; Segura, R.; Tapia, D.; Navarro, F.; Fuenzalida, F.; Aguirre, M.J. Inexpensive and green electrochemical sensor for the determination of Cd(II) and Pb(II) by square wave anodic stripping voltammetry in bivalve mollusks. *Food Chem.* **2020**, *321*, 126682. [CrossRef]
35. Zhang, Y.; Li, C.; Su, Y.; Mu, W.; Han, X. Simultaneous detection of trace Cd(II) and Pb(II) by differential pulse anodic stripping voltammetry using a bismuth oxycarbide/nafion electrode. *Inorg. Chem. Commun.* **2019**, *111*, 107672. [CrossRef]
36. Adamczyk, M.; Grabarczyk, M. Simple, insensitive to environmental matrix interferences method of trace cadmium determination in natural water samples. *Ionics* **2019**, *25*, 1959–1966. [CrossRef]
37. Lisboa, T.P.; de Faria, L.V.; Matos, M.A.C.; Matos, R.C.; de Sousa, R.A. Simultaneous determination of cadmium, lead, and copper in the constituent parts of the illegal cigarettes by Square Wave Anodic Stripping Voltammetry. *Microchem. J.* **2019**, *150*, 104183. [CrossRef]
38. Pandey, S.K.; Sachan, S.; Singh, S.K. Ultra-trace sensing of cadmium and lead by square wave anodic stripping voltammetry using ionic liquid modified graphene oxide. *Mater. Sci. Energy Technol.* **2019**, *2*, 667–675. [CrossRef]
39. Sreekanth, S.; Alodhayb, A.; Assaifan, A.K.; Alzahrani, K.E.; Muthuramamoorthy, M.; Alkhammash, H.I.; Pandiaraj, S.; Alswieleh, A.M.; Van Le, Q.; Mangaiyarkarasi, R.; et al. Multi-walled carbon nanotube-based nanobiosensor for the detection of cadmium in water. *Environ. Res.* **2021**, *197*, 111148. [CrossRef] [PubMed]

40. Dahaghin, Z.; Kilmartin, P.; Mousavi, H.Z. Simultaneous determination of lead(II) and cadmium(II) at a glassy carbon electrode modified with GO@Fe$_3$O$_4$@benzothiazole-2-carboxaldehyde using square wave anodic stripping voltammetry. *J. Mol. Liq.* **2018**, *249*, 1125–1132. [CrossRef]
41. Siriangkhawut, W.; Pencharee, S.; Grudpan, K.; Jakmunee, J. Sequential injection monosegmented flow voltammetric determination of cadmium and lead using a bismuth film working electrode. *Talanta* **2009**, *79*, 1118–1124. [CrossRef] [PubMed]
42. Sipa, K.; Brycht, M.; Leniart, A.; Urbaniak, P.; Nosal-Wiercińska, A.; Pałecz, B.; Skrzypek, S. β–Cyclodextrins incorporated multi-walled carbon nanotubes modified electrode for the voltammetric determination of the pesticide dichlorophen. *Talanta* **2018**, *176*, 625–634. [CrossRef] [PubMed]
43. Mohapatra, D.; Gowthaman, N.; Sayed, M.S.; Shim, J.-J. Simultaneous ultrasensitive determination of dihydroxybenzene isomers using GC electrodes modified with nitrogen-doped carbon nano-onions. *Sens. Actuators B* **2020**, *304*, 127325. [CrossRef]
44. Gugała-Fekner, D.; Nieszporek, J.; Sieńko, D. Adsorption of anionic surfactant at the electrode-NaClO4 solution interface. *Mon. Chemie-Chem. Monthly* **2015**, *146*, 541–545. [CrossRef]

Article

The Influence of Graphene Oxide Composition on Properties of Surface-Modified Metal Electrodes

Natalia Festinger [1], Aneta Kisielewska [2], Barbara Burnat [3], Katarzyna Ranoszek-Soliwoda [2], Jarosław Grobelny [2], Kamila Koszelska [3], Dariusz Guziejewski [3] and Sylwia Smarzewska [3,*]

1. Łukasiewicz Research Network-Lodz Institute of Technology, Maria Skłodowska-Curie 19/27, 90-570 Lodz, Poland
2. Department of Materials Technology and Chemistry, Faculty of Chemistry, University of Lodz, Pomorska 163, 90-236 Lodz, Poland
3. Department of Inorganic and Analytical Chemistry, Faculty of Chemistry, University of Lodz, Tamka 12, 91-403 Lodz, Poland
* Correspondence: sylwia.smarzewska@chemia.uni.lodz.pl

Abstract: The present paper describes the effect of the concentration of two graphene oxides (with different oxygen content) in the modifier layer on the electrochemical and structural properties of noble metal disk electrodes used as working electrodes in voltammetry. The chemistry of graphene oxides was tested using EDS, FTIR, UV–Vis spectroscopy, and combustion analysis. The structural properties of the obtained modifier layers were examined by means of scanning electron and atomic force microscopy. Cyclic voltammetry was employed for comparative electrochemical studies.

Keywords: gold electrode; platinum electrode; graphene oxide; reduced graphene oxide; voltammetry

Citation: Festinger, N.; Kisielewska, A.; Burnat, B.; Ranoszek-Soliwoda, K.; Grobelny, J.; Koszelska, K.; Guziejewski, D.; Smarzewska, S. The Influence of Graphene Oxide Composition on Properties of Surface-Modified Metal Electrodes. *Materials* 2022, *15*, 7684. https://doi.org/10.3390/ma15217684

Academic Editor: Alexander N. Obraztsov

Received: 22 July 2022
Accepted: 29 October 2022
Published: 1 November 2022

Publisher's Note: MDPI stays neutral with regard to jurisdictional claims in published maps and institutional affiliations.

Copyright: © 2022 by the authors. Licensee MDPI, Basel, Switzerland. This article is an open access article distributed under the terms and conditions of the Creative Commons Attribution (CC BY) license (https:// creativecommons.org/licenses/by/ 4.0/).

1. Introduction

Electrochemical methods are based on measurement of the current associated with the molecular properties and interfacial processes of chemical species. The recorded response results from direct transformation of the desired chemical information (concentration, activity) into a current signal (potential, current, resistance, or capacity), according to the selected method. Voltammetry is considered as one of the most sensitive electroanalytical methods, suitable for the determination of trace amounts of many metals and compounds in clinical, industrial, and environmental samples [1–4]. Various voltammetric techniques provide a wealth of chemical, electrochemical, and physical information, such as quantitative analysis, diffusion and reaction rate constants, and number of electrons involved in redox reactions [5,6]. The effectiveness of voltammetric procedures is strongly influenced by the working electrode material [7,8]. The working electrode should provide a high signal-to-noise ratio as well as reproducible signals. Thus, electrode selection depends mainly on the redox behavior of the target analyte, electrical conductivity, surface reproducibility, and background current over the potential window required for measurement. A range of materials have found application as working electrodes in electroanalysis. The most popular types contain mercury, carbon, or noble metals [8]. Among the noble metals, platinum and gold are the most widely used for metallic electrodes, as they offer very favorable electron transfer kinetics and a large available potential range [8]. In contrast, the low hydrogen overvoltage on those electrodes limits the cathodic potential window. Additionally, high background currents associated with the formation of surface oxides or adsorbed hydrogen layers can cause problems. Such films can also strongly alter the kinetics of electrode reaction, leading to irreproducible data [8]. Compared with platinum electrode, the gold one is more inert, and hence, less prone to the formation of stable oxide films or surface contamination. The abovementioned difficulties can be addressed by modifying the surface of platinum and gold electrodes with a specific modifier layer. Still, noble metals used as starting materials for electrodes are well-known for their versatile usage and applications, also in sensing and as catalysts [9,10]. The occurrence of stable oxide films at the electrode surface together with enhanced physical and electrocatalytical properties tuned by the presence of graphene oxides makes this combination a primary choice for electroanalytical purposes.

Graphene, which consists of a one-atom-thick planar sheet containing an sp2-bonded carbon structure with exceptionally high crystalline and electronic quality, is a novel material that has emerged as a rapidly rising star in the field of material science [11,12]. Ever since its discovery in 2004 [13], graphene has been making a profound impact in many areas of science and technology due to its remarkable physicochemical properties. These include a high specific surface area [14], extraordinary electronic properties and electron transport capabilities [15], unprecedented pliability [16] and impermeability, high mechanical strength [17], and excellent thermal and electrical conductivity [18]. One branch of graphene research deals with graphene oxide (GO), which is a precursor in graphene synthesis by either chemical or thermal reduction processes. One of the advantages of graphene oxide is easy dispersibility in water and other organic solvents, as well as in different matrices, due to the presence of oxygen functionalities. This property is very important when it is mixed with other materials with a view to improving their electrical and mechanical properties. On the other hand, in terms of electrical conductivity, graphene oxide is often described as an electrical insulator due to the disruption of its sp2 bonding networks. In order to recover the honeycomb hexagonal lattice—and with it, electrical conductivity—graphene oxide must be reduced. However, once most oxygen groups are removed, the obtained reduced product is more difficult to disperse due to its tendency to aggregate. It is worth noting that graphene oxide and graphene have attracted exceptional attention from the scientific community. Concerning 2022 (data for 15 July 2022), a keyword query in the Scopus database revealed 12,194 and 4258 research papers concerning graphene and graphene oxide, respectively. Those publications can certainly be very inspiring, but also frustrating, if other research teams fail to reproduce the results. A lack of research reproducibility has always been a major issue in the scientific community. With graphene oxide and reduced graphene oxide, the situation is very complicated. While each single carbon layer containing oxygen groups is called graphene oxide, material obtained after GO reduction is called reduced graphene oxide (smaller quantity of oxygen groups after GO reduction is usually confirmed by spectroscopic measurements). Indeed, it cannot be excluded that an oxide synthesized in one laboratory as graphene oxide is structurally similar to reduced graphene oxide obtained by another research team. A similar problem occurs when purchasing graphene oxide from different suppliers. This is attributable to the fact that the precise atomic structure of GO still remains uncertain and perfect stoichiometry has never been achieved [19]. The study of GO structure is derived from the structural analysis of graphite oxide. Over the years, considerable efforts have been directed toward understanding the structure of that compound with the result that several conflicting explanations have been successively proposed. In 1939, Hofmann and Holst [20] developed a simple model in which graphite oxide was thought to consist of planar carbon layers modified with an epoxy (1,2-ether) group, with an overall molecular formula of C_2O. Seven years later, Ruess [21] suggested that the carbon layers were not in fact planar but puckered and that the oxygen-containing groups were hydroxyl and ether-like oxygen bridges, randomly distributed on the carbon skeleton. In order to account for the acidic properties of graphite oxide, Hofmann [22] proposed enol- and keto-type structures, which also contained hydroxyls and ether bridges. In 1969, Scholz and Boehm [23] presented a GO model in which epoxide and ether groups were completely replaced by carbonyl and hydroxyl groups. According to Nakajima et al. [24], GO consisted of two carbon layers linked to each other by sp3 carbon–carbon bonds. Szabó [25] proposed a new structural model that involves a carbon network consisting of trans-linked cyclohexane chairs and ribbons of flat hexagons with C=C double bonds as well as functional groups such as tertiary OH, 1,3-ether, ketone, quinone, and phenol (aromatic diol). One of the few common features of all the published models is the presence of various oxygen-containing functional groups in GO. It is known that oxygenated groups (e.g., hydroxyl, epoxy, carboxyl, carbonyl, phenol, lactone, and quinone) can strongly affect the electronic, mechanical, and electrochemical properties of GO [26–28]. In recent years, research on GO-based materials has been extensive, particularly with respect to their electrochemical applications [29–34]. In such studies, GO was produced by different methods, which were sometimes applied interchangeably as it was assumed that they all led to the same reaction product. Nowadays, it is known that the type and quantity of oxygen groups depends largely on the synthesis method. GO is generally produced by synthesis with concentrated H_2SO_4 along with: (1) sodium nitrate for in situ production of nitric acid in the presence of $KMnO_4$ (Hummers' method); (2) fuming nitric acid and a $KClO_3$ oxidant (Staudenmaier method); (3) concentrated phosphoric acid with $KMnO_4$ (Tour method); or (4) concentrated nitric acid and a $KClO_3$ oxidant (Hoffmann method) [35–38]. However, the obtained oxides differ significantly in the number of oxygen-containing groups (C/O ratio) as well as in terms of the types of oxygenated carbon bonds present. Generally speaking, oxides prepared using $KClO_3$ as an oxidant (Staudenmaier's and Hofmann's methods) exhibit a higher C/O ratio, whereas methods employing $KMnO_4$ (Hummers and Tour) yield a larger proportion of oxygen-containing groups. Oxidative methods employing $KClO_3$ result in oxides containing mostly CO groups (hydroxyl, epoxy), whereas methods using $KMnO_4$ lead to oxides containing large amounts of carbonyl (Hummers) or both carbonyl and carboxyl groups (Tours) [39].

The aim of this study was to compare the properties of noble metal electrodes modified with graphene oxide and to determine how graphene oxide chemical composition and concentration affect their electrochemical and structural (morphology, topography, roughness) properties.

2. Materials and Methods

2.1. Apparatus and Solutions

The surface topography of the studied electrodes was investigated using an atomic force microscope (AFM, Dimension Icon, Bruker Corporation, Billerica, MA, USA). The AFM measurements with the scan size of 5 μm × 5 μm were performed in the tapping mode using silicon scanning probe (TESPA-V2, Bruker AFM Probes) with a nominal spring constant of 42 N/m and resonance frequency of 320 kHz. The roughness parameters Ra and Rq were defined on the basis of AFM topography images (average values taken from 256 surface profiles). The surface morphology and elemental composition of electrodes was investigated with high-resolution scanning electron microscopy (Nova NanoSEM 450, FEI, Hillsboro, OR, USA) equipped in a Schottky field emission electron emitter. The surface morphology measurements were performed using CBS detector, enabling the observation of surface using signal of backscattered electrons (BSE). The images were recorded at 3 kV with electron beam deceleration 4 kV, spot size 2.5, and working distance 6.7 mm. The chemical elemental analysis of GO I- and GO II-modified gold electrodes was determined with EDAX Roentgen spectrometer (EDS) with Octane Pro Silicon Drift Detector (SDD) (AMELTEK, Berwyn, PA, USA). Attenuated total reflection Fourier transform infrared (ATR–FTIR) spectra were recorded on a Nicolet iS50 spectrometer equipped (Thermo Scientific, Madison, WI, USA) with an MCT detector and the GATR accessory with Ge crystal over the spectral region from 600 to 4000 cm^{-1} with a resolution of 4 cm^{-1}. Electroanalytical measurements were carried out using a μAutolab instrument (EcoChemie, The Netherlands) controlled by GPES 4.9 electrochemical software. The three-electrode electrochemical cell employed in the study consisted of a reference electrode, an auxiliary electrode (platinum wire), and a working disk electrode (platinum or gold—please notice that experiments with platinum and gold electrodes are performed with different potential windows).

The potential of the working electrode was measured vs. an Ag/AgCl electrode. Double-distilled water was used throughout the experiments. All the chemicals, including graphene oxide (GO I), reduced graphene oxide (GO II), and hexacyanoferrate system were purchased from Sigma-Aldrich and used as received.

The GO I/GO II suspension was prepared weekly by dispersing an appropriate amount of GO I/GO II powder in DMF (dimethylformamide) in a 5 mL volumetric flask. The resultant solution was kept in a refrigerator at 4 °C in the dark. Spectrophotometric measurements were made using a Cary 100 Bio UV–Vis spectrophotometer (Agilent, Santa Clara, CA, USA).

2.2. Measurement Procedure

The general procedure used to obtain voltammograms was as follows: Ten mL of supporting electrolyte was placed in the voltammetric cell and the solution was purged with argon for 10 min (if necessary). After recording the initial blank, the required volumes of the analyte were added by means of a micropipette. Then, the solution was deoxygenated for 10 s (if necessary), and a voltammogram was recorded. All electrochemical measurements were carried out at ambient temperature. Each measurement was repeated three times and a mean was calculated.

In order to obtain FTIR spectra of both graphene derivatives, DMF solutions of GO I and GO II at a concentration of 20.0 g L^{-1} were prepared. Afterwards, 20 µL of each solution was deposited on alumina foil by sessile drop technique and, subsequently, the samples were allowed to evaporate the solvent. Final spectra were received by the addition of 64 scans at a resolution of 4 cm^{-1}.

2.3. Preparation of Working Electrodes

Modifying solutions containing graphene oxides were prepared with dimethylformamide. Suspensions at concentrations ranging from 1.0 to 50.0 g L^{-1} were prepared for each oxide. Prior to modification with GO I or GO II, the working electrode surface was sonicated in ethanol for 60 s. Next, working electrode surface was mechanically polished with 0.05 µm Al_2O_3 slurry on a polishing cloth to a mirror finish. Then, it was ultrasonically treated in ethanol for 180 s and washed with double-distilled water. The modifier suspension (3 µL) was dropped onto the surface of the cleaned electrode and dried at ambient temperature. A new modifier film was prepared before each series of measurements.

3. Results and Discussion

3.1. Electrochemical Studies

Voltammetric measurements provide a wealth of information concerning the properties and characteristics of electrochemical processes. Cyclic voltammetry is an important and widely used technique determining analyte electrochemical behavior, including the formal redox potential, thermodynamic and transport properties, electron transfer kinetics, and adsorption processes. The hexacyanoferrate(II)/(III) redox couple undergoes a nearly reversible electrode reaction without any complications of preceding or post-chemical reactions; so, it has been a popular choice as a redox standard in CV. Electrochemical measurements were performed at modified working electrodes using the $Fe(CN)_6^{3-}/Fe(CN)_6^{4-}$ redox couple and suspensions of graphene oxides (GO I and GO II) in a concentration range from 1.0 to 50.0 g L^{-1}. Figure 1 shows a voltammogram of the $Fe(CN)_6^{3-}/Fe(CN)_6^{4-}$ redox couple recorded on bare and modified gold (A) and platinum (B) electrodes. Both working electrodes were modified with graphene oxide suspensions at a concentration of 1.0 g L^{-1}.

The strongest signals at both electrodes were observed for the electrode surface-modified with GO I. Significant difference may be observed for GO II suspension, which influence tested working electrodes in the opposite way. As can be seen in Figure 1, model redox system signals recorded on platinum electrode modified with GO II exhibit very good morphology and increased current in comparison with bare electrode. At gold electrodes, such modification led to signals that were weaker even than those recorded at the bare electrode. This suggests that some concentrations of the modifier layer may block the electrode surface causing significant deterioration of analytical parameters. According to this, additional experiments were made for Au electrode and it was concluded that such behavior is observed for GO II suspension concentration 1.5 g L^{-1} and lower. The relationships between hexacyanoferrate anodic peak current and the surface concentration of GO I recorded at both Au and Pt electrodes are shown in Figure 2A. As can be seen, both dependencies have a similar course, with a maximum for electrodes modified with 20.0 g L^{-1} GO I suspension. An analogous study using GO II suspensions also revealed the strongest signals of the Fe(CN)$_6^{3-}$/Fe(CN)$_6^{4-}$ redox couple for the 20.0 g L^{-1} concentration at both Pt and Au electrodes. Figure 2B shows Fe(CN)$_6^{3-}$/Fe(CN)$_6^{4-}$ voltammograms recorded at the GO II-modified platinum electrode.

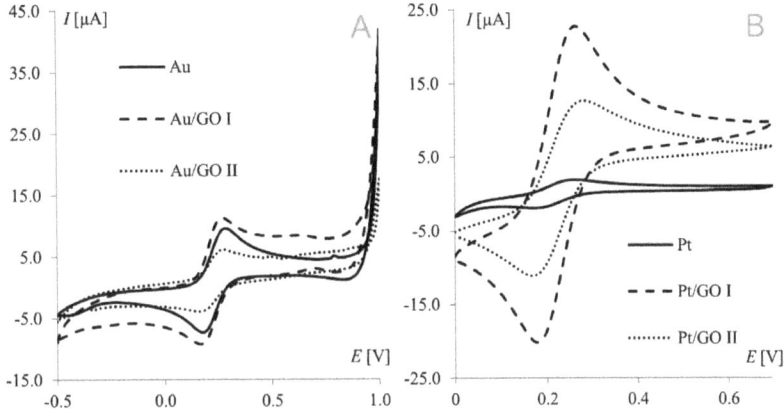

Figure 1. Voltammograms of 1 mM Fe(CN)$_6^{3-}$/Fe(CN)$_6^{4-}$ recorded from bare, graphene oxide I-, and graphene oxide II-modified working electrodes ((**A**) gold electrode, (**B**) platinum electrode); scan rate, 50 mV s^{-1}; supporting electrolyte, 1 M KCl.

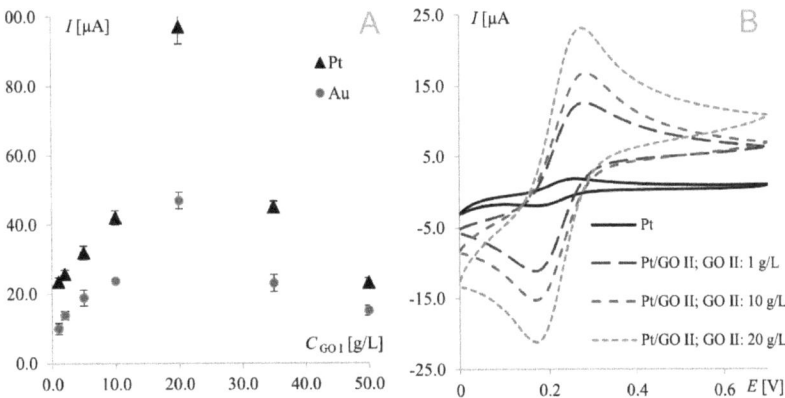

Figure 2. (**A**) Relationship between 1 mM Fe(CN)$_6^{4-}$ anodic peak current recorded at the platinum/gold electrode modified with graphene oxide I suspensions prepared in dimethylformamide and GO I concentration (GO I concentrations from 1 to 50.0 g L^{-1}). (**B**) Voltammograms of 1.0 mM Fe(CN)$_6^{3-}$/Fe(CN)$_6^{4-}$ recorded at the bare platinum electrode and at the electrode modified with graphene oxide II; scan rate, 50 mV s^{-1}; supporting electrolyte, 1 M KCl.

It is worth noting that the highest currents recorded at electrodes modified with 20.0 g L^{-1} GO II (17 µA and 14 µA at Pt and Au, respectively) were much lower than those obtained at electrodes modified with 20.0 g L^{-1} GO I (97 µA and 47 µA for Pt and Au, respectively). Since GO II was purchased as reduced graphene oxide and RGO has fewer oxygen groups (than GO), which improves its electrical conductivity, the opposite results were expected (oxygen content in both GOs will be thoroughly discussed in Section 3.3). However, this should be discussed not only from the point of view of peak current but also the shape of the voltammograms (Figure 3). In voltammetry, the current signal is produced by two different currents: the first one (faradic) corresponds to analyte oxidation or reduction, with its magnitude depending on analyte concentration in solution and all kinetic steps occurring at the electrode (electron-transfer process); the second one (capacitive) is generated by the "electric double layer" at the electrode–solution interface and, as such, is unrelated to the electron transfer process. As it is impossible to completely separate the two types of current, techniques and electrodes minimizing the influence of capacitive current are sought. As can be seen in Figure 3, gold electrodes modified with GO I do not seem to meet these requirements. The characteristic rectangular shape of voltammograms recorded on GO I layer is a favorable and desirable phenomenon in high-performance supercapacitors research but not in electroanalysis [40]. It is worth noting that such capacitive performance is connected with partial restoration of π-conjugation structure and improved electronic conductivity. Both mentioned features are characteristic for reduced grapheme oxide.

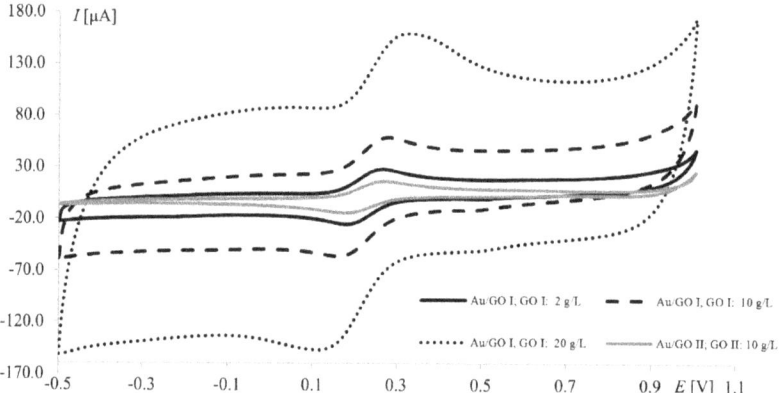

Figure 3. Voltammograms of 1.0 mM Fe(CN)$_6$$^{3-}$/Fe(CN)$_6$$^{4-}$ recorded at the gold electrode modified with graphene oxides I and II; scan rate, 75 mV s^{-1}; supporting electrolyte, 1 M KCl.

The model redox pair Fe(CN)$_6$$^{3-/}$Fe(CN)$_6$$^{4-}$ is a chemically reversible system, in which the oxidized form of the solution species can be regenerated from the reduced form (and vice versa), and both forms are stable on the time scale of the voltammetric experiment. However, after preliminary studies, a negative impact of modifications on electrochemical reversibility could not be excluded. If this was the case, the electron transfer would be so slow that the peak potential would not reflect the equilibrium activity of the redox couple at the electrode surface. To determine whether GO modification had a positive or negative effect, cathodic and anodic signal separation (difference between cathodic and anodic peak potentials) was analyzed. In the theoretical model, the difference between the anodic and cathodic peak potentials (E_a and E_c, respectively) should be equal to 59 mV/n for fully electrochemically reversible systems. However, the majority of redox systems used are quasi-reversible, with anodic/cathodic separation being much higher than 59 mV, even for fast quasi-reversible redox pairs. The separations measured in the present study are shown in Figure 4. To make the figure clearer, the separation obtained for the bare electrode was used as reference (100%).

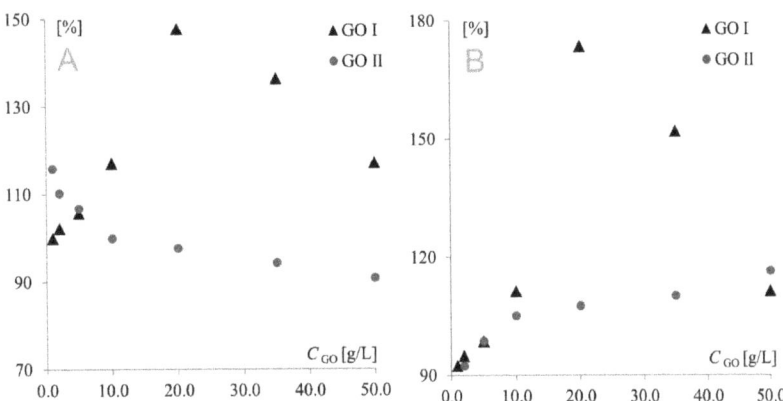

Figure 4. Signal separation measured for Au (**A**) and Pt (**B**) electrodes modified with GO I and GO II.

As can be seen, similar behavior was observed for Au and Pt electrodes modified with GO I. In this case, modifier concentration had a very strong influence on peak current; therefore, it was also the predominant factor affecting separation. This fact precluded investigation of the impact of the remaining factors on separation, and consequently, on the electrochemical reversibility of the system. A different behavior was observed for GO II, where signal intensity (with the highest currents also observed for the 20 g L^{-1} concentration) was only one of several factors influencing peak separation. At the platinum electrode, the lowest separation values were observed for the lowest modifier concentrations, while at the gold electrode, the highest concentrations afforded the best reversibility. This shows again that GO I and GO II act like two totally different compounds. Another important diagnostic characterizing an electrode reaction is peak potential (E_p). At fast electron transfer rates, E_p is independent of the scan rate, indicating a reversible electrode reaction. The influence of the scan rate was tested in the range of 10–500 mV s^{-1}. It was found that the scan rate did not have any effect on peak potential (except for changes resulting from signal increment), indicating a quasi-reversible electrode reaction. The next very important diagnostic tool consists of the relationships log I_p vs. log v and I_p vs. $v^{1/2}$. For both graphene oxides, linear plots of log I_p vs. log v were obtained with slopes of approx. 0.5, indicating an electrode reaction with the rate governed by diffusion of the electroactive species to an electrode surface. A linear relationship (for both electrodes with both graphene oxides) was also observed for I_p vs. $v^{1/2}$, which confirmed that the mass transport rate of the electroactive species to the electrode surface occurred across the concentration gradient. In such a case, the peak current I_p is governed by the Randles–Sevcik equation: $I_p = k\, n^{3/2}\, A\, D^{1/2}\, C^*\, v^{1/2}$, where the constant $k = 2.72 \times 10^5$; n is the number of moles of electrons transferred per mole of the electroactive species; A is electrode area; D is the diffusion coefficient; C^* is solution concentration; and v is the scan rate. The above formula was used to calculate the electroactive area of the Au and Pt electrodes modified with both graphene oxides (Figure 5). From an electrochemical point of view, observed dependence for GO I may be easily combined with observed peak currents—the bigger the electroactive surface, the higher the peak currents. The GO II case is much more complicated. For GO II, the highest current of model redox system was also observed for 20.0 g L^{-1} but the biggest electroactive surface was observed for lower concentrations. This suggest that observed currents are strongly connected with GO II chemical composition and the various oxygen functional groups present in its structure. Detailed studies on structural differences between GO I and GO II are described in Section 3.3.

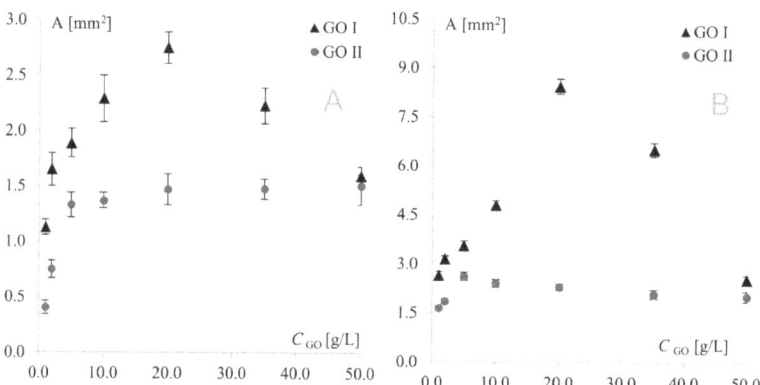

Figure 5. Electroactive surface of Au (**A**) and Pt (**B**) electrodes modified with GO I and GO II.

3.2. Microscopic Analysis

It is well known that the properties of a broad range of materials and the performance of a large variety of devices depend strongly on their surface characteristics [41]. Therefore, surface analyses were performed using scanning electron microscopy and atomic force microscopy to explain observed differences in electrochemical properties of fabricated graphene-oxide-modified electrodes. The first method allows to investigate the surface morphology of the modified electrodes, while the latter one allows to study their surface topography and roughness. Thus, their combination is very often used for surface characterization of the carbon electrodes [42,43]. Scanning electron microscope (SEM) images of GO I- and GO II-modified gold electrodes (suspension concentration 20.0 g L^{-1}) are shown in Figure 6. The comparison of the SEM images do not revealed significant differences in the morphology of samples. In both cases, a lot of irregularly distributed aggregates containing entangled graphene flakes can be observed. The distribution of aggregates on the electrode surfaces is random and homogenous. The size of aggregates is in the range of tens and hundreds of nanometers up to micrometers. The graphene aggregates form a continuous mesoporous layer on the surface of both electrodes, having a pore size in the range of hundreds nm. Moreover, the presence of graphene aggregates results in the high surface roughness and large surface area of both GO I- and GO II-modified gold electrodes.

Figure 6. SEM images of gold electrodes with GO I (**left**) and GO II (**right**).

As mentioned above, the modified electrodes were also characterized by atomic force microscopy (AFM) to give insight into their surface topography. Three-dimensional (3D) views of the graphene-oxide-modified gold electrodes are given in Figure 7. Corresponding two-dimensional (2D) AFM images with cross section profiles are presented in Figure S1. The surface characteristics of the investigated electrodes observed from AFM images are consistent with their SEM images. The applied modifications do not significantly differ in topographies. In both cases, a lot of irregularly distributed aggregates can be observed. There are also free spaces clearly visible between them. Roughness parameters (Rq and Ra) have been calculated from AFM images and are summarized in Table 1. From the presented results, it is clear that the roughness of the GO II-modified electrode is a little bit higher than that for the GO I-modified electrode.

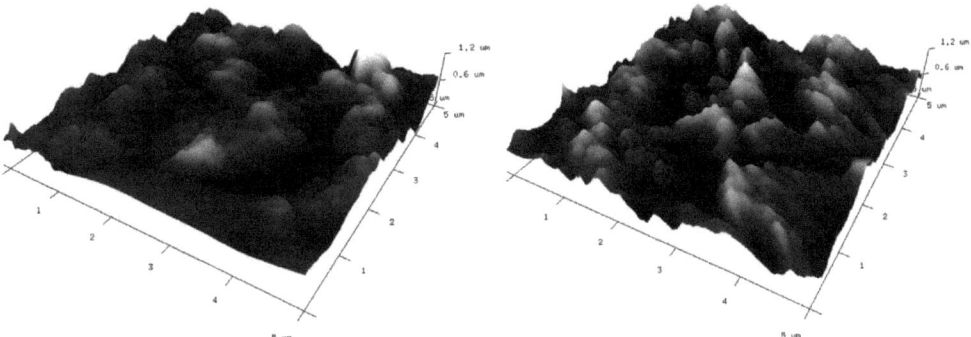

Figure 7. Three-dimensional AFM images of gold electrodes modified with graphene oxides: GO I (**left**) and GO II (**right**) suspensions.

Table 1. Surface roughness parameters calculated from AFM images.

	GO I	GO II
Rq	142.5 ± 5.3	184.0 ± 19.1
Ra	104.0 ± 2.4	144.3 ± 14.4

3.3. GO I and GO II Chemical Composition Analysis

With the aim of determining whether the chemical structure affects the electrochemical properties of the prepared graphene-based electrodes, spectroscopic measurements were conducted. Figure 8 shows the FTIR spectra of GO I and GO II nanostructures. FTIR spectroscopy provided evidence of the presence of various types of oxygen functional groups such as O-H, C=O, C-O, -O-, and C-OH on the GO, which could be located on the basal planes and edges of the GO flakes [44–46]. Small aggregated sharp peaks around 3735 cm^{-1} and a broad peak of very low intensity at 3210 cm^{-1} can be assigned to the stretching mode of the O–H bond, which reveal the presence of hydroxyl groups in graphene oxide [47,48]. Moreover, the peak at 1650 cm^{-1} can be designated to the stretching and bending vibration of water molecules adsorbed on graphene oxide [47]. The FTIR spectra of GO I and II also demonstrate the presence of other oxygenated functional groups with absorption peaks at 1050 cm^{-1} (alkoxy C-O) [44,46,49,50] and 1200 cm^{-1} (epoxide C-O-C or phenolic C-O-H stretch) [44,46,49,50]. It was found that in the case of GO I, absorption bands of all oxygen functional groups were less intense compared with GO II. Further insight into the graphene structure of GO I and GO II revealed other differences. Firstly, more epoxy groups are present on GO II in comparison with GO I. Secondly, the appearance of an absorption band at 1702 cm^{-1} confirms the presence of the C=O from carboxyl and/or carbonyl groups in the GO II structure [44]. The FTIR spectra indicate that GO I is more reduced than GO II.

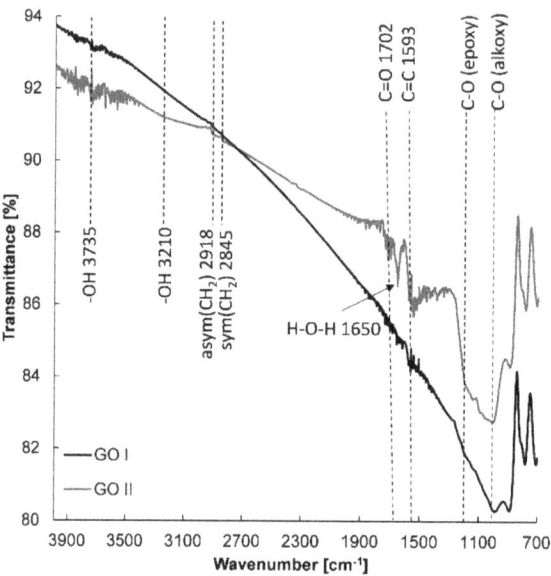

Figure 8. FTIR spectra of GO I and GO II.

This was also confirmed by UV–Vis spectroscopy (Figure S2). UV–Vis spectra of GO I and GO II aqueous suspensions show only one absorption band at 270 nm. This peak is characteristic of reduced graphene oxide and corresponds to the $\pi \to \pi^*$ transition of the C-C bond of the hexagonal carbon ring [51]. Moreover, the presence of this peak indicates the removal of oxygen-containing functional groups by reduction process and the restored electronic conjugation within graphene flakes [51]. However, the difference in the degree of GO I and GO II reduction can be seen in the UV–Vis spectra. Due to the increase in light absorption over the whole spectral region for GO I, it can be assumed that GO I is more reduced than GO II [46]. Another evidence of the higher reduction level of GO I is the lack of an absorption band at around 300 nm, which is attributed to $n \to \pi^*$ transition of C=O bond of edged carboxyl group on graphene oxide [51]. Finally, spectroscopic measurements results were confirmed by combustion analysis (Table 2), which also showed that GO I is more reduced than GO II.

Table 2. Combustion analysis results (mass percent).

Element	GO I	GO II
C	85.44 ± 0.24	81.43 ± 0.13
H	0.380 ± 0.020	0.390 ± 0.059

In the last step, modified electrodes were examined with Energy Dispersive X-ray Spectrometry. The EDS measurements revealed changes in the carbon-to-oxygen ratio in GO I and GO II samples (Table 3). The quantitative EDS element mappings of GO I and GO II are shown in Supplementary Material (Figures S3 and S4). Quantitative analysis shows that GO I had an oxygen content of 11.32 at.% and the atomic ratio of carbon to oxygen was 7.8, while in GO II the oxygen content was 16.62 at.% and the C/O ratio was about 5. These results indicate more oxygen content in the case of GO II compared with GO I and again suggests that GO I is more reduced than GO II.

Table 3. The content of carbon and oxygen in GO I and GO II determined by EDX measurements.

	GO I	GO II
C atomic%	88.68	83.38
C weight%	85.47	79.02
O atomic%	11.32	16.62
O weight%	14.53	20.98

4. Conclusions

In this paper, the electrochemical and structural characteristics of two graphene oxides used as electrodes modifiers were analyzed. AFM and SEM results showed that the topographies of the investigated electrodes/modifications do not significantly differ. However, the EDS elemental analysis revealed changes in the oxygen content. This was also confirmed by combustion analysis, FTIR, and UV–Vis spectroscopy. Unexpectedly, more oxygen content was found in the case of GO II (declared as RGO) compared with GO I (declared as GO). From the electrochemical point of view, the presence of RGO (here, GO II) should improve the electrochemical conductivity; however, the opposite results were obtained. The detailed electrochemical studies with model redox system showed that purchased graphene oxides act as utterly different modifiers. Although basic analytical characteristics are similar—for example, the diffusional nature of registered signals remained unchanged—other more important properties are completely different (peak currents, peak separation, electroactive area, etc.). As a result, we may obtain two completely different sensors that could, but should not, be called uniformly "graphene-oxide-modified".

Based on the obtained results, we can generally conclude that GO I is more reduced than GO II, which is the exact opposite of the general conception of graphene oxide structure. Hence, the following question arises: is there a chance to achieve good reproducibility among the scientific community using the not well-established uniform structure of GO? Undoubtedly, we may find some positive aspects of this graphene oxide case because by using graphene oxides of slightly changed chemical composition we can develop sensors with tailored properties towards a given analyte. On the other hand, we have to be aware of the negative aspect, i.e., the lack of reproducibility between "graphene-oxide-modified sensors" when being developed from diverse research teams.

In conclusion, the authors believe that there is an urgent need to standardize such popular nanomaterials by assigning a CAS Registry Number to them. The orderliness in graphene oxides nomenclature will greatly shorten the search for appropriate graphene oxide and will contribute to improvements in electrochemical analysis.

Supplementary Materials: The following supporting information can be downloaded at https://www.mdpi.com/article/10.3390/ma15217684/s1. Figure S1. 2D AFM images and profiles of gold electrodes modified with graphen oxides: GO I (left) and GO II (right) suspensions in concentration of 20 g L^{-1}; Figure S2. UV-Vis spectra of GO I and GO II aqueous suspensions; Figure S3. The quantitative EDS element mapping of GO I (a); carbon C K (b); oxygen O K (c) and EDS spectrum (d); Figure S4. The quantitative EDS element mapping of GO II (a); carbon C K (b); oxygen O K (c) and EDS spectrum (d).

Author Contributions: N.F., preliminary voltammetric studies—acquisition of data; K.R.-S. and J.G., SEM and EDS analysis—acquisition of data, and analysis and interpretation of data; A.K., D.G. and K.K., spectroscopic analysis—acquisition of data, and analysis and interpretation of data; B.B., AFM studies—acquisition of data, and analysis and interpretation of data; S.S., conception and design of the research, electrochemical analytical studies—acquisition of data, and analysis and interpretation of data, supervision, project administration, and funding acquisition; D.G., K.K. and S.S., writing—original draft preparation with review and editing. All authors have read and agreed to the published version of the manuscript.

Funding: The work was funded partly by NCN under the project no. 2017/01/X/ST4/00092. For the purpose of Open Access, the author has applied a CC-BY public copyright license to any Author Accepted Manuscript (AAM) version arising from this submission.

Institutional Review Board Statement: Not applicable.

Informed Consent Statement: Not applicable.

Data Availability Statement: The data presented in this study are available upon request from the corresponding author.

Conflicts of Interest: The authors declare no conflict of interest.

References

1. Górska, A.; Zambrzycki, M.; Paczosa-Bator, B.; Piech, R. New Electrochemical Sensor Based on Hierarchical Carbon Nanofibers with NiCo Nanoparticles and Its Application for Cetirizine Hydrochloride Determination. *Materials* **2022**, *15*, 3648. [CrossRef]
2. Morawska, K.; Ciesielski, W.; Smarzewska, S. First electroanalytical studies of methoxyfenozide and its interactions with dsDNA. *J. Electroanal. Chem.* **2021**, *882*, 115030. [CrossRef]
3. Kozak, J.; Tyszczuk-Rotko, K.; Wójciak, M.; Sowa, I.; Rotko, M. Electrochemically Pretreated Sensor Based on Screen-Printed Carbon Modified with Pb Nanoparticles for Determination of Testosterone. *Materials* **2022**, *15*, 4948. [CrossRef]
4. Piech, R.; Bugajna, A.; Baś, S.; Kubiak, W.W. Ultrasensitive determination of tungsten(VI) on pikomolar level in voltammetric catalytic adsorptive catechol-chlorate(V) system. *J. Electroanal. Chem.* **2010**, *644*, 74. [CrossRef]
5. Mirceski, V.; Guziejewski, D.; Ciesielski, W. Theoretical Treatment of a Cathodic Stripping Mechanism of an Insoluble Salt Coupled with a Chemical Reaction in Conditions of Square Wave Voltammetry. Application to 6-Mercaptopurine-9-D-Riboside in the Presence of Ni(II). *Electroanalysis* **2011**, *23*, 1365. [CrossRef]
6. Guziejewski, D.; Mirceski, V.; Jadresko, D. Measuring the Electrode Kinetics of Surface Confined Electrode Reactions at a Constant Scan Rate. *Electroanalysis* **2015**, *27*, 67. [CrossRef]
7. Kissinger, P.T.; Heineman, W.R. *Laboratory Techniques in Electroanalytical Chemistry*; Marcel Dekker Inc.: New York, NY, USA, 1996.
8. Wang, J. *Analytical Electrochemistry*; Wiley: Hoboken, NJ, USA, 2006.
9. Ates, A.K.; Er, E.; Celikkan, H.; Erk, N. Reduced graphene oxide/platinum nanoparticles/nafion nanocomposite as a novel 2D electrochemical sensor for voltammetric determination of aliskiren. *New J. Chem.* **2017**, *41*, 15320. [CrossRef]
10. Qiu, X.; Yan, X.; Cen, K.; Sun, D.; Xu, L.; Tang, Y. Achieving Highly Electrocatalytic Performance by Constructing Holey Reduced Graphene Oxide Hollow Nanospheres Sandwiched by Interior and Exterior Platinum Nanoparticles. *ACS Appl. Energy Mater.* **2018**, *1*, 2341. [CrossRef]
11. Geim, A.K.; Novoselov, K.S. The rise of graphene. *Nat. Mater.* **2007**, *6*, 183. [CrossRef]
12. Brumfiel, G. Graphene gets ready for the big time. *Nature* **2009**, *458*, 390. [CrossRef]
13. Novoselov, K.S.; Geim, A.K.; Morozov, S.V.; Jiang, D.; Zhang, Y.; Dubonos, S.V.; Grigorieva, I.V.; Firsov, A.A. Electric field effect in atomically thin carbon films. *Science* **2004**, *306*, 666. [CrossRef]
14. Park, S.; Ruoff, R.S. Chemical methods for the production of graphenes. *Nat. Nanotechnol.* **2009**, *4*, 217. [CrossRef] [PubMed]
15. Novoselov, K.S.; Jiang, Z.; Zhang, Y.; Morozov, S.V.; Stormer, H.L.; Zeitler, U.; Maan, J.C.; Boebinger, G.S.; Kim, P.; Geim, A.K. Room-Temperature Quantum Hall Effect in Graphene. *Science* **2007**, *315*, 1379. [CrossRef]
16. Bunch, J.S.; Verbridge, S.S.; Alden, J.S.; van der Zande, A.M.; Parpia, J.M.; Craighead, H.G.; McEuen, P.L. Impermeable atomic membranes from graphene sheets. *Nano Lett.* **2008**, *8*, 2458. [CrossRef]
17. Lee, C.; Wei, X.; Kysar, J.W.; Hone, J. Measurement of the elastic properties and intrinsic strength of monolayer graphene. *Science* **2008**, *321*, 385. [CrossRef]
18. Bolotin, K.I.; Sikes, K.J.; Jiang, Z.; Klima, M.; Fudenberg, G.; Hone, J.; Kim, P.; Stormer, H.L. Ultrahigh electron mobility in suspended graphene. *Solid State Commun.* **2008**, *146*, 351. [CrossRef]
19. Chen, D.; Feng, H.; Li, J. Graphene Oxide: Preparation, Functionalization, and Electrochemical Applications. *Chem. Rev.* **2012**, *112*, 6027. [CrossRef]
20. Hofmann, U.; Holst, R. Über die Säurenatur und die Methylierung von Graphitoxyd. *Ber. Dtsch. Chem. Ges. B* **1939**, *72*, 754. [CrossRef]
21. Ruess, G. Über das Graphitoxyhydroxyd (Graphitoxyd). *Monatsh. Chem.* **1946**, *76*, 381. [CrossRef]
22. Clauss, A.; Plass, R.; Boehm, H.P.; Hofmann, U. Untersuchungen zur Struktur des Graphitoxyds. *Z. Anorg. Allg. Chem.* **1957**, *291*, 205. [CrossRef]
23. Scholz, W.; Boehm, H.P. Untersuchungen am Graphitoxid. VI. Betrachtungen zur Struktur des Graphitoxids. *Z. Anorg. Allg. Chem.* **1969**, *369*, 327. [CrossRef]
24. Nakajima, T.; Mabuchi, A.; Hagiwara, R. A new structure model of graphite oxide. *Carbon* **1988**, *26*, 357. [CrossRef]
25. Szabó, T.; Berkesi, O.; Forgó, P.; Josepovits, K.; Sanakis, Y.; Petridis, D.; Dékány, I. Evolution of Surface Functional Groups in a Series of Progressively Oxidized Graphite Oxides. *Chem. Mater.* **2006**, *18*, 2740. [CrossRef]
26. Eda, G.; Chhowalla, M. Chemically Derived Graphene Oxide: Towards Large-Area Thin-Film Electronics and Optoelectronics. *Adv. Mater.* **2010**, *22*, 2392. [CrossRef]
27. Li, H.L.; Zhang, G.Y.; Bai, X.D.; Sun, X.M.; Wang, X.R.; Wang, E.G.; Dai, H.J. Highly conducting graphene sheets and Langmuir–Blodgett films. *Nat. Nanotechnol.* **2008**, *3*, 538. [CrossRef]

28. Kim, F.; Cote, L.; Huang, J.X. Graphene Oxide: Surface Activity and Two-Dimensional Assembly. *Adv. Mater.* **2010**, *22*, 1954. [CrossRef]
29. Erdem, A.; Eksin, E.; Isin, D.; Polat, D. Graphene Oxide Modified Chemically Activated Graphite Electrodes for Detection of microRNA. *Electroanalysis* **2017**, *29*, 1350. [CrossRef]
30. Park, M.-O.; Noh, H.-B.; Park, D.-S.; Yoon, J.-H.; Shim, Y.-B. Long-life Heavy Metal Ions Sensor Based on Graphene Oxide-anchored Conducting Polymer. *Electroanalysis* **2017**, *29*, 514. [CrossRef]
31. Smarzewska, S.; Miękoś, E.; Guziejewski, D.; Zieliński, M.; Burnat, B. Graphene oxide activation with a constant magnetic field. *Anal. Chim. Acta* **2018**, *1011*, 35. [CrossRef]
32. Mohamed, M.A.; Atty, S.A.; Salama, N.N.; Banks, C.E. Highly Selective Sensing Platform Utilizing Graphene Oxide and Multi-walled Carbon Nanotubes for the Sensitive Determination of Tramadol in the Presence of Co-Formulated Drugs. *Electroanalysis* **2017**, *29*, 1038. [CrossRef]
33. Festinger, N.; Spilarewicz-Stanek, K.; Borowczyk, K.; Guziejewski, D.; Smarzewska, S. Highly Sensitive Determination of Tenofovir in Pharmaceutical Formulations and Patients Urine—Comparative Electroanalytical Studies Using Different Sensing Methods. *Molecules* **2022**, *27*, 1992. [CrossRef] [PubMed]
34. Smarzewska, S.; Metelka, R.; Festinger, N.; Guziejewski, D.; Ciesielski, W. Comparative Study on Electroanalysis of Fenthion Using Silver Amalgam Film Electrode and Glassy Carbon Electrode Modified with Reduced Graphene Oxide. *Electroanalysis* **2017**, *29*, 1154. [CrossRef]
35. Staudenmaier, L. Verfahren zur Darstellung der Graphitsäure. *Ber. Dtsch. Chem. Ges.* **1898**, *31*, 1481. [CrossRef]
36. Hofmann, U.; Konig, E. Untersuchungen über Graphitoxyd. *Z. Anorg Allg. Chem.* **1937**, *234*, 311. [CrossRef]
37. Hummers, W.S.; Offeman, R.E. Preparation of Graphitic Oxide. *J. Am. Chem. Soc.* **1958**, *80*, 1339. [CrossRef]
38. Marcano, D.C.; Kosynkin, D.V.; Berlin, J.M.; Sinitskii, A.; Sun, Z.Z.; Slesarev, A.; Alemany, L.B.; Lu, W.; Tour, J.M. Improved Synthesis of Graphene Oxide. *ACS Nano* **2010**, *4*, 4806. [CrossRef]
39. Chng, E.L.K.; Pumera, M. The Toxicity of Graphene Oxides: Dependence on the Oxidative Methods Used. *Chem. Eur. J.* **2013**, *19*, 8227. [CrossRef]
40. Yang, J.; Gunasekaran, S. Electrochemically reduced graphene oxide sheets for use in high performance supercapacitors. *Carbon* **2013**, *51*, 36. [CrossRef]
41. Fooladsaz, K.; Negahdary, M.; Rahimi, G.; Habibi-Tamijani, A.; Parsania, S.; Akbari-dastjerdi, H.; Sayad, A.; Jamaleddini, A.; Salahi, F.; Asadi, A. Dopamine Determination with a Biosensor Based on Catalase and Modified Carbon Paste Electrode with Zinc Oxide Nanoparticles. *Int. J. Electrochem. Sci.* **2012**, *7*, 9892–9908.
42. Robak, J.; Burnat, B.; Leniart, B.A.; Kisielewska, A.; Brycht, M.; Skrzypek, S. The effect of carbon material on the electroanalytical determination of 4-chloro-3-methylphenol using the sol-gel derived carbon ceramic electrodes. *Sens. Act. B-Chem.* **2016**, *236*, 318. [CrossRef]
43. Rajawat, D.S.; Kumar, N.; Satsangee, S.P. Trace determination of cadmium in water using anodic stripping voltammetry at a carbon paste electrode modified with coconut shell powder. *J. Anal. Sci. Tech.* **2014**, *5*, 19. [CrossRef]
44. Zhang, H.; Hines, D.; Akins, D.L. Synthesis of a nanocomposite composed of reduced graphene oxide and gold nanoparticles. *Dalton Trans.* **2014**, *43*, 2670. [CrossRef] [PubMed]
45. Spilarewicz-Stanek, K.; Kisielewska, A.; Ginter, J.; Bałuszyńska, K.; Piwoński, I. Elucidation of the function of oxygen moieties on graphene oxide and reduced graphene oxide in the nucleation and growth of silver nanoparticles. *RSC Adv.* **2016**, *6*, 60056. [CrossRef]
46. Spanò, S.F.; Isgrò, G.; Russo, P.; Fragalà, M.E.; Compagnini, G. Tunable properties of graphene oxide reduced by laser irradiation. *Appl. Phys. A* **2014**, *117*, 19. [CrossRef]
47. Das, B.; Kundu, R.; Chakravarty, S. Preparation and characterization of graphene oxide from coal. *Mat. Chem. Phys.* **2022**, *290*, 126597. [CrossRef]
48. Jahan, N.; Roy, H.; Reaz, A.H.; Arshi, S.; Rahman, E.; Firoz, S.H.; Islam, M.S. A comparative study on sorption behavior of graphene oxide and reduced graphene oxide towards methylene blue. *Case Stud. Chem. Environ. Eng.* **2022**, *6*, 100239. [CrossRef]
49. Nawaz, M.; Miran, W.; Jang, J.; Lee, D.S. One-step hydrothermal synthesis of porous 3D reduced graphene oxide/TiO2 aerogel for carbamazepine photodegradation in aqueous solution. *Appl. Catal. B* **2017**, *203*, 85. [CrossRef]
50. Gupta, B.; Kumar, N.; Panda, K.; Kanan, V.; Joshi, S.; Visoly-Fisher, I. Role of oxygen functional groups in reduced graphene oxide for lubrication. *Sci. Rep.* **2017**, *7*, 45030. [CrossRef]
51. Sun, H.; Lin, S.; Peng, T.; Liu, B. Microstructure and Spectral Characteristics of Graphene Oxide during Reduction. *Integr. Ferroelectr.* **2014**, *151*, 21. [CrossRef]

Article

Ratiometric Electrochemical Sensor for Butralin Determination Using a Quinazoline-Engineered Prussian Blue Analogue

Marcio Cristiano Monteiro, João Paulo Winiarski *, Edson Roberto Santana *, Bruno Szpoganicz and Iolanda Cruz Vieira

Department of Chemistry, Federal University of Santa Catarina, Florianópolis 88040-900, SC, Brazil
* Correspondence: joneswnk@gmail.com (J.P.W.); edsonr.santana0@gmail.com (E.R.S.)

Abstract: A ratiometric electrochemical sensor based on a carbon paste electrode modified with quinazoline-engineered ZnFe Prussian blue analogue (PBA-qnz) was developed for the determination of herbicide butralin. The PBA-qnz was synthesized by mixing an excess aqueous solution of zinc chloride with an aqueous solution of precursor sodium pentacyanido(quinazoline)ferrate. The PBA-qnz was characterized by spectroscopic and electrochemical techniques. The stable signal of PBA-qnz at +0.15 V vs. Ag/AgCl, referring to the reduction of iron ions, was used as an internal reference for the ratiometric sensor, which minimized deviations among multiple assays and improved the precision of the method. Furthermore, the PBA-qnz-based sensor provided higher current responses for butralin compared to the bare carbon paste electrode. The calibration plot for butralin was obtained by square wave voltammetry in the range of 0.5 to 30.0 µmol L^{-1}, with a limit of detection of 0.17 µmol L^{-1}. The ratiometric sensor showed excellent precision and accuracy and was applied to determine butralin in lettuce and potato samples.

Keywords: Prussian blue analogue; quinazoline; ratiometric sensor; butralin

Citation: Monteiro, M.C.; Winiarski, J.P.; Santana, E.R.; Szpoganicz, B.; Vieira, I.C. Ratiometric Electrochemical Sensor for Butralin Determination Using a Quinazoline-Engineered Prussian Blue Analogue. *Materials* **2023**, *16*, 1024. https://doi.org/10.3390/ma16031024

Academic Editors: Sławomira Skrzypek, Mariola Brycht and Barbara Burnat

Received: 17 December 2022
Revised: 16 January 2023
Accepted: 20 January 2023
Published: 23 January 2023

Copyright: © 2023 by the authors. Licensee MDPI, Basel, Switzerland. This article is an open access article distributed under the terms and conditions of the Creative Commons Attribution (CC BY) license (https:// creativecommons.org/licenses/by/ 4.0/).

1. Introduction

Prussian blue (PB) was the first polymeric coordination compound recorded in the literature by Diesbach and Dippel in the early 18th century [1]. The different oxidation states between iron atoms coordinated by a cyanide bridge give PB its characteristic blue color due to an intervalence transition around 720 nm [2]. A Prussian blue analogue (PBA) is the result of changes in the chemical composition of PB. When Fe^{2+} and/or Fe^{3+} ions are replaced by other different transition metal centers, such as cobalt, nickel, and zinc [3], it is also possible to change its properties by making small changes in its composition (and consequently in its structure): replacing the metallic centers and/or a CN^- group with other ligands, such as those of quinazoline [4].

Quinazoline (qnz, 1,3-diazanaphthalene) is a heterocyclic hybrid that has the molecular formula $C_8H_6N_2$, and it is an important bicyclic skeleton structure in manifold natural products [5,6]. The quinazoline ring is formed by the union of a benzene ring with a six-membered ring containing 2 N atoms and contains three main isomers, namely, quinoxaline, cinnoline, and phthalazine [5]. Quinazoline and its derivatives have multiple biological activities and show a high affinity for metal ions; they also form all kinds of coordination compounds with sundry transition metals [7]. Regarding the application in electrochemical sensors, introducing nitrogen moieties into the electrode composition has obtained considerable interest, as it leads to the improvement of conductivity and of the electroactive area of the sensor, hence further boosting its electrochemical performance [8]. Therefore, quinazoline is an interesting ligand to be explored in the development of novel PBA for application in the field of electrochemical sensors.

Research on PBA composites and their derivatives has kept growing in the past decade, and they have been applied in energy conversion, energy storage, adsorption, and electrochemical sensors [9]. Regarding the application of electrochemical sensors, for PB/PBA, both oxidized and reduced forms have catalytic activity [3]. In addition, their zeolitic form has a channel diameter of approximately 3.2 Å and a cubic unit cell of 10.2 Å, allowing the diffusion of ions by the structure [3,9]. Furthermore, their high electronic transfer rate is another benefit, which is directly associated to the insertion/disinsertion of small ions [3]. Li et al. [10] developed a PBA-modified glassy carbon electrode for 2-nitrophenol determination. The synthesized PBA ($K_xNi[Fe(CN)_6] \cdot nH_2O$) provided the electrochemical sensor with a higher electrocatalytic performance for the 2-nitrophenol reduction in comparison to bare GCE, which could be attributed to the better intrinsic catalytic nature of Ni, improved conductivity, and larger electroactive area.

Traditional electrochemical sensors usually depend on the precise measurement of a single current intensity, which further leads to low repeatability, reliability, and accuracy, and occasionally false negative results [11]. For this reason, ratiometric electrochemical sensors have recently attracted extensive attention [12–15]. These special sensors quantify the analyte with ratiometric a record of two signals (one is from the analyte and the other is from the inner reference). A peak intensity ratio ($I_{analyte}/I_{inner\ reference}$) is used as the measurement criteria for analytes [13]. Commonly, this ratiometric strategy reduces the intrinsic errors or background electric signals and exhibits a significant ability to further improve the accuracy and precision of the measurements [13,14]. Constant current responses from the internal reference can also indicate that the electrode surface remains homogeneous [14]. Consequently, ratiometric electrochemical sensors are considered more reliable and accurate than common electrochemical sensors [16]. PB has been used as an internal reference for ratiometric electrochemical sensors [15]. However, reports of works using these materials for ratiometric sensors are still limited.

In that regard, the detection of butralin (BTL) is of great importance, since BTL is a dinitroaniline herbicide applied in pre-emergence management of pests in manifold crops such as cotton, sunflower, rice, peanuts, corn, and vegetable crops [17,18]. Dinitroaniline herbicides are slightly soluble in water and moderately persistent in the environment by adsorbing to soil particles, such as organic matter, so it presents an environmental pollution and a potential threat to human health [19–21]. Regarding electrochemical sensors dedicated to the determination of BTL, only two works are found in the literature. Sreedhar and Reddy [22] developed a polarographic method for BTL determination using a dropping mercury electrode, and Gerent et al. [17] used a glassy carbon electrode modified with Co-Ag bimetallic nanoparticles stabilized in poly(vinylpyrrolidone). The electrochemical methods are greatly attractive because of their advantages, such as quick detection, convenient operation, cheap instrumentation, facile integration, and portability [23–25]. Therefore, the development of novel electrochemical tools to detect and supervise the dissipation behavior of BTL in edible raw food and in the environment is relevant.

In this work, the use of the quinazoline ligand and the metals Fe and Zn coordinated by the cyanide bridge was chosen to synthesize a novel PBA. To the best of our knowledge, this is the first report to the use quinazoline ligand to the synthesis of a PBA. Here, the PBA was incorporated into a carbon paste electrode and boosted the conductivity of the system, thus providing greater current intensities and, consequently, greater sensitivity and also serving as an internal reference to improve the precision and accuracy of the novel ratiometric sensor in the determination of BTL.

2. Materials and Methods

2.1. Reagents and Solutions

All reagents used in the experiment were analytical grade and purchased from commercial sources. Acetone, ethanol, sodium iodide, and sodium nitroprusside were purchased from Neon, Cambridge, MA, USA. Chloridric acid, zinc(II) chloride, iron(III) chloride, potassium chloride, sodium chloride, DMSO, butralin, and quinazoline were purchased from Merck, Darmstadt, Germany. The aqueous solutions were prepared with ultrapure water (18.2 MΩ cm), obtained with the Milli-Q system (Millipore, St. Louis, MO, USA).

A stock solution of 10.0 mmol L^{-1} butralin was prepared in acetone and stored at 4 °C. Britton–Robinson (B–R) buffer (H$_3$BO$_3$, CH$_3$COOH, H$_3$PO$_4$) (0.1 mol L^{-1}) was used as the supporting electrolyte. The pH adjustments were performed with 6.0 mol L^{-1} HCl or NaOH.

To build the carbon paste electrode, Acheson 38 graphite powder (Fisher Scientific, Waltham, MA, USA) served as the conductor, and Nujol mineral oil (Merck, Darmstadt, Germany) served as a binding agent.

2.2. Synthesis and Characterization of PBA-qnz

The precursor complex pentacyanido(quinazoline)ferrate (PCF-qnz) was synthesized by solubilizing 0.58 mmol of PCF-amine in 1.0 mL of distilled water and mixing it with 1.0 mL of aqueous quinazoline solution (0.29 mmol). The reaction solution was kept under stirring, out of the reach of light, and in an ice bath for 30 min. After this period, 0.67 mmol of sodium iodide was added to the solution, and then 30 mL of ethanol was slowly added. The precipitated solid was filtered in a vacuum pump, washed with ethanol, and kept in a desiccator until a constant mass was obtained.

The Prussian blue analogue derivative from quinazoline ligand and zinc(II) (PBA-qnz) was synthesized by the direct method, which consists of mixing an excess aqueous solution of zinc chloride (0.40 mmol) with an aqueous solution of PCF-qnz (0.10 mmol) under agitation. After 15 min, the solid was precipitated with acetone and isolated by centrifugation.

The compounds were characterized by UV-Vis spectroscopy using a Lambda 35 spectrometer (Perkin Elmer, Waltham, MA, USA) with quartz cuvettes of 1.0 cm of optical length. FTIR was used to verify the main functional groups of both compounds, using a FTLA 2000 spectrophotometer (Asea Brown Boveri, Zürich, Switzerland). Electron paramagnetic resonance (EPR) spectra were obtained using an EMX micro-9.5/2.7 spectrometer (Bruker, Billerica, MA, USA) with a highly sensitive cylindrical cavity, operating in X-band (9 GHz), at 120 K, with 5 mW microwave power, 5 G modulation amplitude, and 100 kHz modulation frequency. Cyclic voltammetry and electrochemical impedance spectroscopy (EIS) measurements were performed in an Autolab PGSTAT128N potentiostat (Metrohm Autolab B.V., Utrecht, The Netherlands). EIS measurements were performed using the K$_3$[Fe(CN)$_6$]/K$_4$[Fe(CN)$_6$] redox probe (5.0 mmol L^{-1} equimolar mixture) in 0.1 mol L^{-1} KCl. For the EIS measure, the OCP was applied with a perturbation amplitude of 10 mV between the frequencies of 100,000 Hz and 0.1 Hz.

2.3. Construction of Electrochemical Sensor

Studies by our group have described the construction of sensors based on carbon paste [26]. The construction procedure for the sensor involved hand-mixing 18 mg of PBA-qnz (10% w/w) and 135 mg of graphite powder (75% w/w) for twenty minutes. After that, 27 mg (15% w/v) of Nujol was added and hand-mixed for 20 min more in a mortar. The resulting composite was packed firmly into the cavity of a syringe (3.0 mm inner diameter), and a copper wire was inserted to establish electrical contact. For comparison purposes, PCF-qnz/CPE and bare CPE were prepared using a similar procedure.

2.4. Electrochemical Measurements

The electrochemical measurements for the development of the analytical method for BTL were performed using a portable potentiostat PalmSens 4 (Palm Instruments BV, Houten, The Netherlands). The assays were carried out with a system of three electrodes: the proposed sensor (PBA-qnz/CPE) as the working electrode, a platinum plate as the auxiliary electrode, and Ag/AgCl (3.0 mol L^{-1} KCl) as the reference electrode. All assays were carried out at room temperature (25 ± 0.5 °C) in an electrochemical cell containing 10.0 mL of B–R buffer (0.1 mol L^{-1}; pH from 2.0 to 7.0), and successive additions of a standard solution of BTL were carried out using a micropipette. Nitrogen gas was purged to the supporting electrolyte for 10 min before the assays.

2.5. Determination of BTL in Lettuce and Potato Samples

Fresh samples of lettuce (*Lactuca sativa*) and potato (*Solanum tuberosum*) were acquired from a farmers' market in Florianópolis, Brazil. The lettuce and potato samples were prepared as follows: a mixture of 5.0 g of each vegetable with 25.0 mL of acetone was crushed in a blender for 5 min. The extract was filtered (25.0 μm) two times and diluted in acetone in a 50.0 mL volumetric flask for the analysis. For the assays, 500 μL of the samples was added to the electrochemical cell with 9.5 mL of 0.1 mol L^{-1} B–R buffer (pH 2.0).

3. Results and Discussion

3.1. Characterization of PCF-qnz and PBA-qnz

The PCF-qnz complex (Figure 1A) and PBA-qnz (Figure 1B) were first characterized using UV-Vis spectroscopy (Figure 2A). The PCF-qnz complex formed by the exchange of NH$_3$ ligand for qnz ligand exhibits two bands of metal–ligand charge transfer in the visible region (355 and 474 nm, with log ε_{max} equal to 3.33 and 3.40, respectively). One of the characteristic bands of quinazoline [27] has a hypochromic shift when coordinating with the Fe atom, from 271 nm to 290 nm in PCF-qnz and to 280 nm in PBA-qnz.

Infrared spectra (Figure 2B) show that the PCF-qnz complex (curve a) exhibits characteristic bands of benzene (1378–1487 cm^{-1}) and a pyrimidine ring (1580–1617 cm^{-1}) [28]. Furthermore, it is possible to observe the CN^{-} (2047 cm^{-1}) and Fe-CN (568 cm^{-1}) stretches in the complex. Evaluating the FTIR data of PBA-qnz (curve b), it is possible to observe the presence of vibrations, referring to the vibrations of benzene at 1305–1492 cm^{-1} and the pyrimidine ring of qnz at 1592–1619 cm^{-1}. The CN^{-} stretch can be observed at 2094 cm^{-1}, as well as the Fe-CN-Zn stretch at 485 cm^{-1}. Finally, the broadening of the ν(CN^{-}) band in PBA-qnz means a variety of cyanides in the structure [29].

Cyclic voltammetry was used to study the electrochemical behaviors of PCF-qnz complex and PBA-qnz in 0.1 mol L^{-1} KCl (Figure 2C). Pentacyanidoferrates have a well-defined electrochemical process that is influenced by the nature of the ligand. The PCF-qnz complex (curve a) has a half-wave potential (E$_{1/2}$) of 545 mV (I) and 720 mV (II) vs. Ag/AgCl, assigned to the pairs [Fe$^{2+/3+}$(CN)$_5$(qnz)Fe$^{2+/3+}$(CN)$_5$]$^{6-/4-}$. The increase in potential represents a greater difficulty in removing electron density from iron due to the presence of the heterocyclic ligand [29]. Regarding the PBA-qnz (curve b), the E$_{1/2}$ values were 180 mV (III) and 860 mV (IV) vs. Ag/AgCl. The fully reduced form of PBA-qnz, Zn[Fe^{2+}(CN)$_5$(qnz)Fe^{2+}(CN)$_5$] was oxidized at +183 mV vs. Ag/AgCl to form Zn[Fe^{3+}(CN)$_5$(qnz)Fe^{2+}(CN)$_5$]. Due to the presence of quinazoline in its structure, there is an increase in the amount of water coordinated, resulting in a lower σ-donor contribution, consequently resulting in a shift in the oxidation potential to more positive values, compared to traditional Prussian blue [30]. At +906 mV vs. Ag/AgCl, the second metallic center is oxidized, formatting Berlin green (Zn[Fe^{3+}(CN)$_5$(qnz)Fe^{3+}(CN)$_5$]). These processes occurred at more positive potential values than the traditional Prussian blue; in other words, the oxidation of PBA-qnz required a more positive potential value, indicating the coordination of the metal centers with the qnz ligand.

Although PCF compounds present iron atoms with the 2+ oxidation state, a broad signal around g ~2.021 can be observed for PCF-qnz. As it is a metal with six electrons, a value of g greater than g$_e$ is expected [31]. This result occurs due to the magnetic interaction between iron ions, suggesting an Fe-Fe (spin–spin) interaction. The X-band EPR spectrum measured at room temperature reveals a profile similar to the EPR spectrum for PCF-amin (g ~2.2–2.3) presented by Ghobadi et al. [32]. The decrease in the value of g when exchanging the NH$_3$ ligand for qnz suggests that binding with a compound that contributes to a strong field favors the MLCT process Fe^{II}-qnz → Fe^{III}-qnz. When analyzing the EPR spectrum of PBA-qnz (Figure 2D), seven peaks are observed. Zinc atoms fully occupy d orbitals, exhibiting no signs. Thus, the signs suggest a mixture of Fe^{2+} and Fe^{3+} valence states, with g values ranging from ~0.185 to ~0.214 [33]. These results agree with cyclic voltammetry.

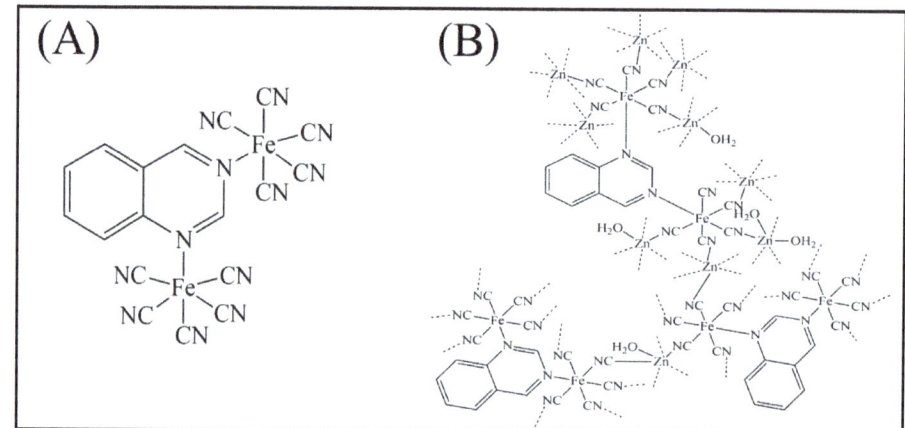

Figure 1. Structures of (**A**) pentacyanidoferrate with quinazoline ligand (PCF-qnz) and (**B**) Prussian blue analogue derivative from quinazoline ligand and zinc(II) (PBA-qnz).

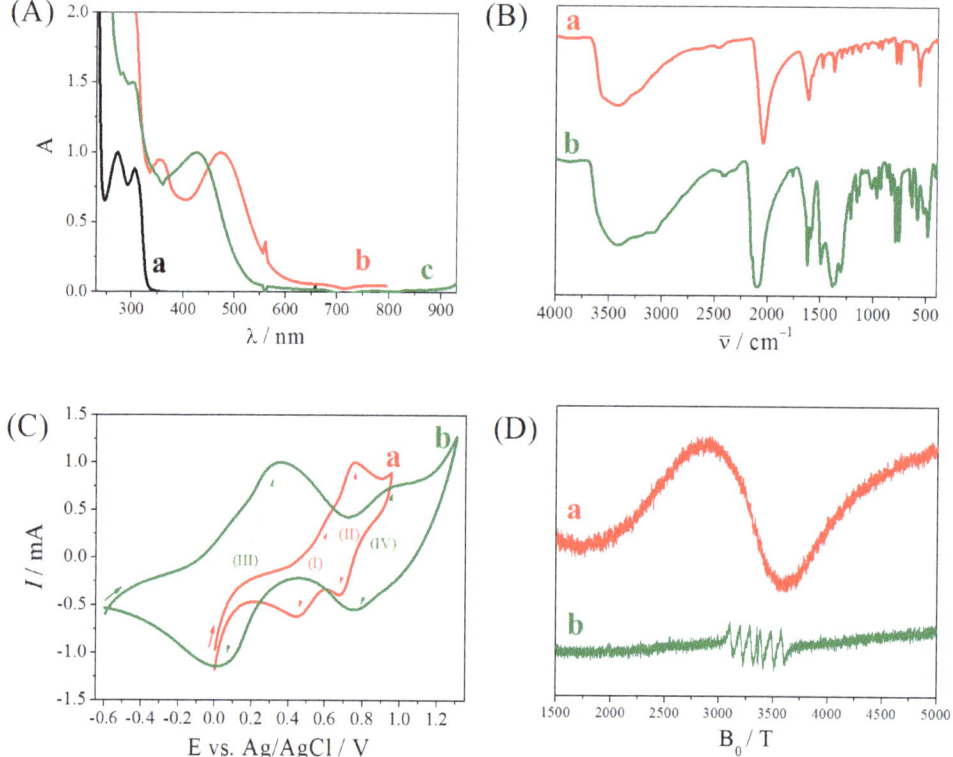

Figure 2. (**A**) UV-Vis spectra of (a) qnz, (b) PCF-qnz, and (c) PBA-qnz in aqueous solution. (**B**) Infrared absorption spectra of (a) PCF-qnz and (b) PBA-qnz. (**C**) Cyclic voltammograms of (a) PCF-qnz/CPE and (b) PBA-qnz/CPE with a scan rate of 25 mV s^{-1} (supporting electrolyte: 0.1 mol L^{-1} KCl). (**D**) EPR spectra in the X-band at room temperature of (a) PCF-qnz and (b) PBA-qnz, revealing the nature of the iron sites.

3.2. Electrochemical Characteristics of PCF-qnz and PBA-qnz

EIS is a useful tool to investigate the interface properties of surface-modified electrodes. Nyquist plots were obtained in $[Fe(CN)_6]^{3-/4-}$ solution in KCl 0.1 mol L^{-1} for the following electrodes: (a) CPE, (b) PCF-qnz/CPE, and (c) PBA-qnz/CPE, which are shown in Figure 3. The charged transfer resistance (R_{ct}) is positively correlated with the semicircle diameter in the high-frequency region of the EIS, and the diffusion process indicates the resistance offered by the mass transfer. Fitting the high-frequency region of the EIS plot, the R_{ct} of CPE is 6.1 kΩ (curve a). A reduced charge transfer resistance value of approximately 1000 Ω was observed for PCF-qnz/CPE (curve b), confirming the improved electrical conductivity based on the complex of Fe(II) and qnz. The R_{ct} for PBA-qnz/CPE is 1484 Ω (curve c), which is significantly lower than two working electrodes. This implies that, due to the polymerization of complex with Zn(II) moiety, the Prussian blue analogue becomes less resistive to charge transfer, increasing the electron transfer pathway between PBA-qnz/CPE and the redox probe. In addition, the introduction of nitrogen moieties into the electrode composition via quinazoline ligands leads to the improvement of conductivity, improving their electrochemical performance [8]. Thus, the PBA-qnz/CPE modified electrode can achieve an electrochemically sensitive determination of BTL.

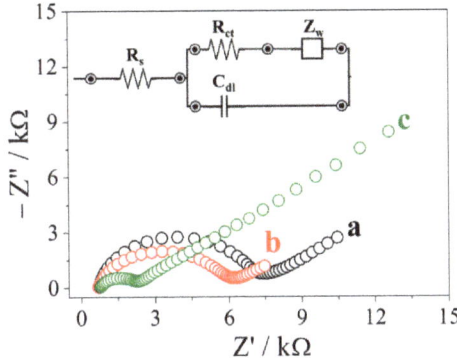

Figure 3. Nyquist plots for 5.0 mmol L^{-1} equimolar mixture of $K_3[Fe(CN)_6]/K_4[Fe(CN)_6]$ in 0.1 mol L^{-1} KCl: (a) CPE, (b) PCF-qnz/CPE, and (c) PBA-qnz/CPE. We inserted the Randles circuit model for the electrodes. R_s: solution resistance; R_{ct}: charge-transfer resistance; Z_w: Warburg impedance; C_{dl}: double-layer capacitance.

3.3. Evaluation of Butralin Ratiometric Sensor Performance

The electrochemical behavior of BTL was studied by square wave voltammetry (SWV) using the bare CPE and the PBA-qnz/CPE ratiometric sensor (Figure 4A). The square wave voltammogram exhibited a peak at −540 mV vs. Ag/AgCl, corresponding to the BTL reduction at bare CPE (curve a). The peak is correlated to the reduction of both nitro groups present in the molecule of BTL [17]. Using the PBA-qnz/CPE in the absence of BTL (curve b), a peak was recorded at +300 mV vs. Ag/AgCl, corresponding to the reduction of iron centers of the complex. Finally, when the BTL was analyzed using the PBA-qnz/CPE (curve c) a three-fold increase in the current intensities of the BTL compared to the performance of the CPE was recorded. This phenomenon can be attributed to the presence of nitrogen moieties in the electrode composition via quinazoline ligands, which leads to the improvement of conductivity and electroactive area of the sensor, boosting their electrochemical performance [8]. Even more important, PBA incorporated in the carbon paste was employed as a promising reference signal for the ratiometric sensor of BTL. The electroactive PBA can be oxidized to Berlin green or reduced to Prussian white at certain potentials and provide stable redox peaks, which can be used as an internal reference [15]. Thus, the measurements provided by the PBA-qnz/CPE ratiometric sensor show two signals (one is from the BTL analyte, and the other is from the PBA internal reference). The constant current responses of PBA-qnz inner reference indicated that the carbon paste was homogeneous, and consequently, the sensor surface was uniform, which contributes to better precision of measurements.

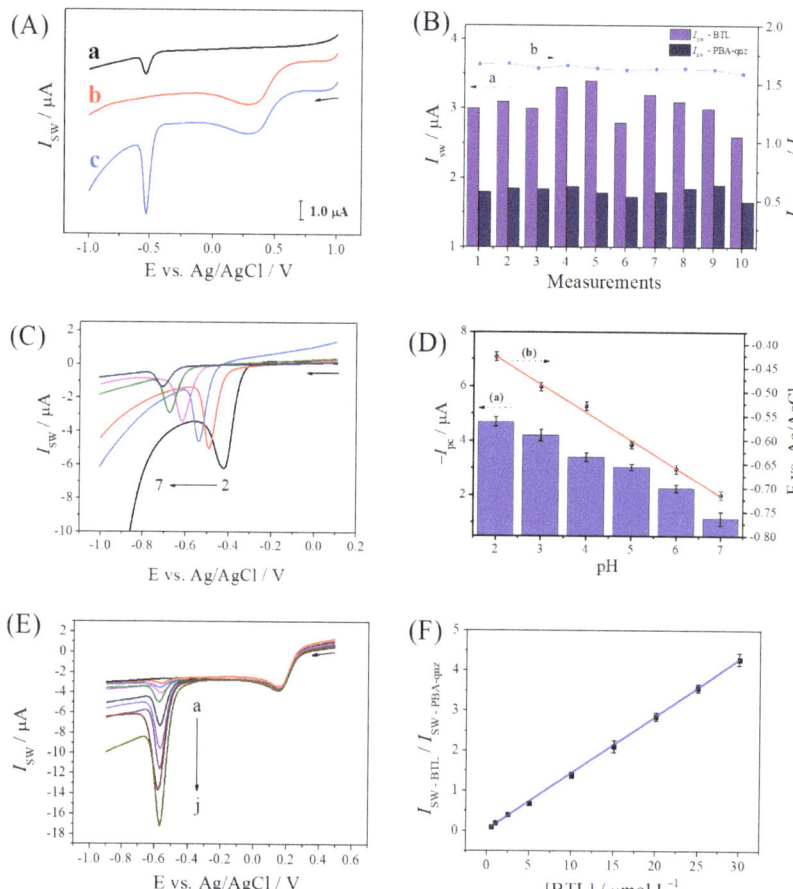

Figure 4. (**A**) Square wave voltammograms in the absence of BTL for (b) PBA-qnz/CPE in buffer solution and in the presence of 10.0 µmol L^{-1} BTL in B–R buffer (pH 2.0) for (a) CPE and (c) PBA-qnz/CPE. SWV parameters: f = 25.0 Hz, ΔE_s = 1.0 mV, a = 50.0 mV (nonoptimized). (**B**) Current responses for I_{SW-BTL} and $I_{SW-PBA\text{-}qnz}$ (axis a) and the $I_{SW-BTL}/I_{SW-PBA\text{-}qnz}$ ratio (axis b) using the same PBA-qnz/CPE on the same day. (**C**) Square wave voltammograms for 10.0 µmol L^{-1} BTL in B–R buffer using the PBA-qnz/CPE at different pH values. (**D**) Current (axis a) and potential (axis b) vs. pH (n = 3). (**E**) Square-wave voltammograms for BTL at PBA-qnz/CPE in 0.1 mol L^{-1} B–R buffer (pH 2.0): (a) blank, (b) 0.5, (c) 1.0, (d) 2.5, (e) 5.0, (f) 10.0, (g) 15.0, (h) 20.0, (i) 25.0, and (j) 30.0 µmol L^{-1} and (**F**) calibration plot (n = 3). SWV parameters: f = 50.0 Hz, ΔE_s = 2.0 mV, a = 60.0 mV (optimized).

The repeatability of responses of the PBA-qnz/CPE sensor for BTL reduction was assessed across 10 consecutive measurements (Figure 4B). The plot shows a variation of current intensities for the BTL (axis a) with a relative standard deviation (RSD) of 8.0%. However, the ratio of $I_{SW-BTL}/I_{SW-PBA\text{-}qnz}$ (axis b) remained nearly unchanged (RSD = 1.5%), which indicates that the developed ratiometric sensing strategy minimized deviations among multiple assays because of the intrinsic built-in correction from the inner reference. This leads to a remarkable enhancement in the precision of the data provided by the sensor.

The effect of the pH of the supporting electrolyte on the electrochemical behavior of BTL was analyzed in the pH range of 2.0 to 7.0 (Figure 4C). Figure 4D (axis a) highlights how the cathodic peak currents vary as a function of the pH of the medium. It can be noted that the current values decreased from pH 2.0 to 7.0. Since hydrogen ions participate in the reduction of aromatic nitroanilines, the peak potential and intensities current of these compounds are pH-dependent. The pH of the medium influences the intensity and direction of the inductive and resonance effects operating in the molecule structure by changing the nature of the substituent [34]. The adsorption of nitroanilines onto the electrode surfaces in an acidic medium occurs because it has three different anchoring sites (the nitro, the amino function, and the aromatic ring system) [35]. Thus, it can be proposed that the adsorption of BTL onto the electrode surface occurs more efficiently at more acidic pH values, which results in higher current intensities. Another point to be considered is the lower stability of PBA-qnz as the pH of the medium increases, due to the affinity of OH^- ions for Fe(III) at pH close to 7.0, breaking the $Fe^{3+}-(CN)-Zn^{2+}$ bridge bond [36]. Thus, in aiming to obtain better sensitivity for the assays of BTL, pH 2.0 was selected for the subsequent analysis.

In addition, a linear shift of the E_p vs. pH plot (Figure 4C, axis b) was reached, with a slope of -58.1 mV pH^{-1}, which was similar to the theoretical value of -59.2 mV pH^{-1} for the Nernst equation. These data indicate that an equal number of mols of electrons and protons are transferred during the reduction of BTL on the surface of the ratiometric sensor. According to the literature, both nitro groups of the molecule were simultaneously reduced, via the one-proton and one-electron mechanism for each nitro group [17]. A proposed reduction reaction for BTL is shown in Figure 5.

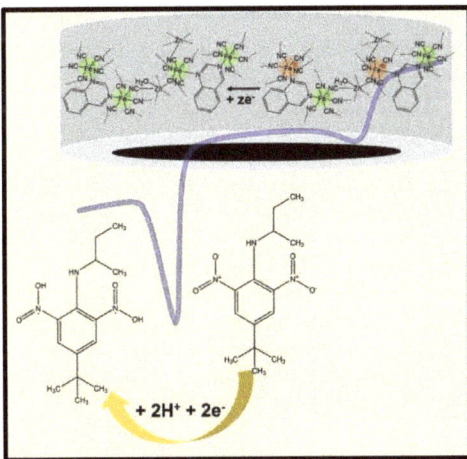

Figure 5. Proposed mechanism of BTL reduction on the surface of the PBA-qnz/CPE ratiometric sensor.

The calibration plot was built under optimized conditions at the PBA-qnz/CPE sensor after successive additions of the BTL standard solution (Figure 4E). The reduction peak of BTL can be observed at -0.57 ± 0.01 V vs. Ag/AgCl, with the current increasing proportionally to the species concentration, while the peak intensities referring to the reduction of the PBA-qnz remained stable at $+0.15$ V. The calibration plot was obtained in the range of 0.5 to 30.0 µmol L^{-1} (r = 0.998) (Figure 4F), and the equation for this plot can be expressed as $I_{SW-BTL}/I_{SW-PBA\text{-}qnz} = 0.05\ (\pm 2.0 \times 10^{-3})\ [BTL]/\mu mol\ L^{-1} + 0.03\ (\pm 2.6 \times 10^{-3})$. The limits of detection and quantification (LOD and LOQ) were calculated according to LOD = 3 × Sb/B and LOQ = 10 × Sb/B, where Sb is the standard deviation of intercept and B is the slope of the calibration plot [37]. The LOD and LOQ values obtained were 0.17 and 0.54 µmol L^{-1}, respectively.

The analytical methods reported for the quantification of BTL are mostly chromatographic (Table 1). Only two works dedicated to the determination of BTL using electrochemical methods were found in the literature [17,22]. In that regard, the LOD obtained with the PBA-qnz/CPE is between the values obtained by other studies. The CPE can be easily prepared and modified, and it can be easily cleaned by manual sanding on a sheet of paper. Furthermore, the use of the PBA modifying agent served to increase the analyte current intensities and as an internal reference for the development of the ratiometric sensor. Thus, in addition to being sensitive, the proposed ratiometric sensor presented excellent precision data.

Table 1. Comparison of the performance of different analytical methods in the quantification of BTL.

Analytical Method	Tools	Matrix	LOD/nmol L^{-1}	Ref.
Chromatographic	HPLC-UV with SPME [a]	Surface water	0.2	[38]
Chromatographic	HPLC-UV-ESI/MS [b]	Tobacco leaf powder	508	[39]
Immunochromatographic	Gold-based strip sensor	Phosphate buffer saline (pH 7.4)	10.4	[20]
Electrochemical	Co-Ag BMNPs-PVP/GCE [c]	B–R buffer (pH 2.0)	32.0	[17]
Electrochemical	Dropping mercury electrode	B–R buffer (pH 4.0)	60.0	[22]
Electrochemical	PBA-qnz/CPE	B–R buffer (pH 2.0)	170	This study

[a] HPLC-UV with SPME—High-performance liquid chromatography with ultraviolet detection with solid phase microextraction fiber coating based on silicone sealant/hollow ZnO@CeO$_2$ composite; [b] HPLC-UV-ESI/MS—High-performance liquid chromatography with ultraviolet detection and electrospray ionization mass spectrometry; [c] Co-Ag BMNPs-PVP/GCE—Glassy carbon electrode modified with Co and Ag bimetallic nanoparticles immobilized in poly(vinylpyrrolidone).

3.4. Interference and Stability Assays

The interference of organic compounds in the electroanalysis of BTL was studied under optimized conditions. The assays were carried out in 0.1 mol L^{-1} B–R buffer (pH 2.0) containing 10.0 µmol L^{-1} BTL (−0.56 V vs. Ag/AgCl) in the presence of 2-nitrophenol (−0.48 V), 3-nitrophenol (−0.45 V), 4-nitrophenol (−0.45 V), and parathion (−0.43 V), which were added at a concentration 10 times higher than that of the BTL. The reduction peak potential of these interferents did not coincide with the reduction peak potential of BTL, and, in addition, the decrease in the $I_{SW-BTL}/I_{SW-PBA\text{-}qnz}$ ratio in the presence of these interferents ranged from −0.5 to −1.5%. The results revealed that the proposed ratiometric sensor is highly selective for BTL quantification in the presence of organic compounds.

The stability of the ratiometric sensor was also inquired by measuring its response to 10.0 µmol L^{-1} BTL over 120 days. After this period, the PBA-qnz/CPE maintained a percentage of I_{SW-BTL} of 90% and 95% of $I_{SW-BTL}/I_{SW-PBA\text{-}qnz}$ in relation to its first response. These results indicated that the ratiometric sensor has excellent stability, evidencing its competence for the quantification of BTL.

3.5. Quantification of BTL in Lettuce and Potato Samples

The quantification of BTL by SWV in fresh samples was performed using the PBA-qnz/CPE ratiometric sensor (Table 2). Assays were carried out in triplicate using the standard addition procedure. The presence of BTL was not detected in any sample. Recovery values were obtained between 94 and 110%. These data confirm the accuracy of the data provided by the analytical method. Furthermore, the slopes of the standard addition plots were similar to those of the calibration plot (Figure 4F), which indicated that there were no influences from the matrix species of fresh samples.

Table 2. Determination of the level of BTL in fresh samples using the ratiometric sensor.

Samples	Determined [a] /μmol L^{-1}	Added/μmol L^{-1}	Found [a]/μmol L^{-1}	Recovery [b]/%
Lettuce (*Lactuca sativa*)	Not detected	1.0 10.0	1.03 10.07	95–110 99–102
Potato (*Solanum tuberosum*)	Not detected	1.0 10.0	1.01 9.97	94–110 97–105

[a] Mean of three measurements under the same conditions by SWV using the PBA-qnz/CPE; [b] Recovery = (amount found − amount determined)/amount added × 100.

4. Conclusions

A ratiometric sensor based on carbon paste modified with Prussian blue analogue derived from quinazoline ligand and zinc(II) was developed for the determination of BTL. This is the first device based on carbon paste dedicated to the electroanalysis of this herbicide. The use of the PBA modifying agent served to increase the BTL current intensities and as an internal reference for the development of the ratiometric sensor. The ratiometric sensor showed excellent precision and accuracy data and adequate selectivity for BTL. All of these capacities indicate the viability of the use of the PBA-qnz/CPE in the determination of BTL.

Author Contributions: Conceptualization, M.C.M., J.P.W. and E.R.S.; methodology, M.C.M., J.P.W. and E.R.S.; validation, J.P.W. and E.R.S.; formal analysis, M.C.M., J.P.W. and E.R.S.; investigation, M.C.M., J.P.W. and E.R.S.; writing—original draft preparation, M.C.M., J.P.W., and E.R.S.; writing—review and editing, M.C.M., J.P.W., E.R.S., B.S. and I.C.V.; supervision, B.S. and I.C.V.; project administration, B.S. and I.C.V.; funding acquisition, B.S. and I.C.V. All authors have read and agreed to the published version of the manuscript.

Funding: This research was funded by Conselho Nacional de Desenvolvimento Científico e Tecnológico (CNPq) and Coordenação de Aperfeiçoamento de Pessoal de Nível Superior (Capes), finance code 001.

Institutional Review Board Statement: Not applicable.

Informed Consent Statement: Not applicable.

Data Availability Statement: Not applicable.

Acknowledgments: The authors are grateful to the Brazilian government agencies CNPq and Capes for scholarships and financial support. This research was also supported by LACBIO (Laboratório de Catálise Biomimética) and the Analysis Center of the Department of Chemistry of the UFSC.

Conflicts of Interest: The authors declare no conflict of interest.

References

1. Song, X.; Song, S.; Wang, D.; Zhang, H. Prussian Blue Analogs and Their Derived Nanomaterials for Electrochemical Energy Storage and Electrocatalysis. *Small Methods* **2021**, *5*, 2001000. [CrossRef] [PubMed]
2. Qiu, S.; Xu, Y.; Wu, X.; Ji, X. Prussian Blue Analogues as Electrodes for Aqueous Monovalent Ion Batteries. *Electrochem. Energy Rev.* **2022**, *5*, 242–262. [CrossRef]
3. Ying, S.; Chen, C.; Wang, J.; Lu, C.; Liu, T.; Kong, Y.; Yi, F.Y. Synthesis and Applications of Prussian Blue and Its Analogues as Electrochemical Sensors. *Chempluschem* **2021**, *86*, 1608–1622. [CrossRef]
4. Pires, B.M.; Jannuzzi, S.V.A.; Formiga, A.L.B.; Bonacin, J.A. Prussian Blue Films Produced by Pentacyanidoferrate(II) and Their Application as Active Electrochemical Layers. *Eur. J. Inorg. Chem.* **2014**, *2014*, 5812–N5819. [CrossRef]
5. Chen, J.; Wang, Y.; Luo, X.; Chen, Y. Recent Research Progress and Outlook in Agricultural Chemical Discovery Based on Quinazoline Scaffold. *Pestic. Biochem. Physiol.* **2022**, *184*, 105122. [CrossRef]
6. Cheke, R.S.; Shinde, S.D.; Ambhore, J.P.; Chaudhari, S.R.; Bari, S.B. Quinazoline: An Update on Current Status against Convulsions. *J. Mol. Struct.* **2022**, *1248*, 131384. [CrossRef]
7. Chai, L.Q.; Chai, Y.M.; Li, C.G.; Zhou, L. Two Mono- and Dinuclear Cu (II) Complexes Derived from 3-Ethoxy Salicylaldehyde: X-ray Structures, Spectroscopic, Electrochemical, Antibacterial Activities, Hirshfeld Surfaces Analyses, and Time-Dependent Density Functional Theory Studies. *Appl. Organomet. Chem.* **2022**, *36*, e6475. [CrossRef]

8. Yang, Z.; Wang, T.; Chen, H.; Suo, X.; Halstenberg, P.; Lyu, H.; Jiang, W.; Mahurin, S.M.; Popovs, I.; Dai, S. Surpassing the Organic Cathode Performance for Lithium-Ion Batteries with Robust Fluorinated Covalent Quinazoline Networks. *ACS Energy Lett.* **2021**, *6*, 41–51. [CrossRef]
9. Wu, X.; Ru, Y.; Bai, Y.; Zhang, G.; Shi, Y.; Pang, H. PBA Composites and Their Derivatives in Energy and Environmental Applications. *Coord. Chem. Rev.* **2022**, *451*, 214260. [CrossRef]
10. Li, J.; He, L.; Jiang, J.; Xu, Z.; Liu, M.; Liu, X.; Tong, H.; Liu, Z.; Qian, D. Facile Syntheses of Bimetallic Prussian Blue Analogues ($K_xM[Fe(CN)_6]\cdot nH_2O$, M=Ni, Co, and Mn) for Electrochemical Determination of Toxic 2-Nitrophenol. *Electrochim. Acta* **2020**, *353*, 136579. [CrossRef]
11. Zhang, W.; Wen, J.; Wang, J.; Yang, K.; Sun, S. Recent Development and Application of Ratiometric Electrochemical Biosensor. *J. Electroanal. Chem.* **2022**, *921*, 116653. [CrossRef]
12. Xu, Z.; Li, P.; Liu, X.; Zhu, X.; Liu, M.; Zhang, Y.; Yao, S. Dual-Signal Intrinsic Self-Calibration Ratio Electrochemical Sensor for Glutathione Based on Silver Nanoparticle Decorated Prussian Blue Analog. *Electrochim. Acta* **2022**, *434*, 141273. [CrossRef]
13. Jin, H.; Gui, R.; Yu, J.; Lv, W.; Wang, Z. Fabrication Strategies, Sensing Modes and Analytical Applications of Ratiometric Electrochemical Biosensors. *Biosens. Bioelectron.* **2017**, *91*, 523–537. [CrossRef] [PubMed]
14. Spring, S.A.; Goggins, S.; Frost, C.G. Ratiometric Electrochemistry: Improving the Robustness, Reproducibility and Reliability of Biosensors. *Molecules* **2021**, *26*, 2130. [CrossRef]
15. Liu, C.; Wei, X.; Wang, X.; Shi, J.; Chen, Z.; Zhang, H.; Zhang, W.; Zou, X. Ratiometric Electrochemical Analysis on a Flexibly-Fabricated Vibratory Electrode Module for Reliable and Selective Determination of Imidacloprid. *Sens. Actuators B Chem.* **2021**, *329*, 129228. [CrossRef]
16. Yang, T.; Yu, R.; Yan, Y.; Zeng, H.; Luo, S.; Liu, N.; Morrin, A.; Luo, X.; Li, W. A Review of Ratiometric Electrochemical Sensors: From Design Schemes to Future Prospects. *Sens. Actuators B Chem.* **2018**, *274*, 501–516. [CrossRef]
17. Gerent, G.G.; Santana, E.R.; Martins, E.C.; Spinelli, A. A Non-Mercury Electrode for the Voltammetric Determination of Butralin in Foods. *Food Chem.* **2021**, *343*, 128419. [CrossRef]
18. Yang, L.; Song, X.; Zhou, X.; Zhou, Y.; Zhou, Y.; Gong, D.; Luo, H.; Deng, Y.; Yang, D.; Chen, L. Residual Behavior and Risk Assessment of Butralin in Peanut Fields. *Environ. Monit. Assess.* **2020**, *192*, 62. [CrossRef]
19. Ghatge, S.; Yang, Y.; Moon, S.; Song, W.Y.; Kim, T.Y.; Liu, K.H.; Hur, H.G. A Novel Pathway for Initial Biotransformation of Dinitroaniline Herbicide Butralin from a Newly Isolated Bacterium Sphingopyxis Sp. Strain HMH. *J. Hazard. Mater.* **2021**, *402*, 123510. [CrossRef]
20. Xu, X.; Guo, X.; Song, S.; Wu, A.; Xu, C.; Kuang, H.; Liu, L. Gold-Based Strip Sensor for the Rapid and Sensitive Detection of Butralin in Tomatoes and Peppers. *Food Addit. Contam.—Part A Chem. Anal. Control Expo. Risk Assess.* **2022**, *39*, 1255–1264. [CrossRef]
21. Wang, X.; You, Q.; Hou, Z.; Yu, X.; Gao, H.; Gao, Y.; Wang, L.; Wei, L.; Lu, Z. Establishing the HPLC-MS/MS Method for Monitoring the Residue and Degradation of Butralin in Ginseng during Field and Risk Assessments. *Agronomy* **2022**, *12*, 2675. [CrossRef]
22. Sreedhar, M.; Reddy, S.J. Electrochemical Reduction and Differential Pulse Polarographic Determination of Butralin and Isopropalin in Environmental Samples at a Mercury Electrode. *Bull. Chem. Soc. Jpn.* **2002**, *75*, 2155–2159. [CrossRef]
23. Deng, X.; Lin, X.; Zhou, H.; Liu, J. Equipment of Vertically-Ordered Mesoporous Silica Film on Electrochemically Pretreated Three-Dimensional Graphene Electrodes for Sensitive Detection of Methidazine in Urine. *Nanomaterials* **2023**, *13*, 239. [CrossRef] [PubMed]
24. Gong, J.; Zhang, T.; Chen, P.; Yan, F.; Liu, J. Bipolar Silica Nanochannel Array for Dual-Mode Electrochemiluminescence and Electrochemical Immunosensing Platform. *Sens. Actuators B Chem.* **2022**, *368*, 132086. [CrossRef]
25. Gong, J.; Tang, H.; Wang, M.; Lin, X.; Wang, K.; Liu, J. Novel Three-Dimensional Graphene Nanomesh Prepared by Facile Electro-Etching for Improved Electroanalytical Performance for Small Biomolecules. *Mater. Des.* **2022**, *215*, 110506. [CrossRef]
26. Franzoi, A.C.; Peralta, R.A.; Neves, A.; Vieira, I.C. Biomimetic Sensor Based on $Mn^{III}Mn^{II}$ Complex as Manganese Peroxidase Mimetic for Determination of Rutin. *Talanta* **2009**, *78*, 221–226. [CrossRef]
27. Achelle, S.; Rodríguez-López, J.; Robin-Le Guen, F. Synthesis and Photophysical Studies of a Series of Quinazoline Chromophores. *J. Org. Chem.* **2014**, *79*, 7564–7571. [CrossRef]
28. Jannuzzi, S.A.V.; Martins, B.; Felisberti, M.I.; Formiga, A.L.B. Supramolecular Interactions between Inorganic and Organic Blocks of Pentacyanoferrate/Poly(4-Vinylpyridine) Hybrid Metallopolymer. *J. Phys. Chem. B* **2012**, *116*, 14933–14942. [CrossRef]
29. Monteiro, M.C.; Toledo, K.C.F.; Pires, B.M.; Wick, R.; Bonacin, J.A. Improvement in Efficiency of the Electrocatalytic Reduction of Hydrogen Peroxide by Prussian Blue Produced from the $[Fe(CN)_5(Mpz)]^{2-}$Complex. *Eur. J. Inorg. Chem.* **2017**, *2017*, 1979–1988. [CrossRef]
30. Moore, K.J.; Lee, L.; Figard, J.E.; Gelroth, J.A.; Stinson, A.J.; Wohlers, H.D.; Petersen, J.D. Photochemistry of Mixed-Metal Bimetallic Complexes Containing Pentacyanoferrate(II) or Pentaammineruthenium(II) Metal Centers. Evidence for Some Intramolecular Energy-Transfer Reactions. *J. Am. Chem. Soc.* **1983**, *105*, 2274–2279. [CrossRef]
31. Gatteschi, D. *NMR, NQR, EPR and Mössbauer Spectroscopy in Inorganic Chemistry*, 1st ed.; Ellis Horwood Limited: West Sussex, UK, 1990.
32. Ghobadi, T.G.U.; Ghobadi, A.; Demirtas, M.; Buyuktemiz, M.; Ozvural, K.N.; Yildiz, E.A.; Erdem, E.; Yaglioglu, H.G.; Durgun, E.; Dede, Y.; et al. Building an Iron Chromophore Incorporating Prussian Blue Analogue for Photoelectrochemical Water Oxidation. *Chem.—A Eur. J.* **2021**, *27*, 8966–8976. [CrossRef] [PubMed]

33. Li, J.; Chu, Y.; Zhang, C.; Zhang, X.; Wu, C.; Xiong, X.; Zhou, L.; Wu, C.; Han, D. CoFe Prussian Blue Decorated BiVO$_4$ as Novel Photoanode for Continuous Photocathodic Protection of 304 Stainless Steel. *J. Alloys Compd.* **2021**, *887*, 161279. [CrossRef]
34. Glicksman, R.; Morehouse, C.K. Investigation of the Electrochemical Properties of Organic Compounds. I. Aromatic Nitro Compounds. *J. Electrochem. Soc.* **1958**, *105*, 299. [CrossRef]
35. Kumar, S.A.; Chen, S. Myoglobin/Arylhydroxylamine Film Modified Electrode: Direct Electrochemistry and Electrochemical Catalysis. *Talanta* **2007**, *72*, 831–838. [CrossRef]
36. Bonacin, J.A.; Dos Santos, P.L.; Katic, V.; Foster, C.W.; Banks, C.E. Use of Screen-Printed Electrodes Modified by Prussian Blue and Analogues in Sensing of Cysteine. *Electroanalysis* **2018**, *30*, 170–179. [CrossRef]
37. Gumustas, M.; Ozkan, S.A. The Role of and the Place of Method Validation in Drug Analysis Using Electroanalytical Techniques. *Open Anal. Chem. J.* **2011**, *5*, 1–21. [CrossRef]
38. Wang, X.; Huang, L.; Yuan, N.; Huang, P.; Du, X.; Lu, X. Facile Fabrication of a Novel SPME Fiber Based on Silicone Sealant/Hollqow ZnO@CeO$_2$ Composite with Super-Hydrophobicity for the Enhanced Capture of Pesticides from Water. *Microchem. J.* **2022**, *183*, 108118. [CrossRef]
39. Liu, H.; Ding, C.; Zhang, S.; Liu, H.; Liao, X.; Qu, L.; Zhao, Y.; Wu, Y. Simultaneous Residue Measurement of Pendimethalin, Isopropalin, and Butralin in Tobacco Using High-Performance Liquid Chromatography with Ultraviolet Detection and Electrospray Ionization/Mass Spectrometric Identification. *J. Agric. Food Chem.* **2004**, *52*, 6912–6915. [CrossRef]

Disclaimer/Publisher's Note: The statements, opinions and data contained in all publications are solely those of the individual author(s) and contributor(s) and not of MDPI and/or the editor(s). MDPI and/or the editor(s) disclaim responsibility for any injury to people or property resulting from any ideas, methods, instructions or products referred to in the content.

Article

Incorporation of Bismuth(III) Oxide Nanoparticles into Carbon Ceramic Composite: Electrode Material with Improved Electroanalytical Performance in 4-Chloro-3-Methylphenol Determination

Mariola Brycht, Andrzej Leniart, Sławomira Skrzypek and Barbara Burnat *

University of Lodz, Faculty of Chemistry, Department of Inorganic and Analytical Chemistry, Tamka 12, 91-403 Lodz, Poland; mariola.brycht@chemia.uni.lodz.pl (M.B.); andrzej.leniart@chemia.uni.lodz.pl (A.L.); slawomira.skrzypek@chemia.uni.lodz.pl (S.S)
* Correspondence: barbara.burnat@chemia.uni.lodz.pl

Citation: Brycht, M.; Leniart, A.; Skrzypek, S.; Burnat, B. Incorporation of Bismuth(III) Oxide Nanoparticles into Carbon Ceramic Composite: Electrode Material with Improved Electroanalytical Performance in 4-Chloro-3-Methylphenol Determination. *Materials* 2024, 17, 665. https://doi.org/10.3390/ma17030665

Academic Editor: Alessandro Dell'Era

Received: 20 December 2023
Revised: 21 January 2024
Accepted: 26 January 2024
Published: 29 January 2024

Copyright: © 2024 by the authors. Licensee MDPI, Basel, Switzerland. This article is an open access article distributed under the terms and conditions of the Creative Commons Attribution (CC BY) license (https://creativecommons.org/licenses/by/4.0/).

Abstract: In this study, a carbon ceramic electrode (CCE) with improved electroanalytical performance was developed by bulk-modifying it with bismuth(III) oxide nanoparticles (Bi-CCE). Characterization of the Bi-CCE was conducted employing atomic force microscopy, scanning electron microscopy with energy-dispersive X-ray spectroscopy, cyclic voltammetry (CV), and electrochemical impedance spectroscopy. Comparative analysis was conducted using an unmodified CCE. The findings proved that the incorporation of Bi_2O_3 nanoparticles into the CCE significantly altered the morphology and topography of the ceramic composite, and it improved the electrochemical properties of CCE. Notably, the Bi-CCE demonstrated a prolonged operational lifespan of at least three months, and there was a high reproducibility of the electrode preparation procedure. The developed Bi-CCE was effectively employed to explore the electrochemical behavior and quantify the priority environmental pollutant 4-chloro-3-methylphenol (PCMC) using CV and square-wave voltammetry (SWV), respectively. Notably, the developed SWV procedure utilizing Bi-CCE exhibited significantly enhanced sensitivity (0.115 $\mu A\ L\ mol^{-1}$), an extended linearity (0.5–58.0 $\mu mol\ L^{-1}$), and a lower limit of detection (0.17 $\mu mol\ L^{-1}$) in comparison with the unmodified electrode. Furthermore, the Bi-CCE was utilized effectively for the detection of PCMC in a river water sample intentionally spiked with the compound. The selectivity toward PCMC determination was also successfully assessed.

Keywords: bulk modification; electrochemical characterization; surface characterization; effective surface area; pollutant determination

1. Introduction

Over the past several years, significant attention has been devoted to advancing a diverse range of carbon-based electrode materials. These electrodes offer numerous advantages, including straightforward preparation, extended durability, a wide potential window, and facile surface renewability [1]. Carbon-based electrodes find extensive use in electrochemical sensors, facilitating the identification of diverse substances such as pollutants, biomolecules, and drugs [1–4]. Additionally, carbon-based materials are crucial in biosensors, contributing significantly to the detection of biological molecules. This makes them invaluable in applications such as medical diagnostics, environmental monitoring, and ensuring food safety [2,3,5–7]. Furthermore, these electrodes are utilized in electroanalytical techniques, aiding in the precise determination of trace elements and the exploration of reaction mechanisms [8,9].

Among the spectrum of carbon-based electrodes, carbon ceramic electrodes (CCEs) pioneered by Tsionsky et al. in 1994 [10] occupy a prominent position. Fabricated through a sol–gel approach integrating carbon powder (graphite) into a silica sol–gel matrix [11],

CCEs not only showcase the above-mentioned features of carbon-based electrodes but also exhibit exceptional mechanical resistance and robustness [12]. The versatility of CCEs for modification through diverse techniques stands as a crucial advantage, enhancing their electroanalytical capabilities. Surface modifications of CCEs achieved via electrodeposition [13–15] or drop-casting procedure [16–18] contribute to improved electrode sensitivity. However, these approaches pose challenges related to surface renewal, demanding the preparation of new modification layers and potentially leading to inconsistent results. Moreover, these modification techniques encounter limitations in controlling electrode surface film thickness [12]. An efficient alternative to the surface modification procedure, ensuring greater result reproducibility, involves the bulk modification of the electrode material. In the case of CCEs, this approach includes the partial [19–23] or complete [24–27] replacement of the original carbon material (graphite) in the silicon matrix with other carbon-based materials, such as carbon nanotubes [19,24–27], graphene oxide [20], carbon black [21], and non-carbon materials like nanoparticles [23,28], zeolites [22], and Prussian blue [29]. The advantage of this approach lies in obtaining a carbon–ceramic composite with an evenly distributed modifier throughout the entire volume of the electrode material. This allows for straightforward surface renewal through mechanical polishing while maintaining surface reproducibility after each polishing step [12].

Metal oxide nanoparticles have gathered significant attention in the electrochemistry due to their unique properties, making them ideal for fabricating electrochemical sensors used in electroanalysis [30]. Bismuth(III) oxide (Bi_2O_3) nanoparticles (Bi_2O_3NPs) exhibit promising electronic characteristics, including a low-energy bandgap, large surface area, electrochemical stability, and catalytic behavior, making them well-suited for various applications [30–32]. Although Bi_2O_3NPs have primarily functioned as surface modifiers for glassy carbon electrodes and screen-printed electrodes, their utilization as a bulk modifier has been limited to carbon paste electrodes. It is envisioned that employing Bi_2O_3NPs as a bulk modifier of CCEs will result in electrode materials with significantly enhanced surface properties, an enlarged electroactive surface area, and improved electroanalytical performance.

Phenolic compounds have been designated as high-priority pollutants by both the United States Environmental Protection Agency and the European Union due to their known toxicity. These chemicals exhibit significant short- and long-term adverse effects on both human health and animal well-being. The existence of phenolic compounds in the aquatic environments is not only undesirable but also poses a significant threat to human health and wildlife. Consequently, various wastewater treatment methods have been developed and implemented to eliminate phenolic compounds from industrial, domestic, and municipal wastewater streams before their release into water bodies [33]. Moreover, numerous procedures based on chromatographic and electroanalytical methods have been developed for the quantitative determination of phenolic compounds in water. One notable phenolic priority environmental pollutant is 4-chloro-3-methylphenol (synonym: 4-chloro-m-cresol, PCMC), which is used as a medicinal or non-medicinal ingredient in final pharmaceuticals, disinfectants, veterinary drug products, and cosmetics, and it is used as an active ingredient in registered pest control products. Due to the wide use of PCMC, it is necessary to monitor its content in the aquatic environment to prevent its destructive effects on both human and aquatic lives. In the literature, only a few reports on electroanalytical procedures for PCMC determination using carbon-based electrodes can be found [24,34–38]. One of these reports is from our previous study, in which a voltammetric procedure for PCMC determination involving CCE modified with carbon nanotubes was detailed [24]. In that study, a linear range of 3–32 $\mu mol\ L^{-1}$ and a limit of detection (LOD) of 0.71 $\mu mol\ L^{-1}$ was achieved. Considering the potential for improvement, we hypothesize that using Bi_2O_3NPs to prepare a modified CCE could enhance the performance of such a voltammetric procedure. To our knowledge, this specific electrode type has not been employed for PCMC determination previously.

Our study aims to prepare a bulk-modified CCE by incorporating Bi$_2$O$_3$NPs (Bi-CCE), replacing a portion of graphite within a ceramic composite. The investigation focuses on assessing how the incorporation of Bi$_2$O$_3$NPs influences the surface characteristics, electrochemical properties toward the ferro/ferricyanide ([Fe(CN)$_6$]$^{4-}$/[Fe(CN)$_6$]$^{3-}$) redox model system, and the overall electroanalytical performance of the Bi-CCE toward the priority environmental pollutant PCMC.

2. Materials and Methods

2.1. Materials

For the CCEs preparation, methyltrimethoxysilane (MTMS, 98%, Sigma Aldrich, Warsaw, Poland), methanol (CH$_3$OH, 99.8%, Avantor Performance, Gliwice, Poland), and hydrochloric acid (HCl, pure p.a., 36–38%, Avantor Performance, Gliwice, Poland) were employed without undergoing any further purification. Graphite flakes (99%, 7–10 micron, Alfa Aesar, Karlsruhe, Germany) served as the carbon material for the CCEs. Bi$_2$O$_3$NPs (99.999%, particle size of 90–210 nm, Sigma Aldrich, Warsaw, Poland) were utilized as the modifier in the Bi-CCE preparation.

All chemicals employed were of analytical reagent grade, and the solutions were prepared using triply distilled water. The electrochemical characterization of the CCEs involved the use of 1.0 mol L^{-1} potassium chloride (KCl, 99%, Avantor Performance, Gliwice, Poland) solution and 1.0 mmol L^{-1} potassium ferricyanide (K$_3$[Fe(CN)$_6$], 99%, Avantor Performance, Gliwice, Poland) solution. For the preparation of a 1.0 mmol L^{-1} PCMC stock solution, PCMC (99%, Sigma Aldrich, Warsaw, Poland) was dissolved in water and stored in a glass flask in a refrigerator when not in use. For the preparation of the Britton–Robinson buffer (BRB) across a pH range of 2.0–12.0, the following reagents were utilized: phosphoric acid (H$_3$PO$_4$, 85.0%, Avantor Performance, Gliwice, Poland), boric acid (H$_3$BO$_3$, pure p.a., Avantor Performance, Gliwice, Poland), and acetic acid (CH$_3$COOH, pure p.a., 99.5%, Avantor Performance, Gliwice, Poland), all at a concentration of 40.0 mmol L^{-1}. Additionally, sodium hydroxide (NaOH, pure p.a., Avantor Performance, Gliwice, Poland) at a concentration of 0.20 mol L^{-1} was employed to adjust the pH values of the buffer solutions. River water samples, collected from the Rudawa River at coordinates 50.057040, 19.906389, were utilized for the analytical purposes. These samples, spiked with the known concentration of PCMC, underwent examination without any pretreatment or filtration except for dilution. The samples were stored in a refrigerator before experiments and analyzed within one week of collection. The concentration of stock solutions of interferents (Cd^{2+}, Ni^{2+}, Cu^{2+}, HCO$_3^-$, and SO$_4^{2-}$; all from Avantor Performance, Gliwice, Poland) was 1.0 mmol L^{-1}.

2.2. Preparation of the CCEs

To prepare Bi-CCE, the sol–gel method was employed. Initially, a mixture containing 750 µL of CH$_3$OH as the solvent, 500 µL of MTMS serving as the silica matrix precursor, and 50 µL of concentrated HCl as the catalyst was stirred for 5 min using a magnetic stirrer set at 450 rpm. Next, 600 mg of the activated graphite (prepared following the procedure outlined in [39]) was combined with 150 mg of Bi$_2$O$_3$NPs in the silica sol solution. This mixture was thoroughly stirred using a spatula and then promptly transferred into a Teflon tube (measuring 5 mm in length and 3 mm in inner diameter), and silver-painted copper wire (1 mm in diameter) was used as the electrical contact. The unmodified CCE was prepared analogously with the only difference being the addition of 750 mg of activated graphite into the silica sol solution. The resulting CCEs were then air-dried for 48 h at room temperature. The CCEs were polished using 2000 grit polishing paper, which was followed by a cleansing with water and drying using argon. This process was performed before their initial use and repeated before each measurement series.

2.3. Apparatus

An atomic force microscope (AFM, Dimension Icon, Bruker Corporation, Santa Barbara, CA, USA) and field emission scanning electron microscope (SEM, Nova NanoSEM 450, FEI, USA) with an energy-dispersive X-ray analyzer (EDX, Ametek Inc, Berwyn, PA, USA) were employed to analyze both CCEs. AFM measurements were carried out in tapping mode, employing TESPA (NanoWorld, Neuchâtel, Switzerland) probes, featuring a nominal spring constant of 42 N m^{-1} and a resonance frequency of 320 kHz. The AFM images were captured with a scan size of 5 μm × 5 μm in randomly chosen places on the CCE surfaces. AFM topography images were used to determine roughness parameters, including root mean square average roughness (R_q) and surface area difference (SAD), using NanoScope Analysis software (version 1.4, Bruker, Santa Barbara, CA, USA). SEM measurements were performed with an accelerating voltage of 10 kV using a through-the-lens detector (TLD). Elemental surface composition was obtained from EDX spectra, which were acquired through area analysis.

Cyclic voltammetry (CV) measurements were undertaken using an μAutolab type II potentiostat-galvanostat (Eco Chemie, Utrecht, the Netherlands) under the control of GPES software (version 4.9), whereas electrochemical impedance spectroscopy (EIS) experiments were conducted employing an AUTOLAB N128 electrochemical analyzer with FRA2 module (Eco Chemie, Utrecht, the Netherlands) operated by FRA software (version 4.9). Both potentiostats were linked to an M164 electrode stand (MTM Anko Instruments, Cracow, Poland). For these experiments, a three-electrode electrochemical cell configuration was utilized, consisting of Ag|AgCl|3 mol L^{-1} KCl (Mineral, Łomianki-Sadowa, Poland) as the reference electrode, Pt wire (99.99%, The Mint of Poland, Warsaw, Poland) as the counter electrode, and laboratory-made CCEs as the working electrodes.

2.4. Electrochemical Procedures

The electrochemical characterization of both CCEs was implemented in 1.0 mol L^{-1} KCl (to assess the potential window width) and 1.0 mmol L^{-1} K$_3$[Fe(CN)$_6$] solutions (to evaluate the reversibility of the model redox process) using CV and EIS. The cyclic voltammograms in KCl solution were recorded at a scan rate of 100 mV s^{-1}, while those of K$_3$[Fe(CN)$_6$] solution were registered over a scan rate range from 5 to 400 mV s^{-1}. The EIS spectra were captured within the frequency ranging from 10,000 to 0.01 Hz (amplitude 10 mV, 50 measuring points).

The electrochemical behavior of PCMC was investigated on the Bi-CCE using CV. Cyclic voltammograms were recorded in BRB at an optimized pH value of 5.0 with a PCMC concentration of 50.0 μmol L^{-1} across the potential range from −0.3 to +1.3 V, employing scan rates ranging from 5 to 400 mV s^{-1}. To quantitatively determine PCMC on the Bi-CCE, the SWV was utilized. The PCMC stock solution (1.0 mmol L^{-1}) was successively added to the cell containing BRB at pH 5.0, covering concentrations ranging from 0.5 to 58.0 μmol L^{-1}. SW voltammograms were recorded with a potential ranging from +0.25 to +1.25 V, using optimized SWV parameters: amplitude of 50 mV, frequency of 60 Hz, and step potential of 5 mV. SWV signals were measured after the baseline correction. Comparative analysis was conducted using the unmodified CCE.

The real sample analysis was conducted in spiked river water samples using the standard addition method. Initially, to study the possible interferences caused by river water components, a blank SW voltammogram was registered for a solution comprising 9.0 mL of BRB at pH 5.0 and 1 mL of unspiked river water. Subsequently, SW voltammogram for the river water sample spiked with PCMC (the cell contained 9.0 mL of BRB at pH 5.0, 1.0 mL of river water spiked with the PCMC stock solution, resulting in a PCMC concentration of 6.0 μmol L^{-1}) was recorded. Following this, SW voltammograms were registered for three consecutive additions of the PCMC stock solution (each 60 μL) to the cell containing the river water sample that was spiked with PCMC and diluted with BRB at pH 5.0. SWV signals were measured after the baseline correction.

The effect of interferents (ions) that might be possibly present in river water, i.e., Cd^{2+}, Ni^{2+}, Cu^{2+}, HCO_3^-, and SO_4^{2-}, on the PCMC SWV signal was investigated. The ratio of the PCMC concentration to the interferents concentrations in the voltammetric cell was equal to 1:0.5, 1:1, 1:2, 1:10, 1:50, and 1:100.

3. Results and Discussion

3.1. Surface Characterization of CCEs

The morphology and topography of both the unmodified CCE and the CCE bulk-modified with the Bi_2O_3NPs were examined using SEM and AFM techniques. Both SEM and AFM images of the unmodified CCE, depicted in Figure 1A,B, illustrate flatly arranged graphite flakes covered by a silicon matrix. Elemental analysis shows only C, O, and Si elements, as depicted in Figure 1C. Between individual flakes, noticeable gaps were observed, which is a common feature in such electrodes [12,24,25,27]. In this instance, the observed gaps measure approximately 100 nm deep (Figure 1D). Conversely, SEM and AFM images of the Bi-CCE (Figure 1E,F) show changes in the surface morphology due to the incorporation of Bi_2O_3NPs into the ceramic composite. Notably, the morphology of the Bi-CCE differs, presenting a more compact surface with graphite flakes decorated with the Bi_2O_3NPs primarily organized in larger agglomerates (Figure 1E). Their presence was confirmed by EDX results indicating ca. 13 wt.% (ca. 0.99 at.%) of Bi_2O_3NPs in the ceramic composite (Figure 1G). The AFM results prove a more compact surface of the Bi-CCE (Figure 1F), and the cross-section (Figure 1H) demonstrates that no pinholes are visible. While individual graphite flakes are also distinguished in the AFM image, identifying the Bi_2O_3NPs used for modification remains challenging. This difficulty arises from their substantial coverage by the ceramic composite, complicating their distinct visualization in the AFM image. Calculated surface roughness parameter Rq values (88.5 nm for the unmodified CCE, 5.4 nm for the Bi-CCE) indicate reduced roughness after modification. Significant differences in surface morphology are evident upon comparison of the surface area difference (SAD) parameter values between prepared CCEs derived from AFM images shown in Figure 1B,F (16.0% for the unmodified CCE, and 0.5% for the Bi-CCE). These SAD values confirm that the bulk modification of CCE with Bi_2O_3NPs leads to a smoother electrode surface with a smaller surface area. Nevertheless, a reduced surface area does not necessarily imply a smaller electroactive surface area, as will be demonstrated by results from subsequent electrochemical measurements. Additionally, from a practical perspective, a smoother electrode surface poses fewer challenges in laboratory applications compared to porous electrodes.

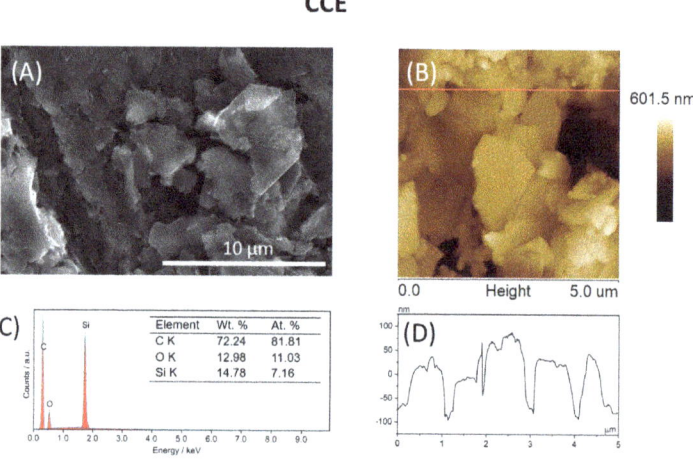

Figure 1. Cont.

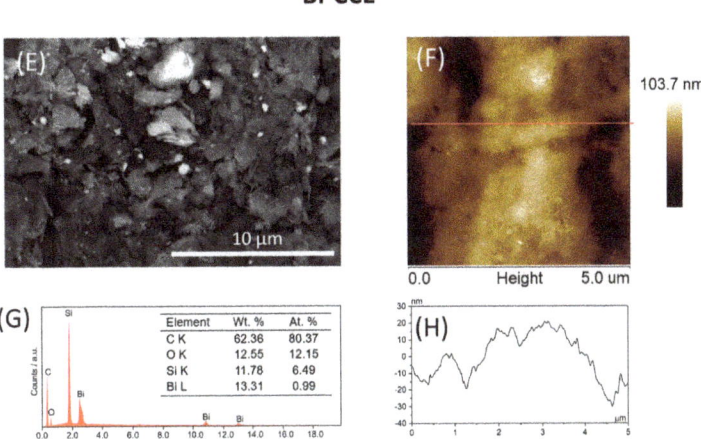

Figure 1. Surface morphology, composition, and topography of the unmodified CCE and Bi-CCE: (**A**,**E**) SEM images, (**C**,**G**) EDX spectra, (**B**,**F**) 2D AFM images, (**D**,**H**) cross-sections. SEM imaging: HV 10 kV, TLD detector, mag. 10,000×. AFM imaging: tapping mode, scan size: 5 μm × 5 μm.

3.2. Electrochemical Characterization of CCEs

The initial step in the electrochemical characterization of the prepared CCEs involved determining their working potential window, which is a critical factor indicating the applicability of the working electrodes and providing valuable insights into their surface properties. To assess the potential window width of the Bi-CCE, CV measurements were conducted in a 1.0 mol L^{-1} KCl solution, which is commonly used for characterizing electrode potential windows. The obtained results were then compared to the corresponding curve acquired for the unmodified CCE. As depicted in Figure 2A, the introduction of Bi$_2$O$_3$NPs into the CCE had an insignificant impact on the potential window width. Both CCEs displayed the same accessible cathodic potential limit (at −0.25 V), while a slightly higher anodic limit was observed in the Bi-CCE (at +1.25 V) compared to the unmodified CCE (at +1.2 V). Significantly, the incorporation of Bi$_2$O$_3$NPs within the ceramic matrix notably and positively influenced the background current; the Bi-CCE demonstrated a considerably lower capacitive current compared to the unmodified CCE. The heightened background current in the unmodified CCE may be attributed to its more porous surface in contrast to the Bi-CCE. Conversely, the reduced background current observed in the Bi-CCE could be credited to the exceptional properties of Bi$_2$O$_3$NPs incorporated into the ceramic composite. Moreover, due to the lower background current, it is anticipated that the Bi-CCE will exhibit a lower detection limit than the unmodified CCE.

The subsequent step in the electrochemical characterization of the prepared CCEs involved assessing the reversibility of a model redox marker using CV and the rate of the electron transfer process using EIS. The representative cyclic voltammograms of 1.0 mmol L^{-1} [Fe(CN)$_6$]$^{4-}$/[Fe(CN)$_6$]$^{3-}$ recorded at a scan rate (v) of 100 mV s^{-1} on both CCEs are depicted in the inset of Figure 2A. While both electrodes exhibited a well-defined redox peak pair, those observed on the Bi-CCE appeared higher than those on the unmodified CCE. Moreover, a peak potential separation (ΔE_p) value closer to the theoretical value of 0.059 V was obtained for the Bi-CCE ([Fe(CN)$_6$]$^{4-}$/[Fe(CN)$_6$]$^{3-}$ redox probe displayed a ΔE_p value of 0.080 V for the Bi-CCE and 0.113 V for the unmodified CCE), suggesting a more reversible electrode process on the Bi-CCE compared to the unmodified CCE. Furthermore, cyclic voltammograms of the [Fe(CN)$_6$]$^{4-}$/[Fe(CN)$_6$]$^{3-}$ redox marker were recorded across a scan rate range of 10–400 mV s^{-1} for the Bi-CCE (Figure 2B) as well as for the unmodified CCE (results not shown). In case of both electrodes, a linear dependence of the I_p vs. $v^{1/2}$ was observed. Additionally, the I_a/I_c ratio ranged from 0.99 to 1.09 for the Bi-CCE and from 1.03 to 1.12 for the unmodified CCE, closely aligning with the theoretical value of 1. These results indicate the enhanced electron transfer process with improved reversibility of the $[Fe(CN)_6]^{4-}/[Fe(CN)_6]^{3-}$ redox marker at the Bi-CCE when compared to the unmodified CCE. The EIS spectra (Nyquist plot of real impedance (Z') vs. imaginary impedance (Z'') fitted by a Randles equivalent circuit) for 1.0 mmol L^{-1} $[Fe(CN)_6]^{4-}/[Fe(CN)_6]^{3-}$ recorded on both CCEs are depicted in the inset of Figure 2C. The EIS results confirmed a reduction in charge transfer resistance for the Bi-CCE (275.7 Ω cm^2) when compared to the unmodified CCE (396.3 Ω cm^2), indicating a more facile redox reaction on this electrode material. All the aforementioned results affirm the beneficial impact of incorporating Bi_2O_3NPs into the ceramic composite on the superior electrochemical properties of the modified CCE.

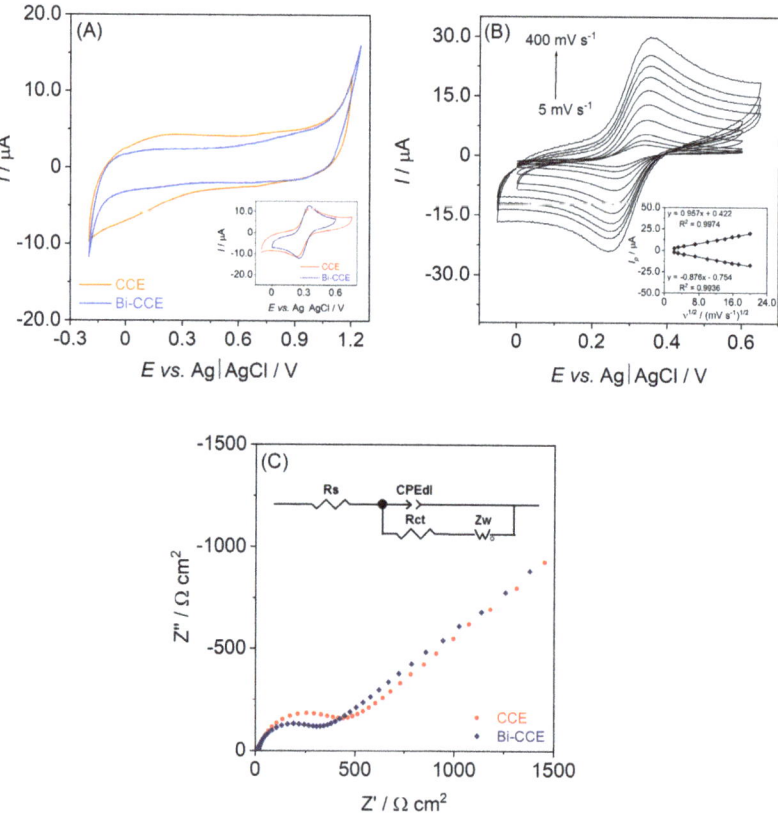

Figure 2. (**A**) Cyclic voltammograms (a scan rate of 100 mV s^{-1}) registered in 1.0 mol L^{-1} KCl solution for the unmodified CCE (orange line) and the Bi-CCE (blue line). Inset shows cyclic voltammograms (a scan rate of 100 mV s^{-1}) recorded in 1.0 mmol L^{-1} $K_3[Fe(CN)_6]$ solution at the unmodified CCE (red line) and the Bi-CCE (navy line); (**B**) Cyclic voltammograms recorded in 1.0 mmol L^{-1} $K_3[Fe(CN)_6]$ solution at the Bi-CCE in the scan rates range of 5–400 mV s^{-1}. Inset illustrates the relationship between the peaks current (I_p) and the square root of the scan rate ($v^{1/2}$); (**C**) EIS spectra captured for 1.0 mmol L^{-1} $[Fe(CN)_6]^{4-}/[Fe(CN)_6]^{3-}$ in 1.0 mol L^{-1} KCl solution at the unmodified CCE (•) and the Bi-CCE (♦). Frequency range of 10,000–0.01 Hz. Inset exhibits the Randles equivalent circuit consisted of the solution resistance (R_s), constant phase element describing double-layer capacitance (CPE_{dl}), charge transfer resistance (R_{ct}), and Warburg impedance related with diffusion (Z_w).

Subsequently, cyclic voltammograms of the $[Fe(CN)_6]^{4-}/[Fe(CN)_6]^{3-}$ redox marker registered in the v range of 5–400 mV s^{-1} were utilized to evaluate the effective surface area (A_{eff}) values for both electrodes. A_{eff} were determined using the Randles–Sevcik equation [40]: $I_p = (2.69 \times 10^5) n^{3/2} A_{eff} D^{1/2} v^{1/2} c_0$, where I_p represents peak current (A), n denotes the number of transferred electrons (1), D signifies the diffusion coefficient of $[Fe(CN)_6]^{4-}/[Fe(CN)_6]^{3-}$ (7.6 × 10^{-6} cm^2 s^{-1}), v denotes the scan rate (V s^{-1}), and c_0 indicates the redox marker concentration (1.0 × 10^{-6} mol cm^{-3}). An increased A_{eff} value (1.6-times higher) was calculated for the Bi-CCE (A_{eff} of 4.09 mm^2) compared to the unmodified CCE (A_{eff} of 2.59 mm^2), indicating the advantageous impact of incorporating Bi$_2$O$_3$NPs in boosting the number of electroactive sites within the ceramic–carbon composite. Furthermore, it is worth noting that the A_{eff} values smaller than the geometric area value (7.07 mm^2) were calculated for both CCEs, implying the existence of electrochemically inactive sites on the electrode surface. These electrochemical results confirm the earlier statement that the electroactive surface area is determined not solely by the electrode surface morphology but predominantly by the ratio of electroactive to electrochemically inactive sites on the electrode surface. The incorporation of electroactive Bi$_2$O$_3$NPs within the electrode material simplifies the process of renewing the electrode surface while preserving its enhanced electroactivity.

3.3. Long-Term Stability and Reproducibility Study of the Bi-CCE

The long-term stability over time of the developed Bi-CCE was assessed by conducting CV experiments in 1.0 mmol L^{-1} K$_3$[Fe(CN)$_6$] solution over an extended period. The measurements were systematically performed at regular intervals (every 10 days) for three months (each time involving the renewal of the electrode surface by polishing). The recorded cyclic voltammograms exhibited negligible changes in the I_p values (Figure 3A) indicated by a low relative standard deviation (RSD) value. Specifically, the RSD of I_p remained below 4% for 50 days and below 8% for 90 days. This consistent performance suggests that the Bi-CCE can be considered stable over time and easily renewable through a straightforward polishing procedure.

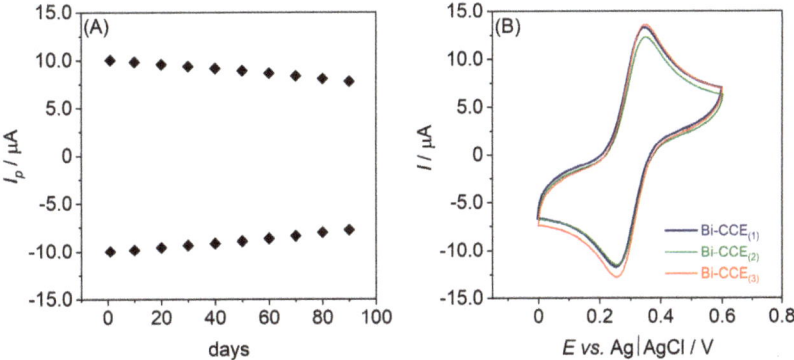

Figure 3. (**A**) The plot representing the changes in the peak current (I_p) over three months (90 days) of the developed Bi-CCE; (**B**) cyclic voltammograms (scan rate of 100 mV s^{-1}) in 1.0 mmol L^{-1} K$_3$[Fe(CN)$_6$] solution at three separate Bi-CCEs.

The reproducibility was assessed by conducting measurements in 1.0 mmol L^{-1} K$_3$[Fe(CN)$_6$] solution employing three separate Bi-CCEs, each prepared on different days (Figure 3B). RSD lower than 5% was observed, affirming the high reproducibility and reliability of the electrode preparation method.

3.4. Electroanalytical Performance of the Bi-CCE

The impact of incorporating Bi_2O_3NPs into the CCE on the electroanalytical performance of the Bi-CCE was evaluated using PCMC. Initially, cyclic voltammograms were recorded at a scan rate of 50 mV s^{-1} within the potential range from −0.25 to +1.35 V in the BRB solution at pH 5.0 both in the absence and presence of 50.0 µmol L^{-1} PCMC. As depicted in Figure 4A, PCMC displays a single oxidation peak on both CCEs, which was observed at +0.883 V for the unmodified CCE and +0.865 V for the Bi-CCE. Since no cathodic peak is evident in the cyclic voltammograms when PCMC is present, the electrochemical oxidation of PCMC on both CCEs can be deemed irreversible. Notably, a better-shaped and higher oxidation peak for PCMC (1.6-fold increase in current response) was observed on the Bi-CCE (I_p = 2.43 µA) compared to the unmodified CCE (I_p = 1.52 µA). These initial findings suggest a positive influence resulting from the addition of Bi_2O_3NPs on the electroanalytical performance of the CCE in determining PCMC.

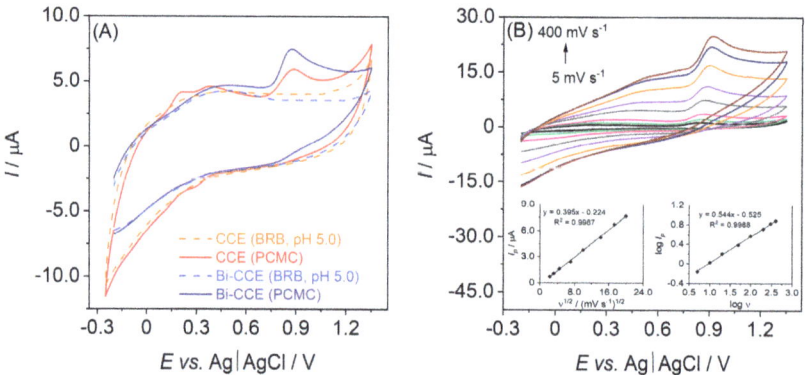

Figure 4. (**A**) Cyclic voltammograms (a scan rate of 50 mV s^{-1}) registered in the BRB solution at pH 5.0 (dashed lines) and in the BRB solution containing 50.0 µmol L^{-1} PCMC (solid lines) on the unmodified CCE and the Bi-CCE; (**B**) cyclic voltammograms recorded across scan rates ranging from 5 to 400 mV s^{-1} in the BRB solution at pH 5.0 containing 50.0 µmol L^{-1} PCMC on the Bi-CCE. Insets exhibit the relationship between I_p and $v^{1/2}$ (**left graph**) and the dependence of logarithm of I_p vs. logarithm of v (**right graph**).

The impact of varying scan rates on the electrochemical behavior of PCMC on the Bi-CCE was investigated across the range of 5–400 mV s^{-1}. As depicted in Figure 4B, the I_p amplifies with the rising v, and the peak potential (E_p) shifts toward more positive values (from +0.84 to +0.92 V). These patterns further affirm the irreversibility of the PCMC oxidation process. Additionally, while a clear linear relationship exists between I_p and $v^{1/2}$ (R^2 = 0.9987, as illustrated in the left inset of Figure 4B), the y-intercept deviates from zero, which is contrary to what is expected in a completely diffusion-controlled process where the intercept is typically zero [41]. Moreover, plotting the linear relationship between log I_p and log v revealed a slope of 0.544 (R^2 = 0.9988; displayed in the right inset of Figure 4B), affirming the earlier hypothesis that the electrochemical oxidation process of PCMC is not exclusively controlled by diffusion. Possibly, another mechanism beyond diffusion significantly contributes to the PCMC oxidation process. This observation aligns with typical behaviors observed in phenolic-type compounds [21,42].

Based on the results obtained, it can be stated that the presence of the diffusion-controlled process is beneficial for the analysis being conducted. Thanks to the absence of an adsorption process in the electrooxidation of PCMC, the bulk-modified CCEs were easily reusable after each experiment. To refresh the electrode's surface, a simple surface-polishing technique was performed only after a series of measurements. In the case of an adsorption process, polishing would be necessary after each scan due to the accumulation of oxidation products on the CCEs surface. On the other hand, an adsorption-controlled process would potentially lead to a decrease in the LOD values in the determination of PCMC. Nevertheless, the presence of the diffusion-controlled process does not limit the electroanalytical application of the sensors and could result in satisfactory validation parameters.

Furthermore, the quantification of PCMC at the Bi-CCE was established through SWV, which is a highly sensitive technique widely employed in electroanalytical applications [8,9]. To ensure an optimally developed PCMC oxidation peak, coupled with a sufficient I_p for precise quantitative analysis, the developed analytical procedure involved optimizing specific conditions. This encompassed determining the ideal pH value of the supporting electrolyte (BRB) and optimizing the SWV parameters such as amplitude, frequency, and step potential.

Initially, the impact of pH of the BRB on the current response of PCMC was investigated within a pH range spanning from 2.0 to 12.0 (as depicted in Figure 5). An increase in I_p was noted with rising pH values until reaching 5.0, beyond which a decline in I_p was observed as pH values further increased (as demonstrated in the left inset of Figure 5). The Bi-CCE exhibited the highest PCMC response in the BRB at pH 5.0; thus, this pH value was picked for successive measurements. Furthermore, Figure 5 shows that the E_p of PCMC shifts toward more negative potentials with increasing pH within the 2.0–12.0 pH range, suggesting a pH-dependent oxidation process. Notably, the relationship between E_p and pH is linear ($R^2 = 0.9985$; presented in the right inset of Figure 5), with a slope of -0.0578 V pH^{-1}, closely resembling the Nernstian value of 0.059 V pH^{-1}. This indicates the exchange of an equivalent number of protons and electrons in the electrochemical reaction [43] of PCMC. Based on the obtained results and a literature overview, the electrochemical oxidation mechanism has been proposed, and it can be concluded that one electron and one proton are involved in the electrooxidation of the –OH group present in the PCMC structure [35,36].

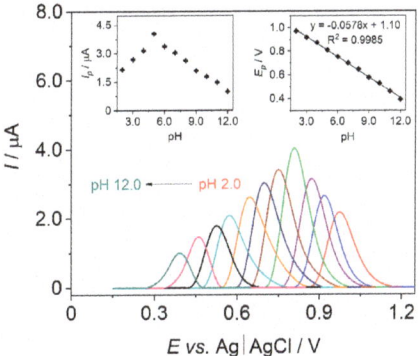

Figure 5. Baseline corrected SW voltammograms of PCMC (50.0 µmol L^{-1}) registered on the Bi-CCE in the BRB solutions in the pH range of 2.0–12.0. Insets exhibit the relationship between I_p and pH (**left graph**) with SD error bars ($n = 3$) and the relationship between E_p and pH (**right graph**). SWV parameters: amplitude of 30 mV, frequency of 20 Hz, step potential of 5 mV.

Subsequently, the optimization of SWV parameters was conducted across specific ranges: amplitude from 10 to 80 mV, frequency from 10 to 90 Hz, and step potential from 1 to 9 mV. The investigation revealed a gradual increase in the peak current of PCMC with rising SW amplitude, stabilizing around 50 mV. Moreover, the oxidation peak of PCMC broadened notably at amplitudes higher than 50 mV. Across the entire studied SW frequency range, the peak current of PCMC consistently exhibited an increase. Additionally, the investigation highlighted a direct correlation between the increment of step potential value and the increase in the PCMC oxidation peak current. However, the application of step potential values exceeding 5 mV resulted in observable peak distortion. Upon careful evaluation of peak height and shape for PCMC, the optimal experimental conditions were identified as an amplitude of 50 mV, a frequency of 60 Hz, and a step potential of 5 mV.

Following optimization, the SW voltammograms (Figure 6A) enabled the construction of a calibration curve on the Bi-CCE that facilitated the determination of various essential validation parameters for assessing the developed analytical procedure. To evaluate the impact of the Bi_2O_3NPs on the electroanalytical performance of the CCE, similar assessments were conducted on both CCE types, and the resulting validation parameters are detailed in Table 1. The data comparison clearly highlights the beneficial impact of Bi_2O_3NPs on the analytical (validation) parameters in the SWV procedure for PCMC determination. The presence of Bi_2O_3NPs considerably widened the LDR at the Bi-CCE and extended the linear response limit. Additionally, the Bi-CCE exhibited over an 11-fold increase in sensitivity, which was determined from the slope of the calibration curve, along with notably reduced LOD and LOQ values (nearly 4.5-times lower compared to the unmodified CCE). Furthermore, the Bi-CCE demonstrated improved precision (expressed as RSD of the lowest concentration within the linear range) and accuracy (determined by the recovery of the lowest concentration within the linear range) in comparison with the unmodified CCE. It is noteworthy that both CCEs met the acceptance criteria outlined in the literature [44] for these parameters.

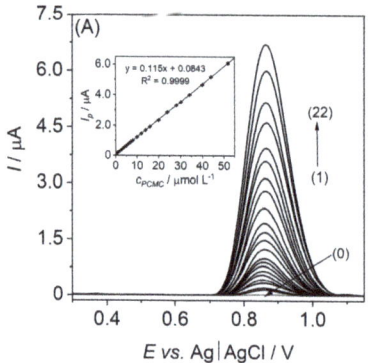
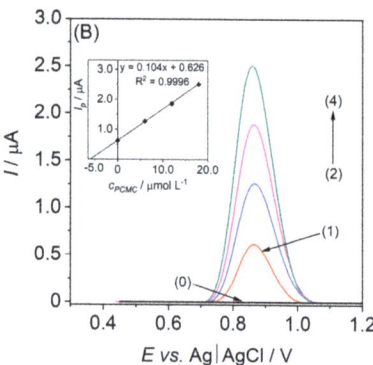

Figure 6. (**A**) Baseline-corrected SW voltammograms recorded on the Bi-CCE in BRB at pH 5.0 containing increasing concentration of the PCMC: (0) blank, (1) 0.5, (2) 1.0, (3) 2.0, (4) 3.0, (5) 4.0, (6) 5.0, (7) 6.0, (8) 7.0, (9) 8.0, (10) 10.0, (11) 12.0, (12) 14.0, (13) 16.0, (14) 20.0, (15) 24.0, (16) 28.0, (17) 30.0, (18) 34.0, (19) 40.0, (20) 44.0, (21) 52.0, and (22) 58.0 µmol L^{-1}. SWV parameters: amplitude of 50 mV, frequency of 60 Hz, and step potential of 5 mV. Inset: calibration curve with SD error bars ($n = 3$). (**B**) Baseline-corrected SW voltammograms recorded on the Bi-CCE in the river water sample employing the standard addition method: (0) unspiked river water sample, (1) spiked river water sample, (2) same as (1) + 6.0 µmol L^{-1} PCMC, (3) same as (1) + 12.0 µmol L^{-1} PCMC, (4) same as (1) + 18.0 µmol L^{-1} PCMC. SWV parameters: amplitude of 50 mV, frequency of 60 Hz, and step potential of 5 mV. The inset displays a corresponding standard addition plot with SD error bars ($n = 3$).

The validation parameters acquired for the Bi-CCE were contrasted with data for various carbon-based electrodes utilized in PCMC determination (Table 2). Overall, the Bi-CCE demonstrates superior electroanalytical performance in terms of both LDR and LOD values compared to most other carbon-based electrodes [24,34,36–38]. Remarkably, slightly better, however, still comparable results were achieved using the anodically pretreated boron-doped diamond electrode with a B/C ratio of 2000 ppm [35]. However, it is important to note that the preparation of a boron-doped diamond electrode is an expensive and intricate process that necessitates sophisticated equipment operated by trained personnel. Therefore, this comparison suggests that the developed Bi-CCE in our study exhibits notably low LOD and a wide linear range, indicating its highly efficient performance in PCMC determination.

Table 1. Validation parameters in the SWV procedure of PCMC determination on the unmodified CCE and the Bi-CCE.

Parameter	CCE	Bi-CCE
Linear range (μmol L^{-1})	4.0–24.0	0.5–58.0
Sensitivity (μA L μmol^{-1})	0.0102	0.115
LOD [a] (μmol L^{-1})	0.73	0.17
LOQ [a] (μmol L^{-1})	2.21	0.50
Precision [b] (%)	2.6	0.8
Accuracy [b] (%)	86.3	101.2

[a] LOD = (3.3 × SD$_a$)/b; LOQ = (10 × SD$_a$)/b, where SD$_a$ is the standard deviation of the slope, b is the intercept value; [b] precision and accuracy calculated for 3 consecutive measurements for PCMC concentration of 4.0 μmol L^{-1} at the unmodified CCE and 0.5 μmol L^{-1} at the Bi-CCE.

Table 2. Comparison of PCMC determination on various carbon-based electrodes.

Electrode	Linear Range (μmol L^{-1})	LOD (μmol L^{-1})	Ref.
GCE	21.0–210.4	9.3	[37]
MWCNTs-GCE	14.0–137.5	8.8	[34]
UiO-66-NH2@PEDOT/GA-GCE	0.6–18.0	0.20	[36]
APT-BDDE (B/C = 2000 ppm)	0.5–100.0	0.11	[35]
SNG–C–PANI	0.7–7.0	0.69	[38]
MWCNTs-CCE	3.0–32.0	0.71	[24]
Bi-CCE	0.5–58.0	0.17	This work

GCE—glassy carbon electrode; MWCNTs-GCE—glassy carbon electrode modified with multi-walled carbon nanotubes; APT-BDDE—anodically pretreated boron-doped diamond electrode; MWCNTs-CCE—carbon ceramic electrode modified with multi-walled carbon nanotubes; UiO-66-NH2@PEDOT/GA-GCE—glassy carbon electrode modified with Zr-based metal–organic framework (UiO-66-NH$_2$), poly-(3,4-ethylenedioxythiophene) and graphene aerogel; SNG–C–PANI—sonogel carbon poly-aniline electrode; Bi-CCE—carbon ceramic electrode bulk-modified with Bi$_2$O$_3$ nanoparticles.

The efficacy of the optimized and validated SWV procedure for PCMC determination using the Bi-CCE was authenticated through real sample analysis. Initial screening of the samples showed no detectable PCMC. To ascertain its presence, spiking experiments were conducted, and PCMC concentrations were assessed via the standard addition method. Figure 6B displays SW voltammograms derived from successive PCMC additions to the river water sample. The quantification of PCMC content in the river water sample, determined through the linear correlation of I_p vs. PCMC concentration (shown in the inset of Figure 6B), was successfully achieved. The calculated PCMC concentration in the tested sample was 6.21 μmol L^{-1}, closely matching the added spiked value of 6.0 μmol L^{-1}. Moreover, the RSD values obtained at each concentration level (n = 3) did not exceed 3.1%, indicating highly reproducible measurements using the Bi-CCE. The estimated PCMC recovery values ranged from 99.8% to 100.3%, affirming the accuracy of the proposed methodology. Notably, there was no noticeable matrix effect from the analyzed samples on the performance of PCMC. These results affirm the suitability of the Bi-CCE for precise quantitative analysis of PCMC in real samples.

Subsequently, to assess method selectivity, various ions commonly present in river water, i.e., Cd^{2+}, Ni^{2+}, Cu^{2+}, HCO_3^-, and SO_4^{2-}, were examined (Figure 7). It was observed that the presence of Cd^{2+}, Ni^{2+}, Cu^{2+}, and HCO_3^- ions did not notably interfere with the PCMC signal. However, the presence of SO_4^{2-} ions resulted in a slight alteration of the PCMC signal. These findings demonstrate a generally favorable selectivity of the developed method.

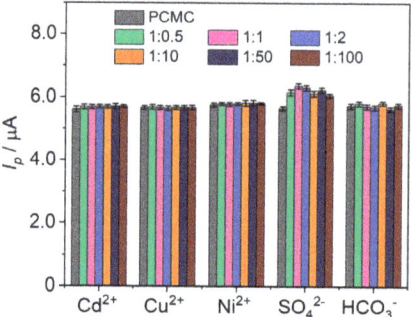

Figure 7. A bar graph depicting the peak current (I_p) of PCMC in the presence of interferents.

4. Conclusions

In this work, the Bi-CCE was prepared by bulk modification with Bi_2O_3NPs and thoroughly investigated using microscopic and electrochemical methods. The results revealed that the incorporation of Bi_2O_3NPs into the ceramic matrix significantly altered the morphology and topography of the ceramic composite, resulting in a more compact electrode material when compared to the more porous unmodified CCE, which exhibited visible pinholes. Moreover, the enhancement of the electrochemical properties of the CCE following modification with Bi_2O_3NPs was confirmed in the presence of the redox marker. Additionally, the Bi-CCE was successfully verified as an outstanding sensing tool for the reliable, sensitive, and selective determination of the priority environmental pollutant PCMC. Importantly, the exceptional features of the Bi-CCE enabled the development of a direct, simple, and rapid protocol for PCMC determination, demonstrating the possibility of avoiding complex and tedious modification procedures as well as accumulation steps. In summary, all analyses affirm the positive impact of Bi_2O_3NPs on the overall performance of the CCE.

As a result, the Bi-CCE developed in this study presents a cheap, prospective, and promising carbon-based electrode material, offering utility as an effective analytical tool for applications in pharmaceutical, clinical, food, and environmental analyses. However, it is important to acknowledge that for the detection of organic compounds in more intricate biological or environmental samples, where concentrations typically fall in the nmol L^{-1} range, additional modifications may be required to further enhance the already superior electroanalytical performance of the Bi-CCE.

Author Contributions: Conceptualization, B.B.; methodology, B.B.; validation, B.B. and M.B.; formal analysis, B.B., M.B. and A.L.; investigation, B.B., M.B. and A.L.; resources, B.B.; data curation, B.B. and M.B.; writing—original draft preparation, B.B. and M.B.; writing—review and editing, B.B., M.B., A.L. and S.S.; visualization, B.B. and M.B.; supervision, B.B.; funding acquisition, S.S. All authors have read and agreed to the published version of the manuscript.

Funding: This research received no external funding.

Institutional Review Board Statement: Not applicable.

Informed Consent Statement: Not applicable.

Data Availability Statement: Data will be made available upon request.

Acknowledgments: B.B. thanks Justyna Węgiel for stimulating discussion.

Conflicts of Interest: The authors declare no conflicts of interest.

References

1. Jiwanti, P.K.; Wardhana, B.Y.; Sutanto, L.G.; Chanif, M.F. A Review on Carbon-Based Electrodes for Electrochemical Sensor of Quinolone Antibiotics. *ChemistrySelect* **2022**, *7*, e202103997. [CrossRef]
2. Laghlimi, C.; Moutcine, A.; Chtaini, A.; Isaad, J.; Soufi, A.; Ziat, Y.; Amhamdi, H.; Belkhanchi, H. Recent Advances in Electrochemical Sensors and Biosensors for Monitoring Drugs and Metabolites in Pharmaceutical and Biological Samples. *ADMET DMPK* **2023**, *11*, 151–173. [CrossRef] [PubMed]
3. Boumya, W.; Taoufik, N.; Achak, M.; Barka, N. Chemically Modified Carbon-Based Electrodes for the Determination of Paracetamol in Drugs and Biological Samples. *J. Pharm. Anal.* **2021**, *11*, 138–154. [CrossRef]
4. Hasoň, S.; Daňhel, A.; Schwarzová-Pecková, K.; Fojta, M. Carbon Electrodes in Electrochemical Analysis of Biomolecules and Bioactive Substances: Roles of Surface Structures and Chemical Groups. In *Nanotechnology and Biosensors*; Elsevier: Amsterdam, The Netherlands, 2018; pp. 51–111. [CrossRef]
5. Michalkiewicz, S.; Skorupa, A.; Jakubczyk, M. Carbon Materials in Electroanalysis of Preservatives: A Review. *Materials* **2021**, *14*, 7630. [CrossRef]
6. Kaur, H.; Siwal, S.S.; Chauhan, G.; Saini, A.K.; Kumari, A.; Thakur, V.K. Recent Advances in Electrochemical-Based Sensors Amplified with Carbon-Based Nanomaterials (CNMs) for Sensing Pharmaceutical and Food Pollutants. *Chemosphere* **2022**, *304*, 135182. [CrossRef] [PubMed]
7. Zhang, C.; Du, X. Electrochemical Sensors Based on Carbon Nanomaterial Used in Diagnosing Metabolic Disease. *Front. Chem.* **2020**, *8*, 651. [CrossRef]
8. Mirceski, V.; Gulaboski, R.; Lovric, M.; Bogeski, I.; Kappl, R.; Hoth, M. Square-Wave Voltammetry: A Review on the Recent Progress. *Electroanalysis* **2013**, *25*, 2411–2422. [CrossRef]
9. Mirceski, V.; Gulaboski, R. Recent Achievements in Square-Wave Voltammetry a Review. *Maced. J. Chem. Chem. Eng.* **2014**, *33*, 1–12.
10. Tsionsky, M.; Gun, G.; Glezer, V.; Lev, O. Sol-Gel-Derived Ceramic-Carbon Composite Electrodes: Introduction and Scope of Applications. *Anal. Chem.* **1994**, *66*, 1747–1753. [CrossRef]
11. Lev, O.; Tsionsky, M.; Rabinovich, L.; Glezer, V.; Sampath, S.; Pankratov, I.; Gun, J. Organically Modified Sol-Gel Sensors. *Anal. Chem.* **1995**, *67*, 22A–30A. [CrossRef]
12. Boumya, W.; Charafi, S.; Achak, M.; Bessbousse, H.; Elhalil, A.; Abdennouri, M.; Barka, N. Modification Strategies of Sol–Gel Carbon Ceramic Electrodes and Their Electrochemical Applications. *Results Chem.* **2022**, *4*, 100623. [CrossRef]
13. Shamsipur, M.; Karimi, Z.; Amouzadeh Tabrizi, M. A Novel Electrochemical Cyanide Sensor Using Gold Nanoparticles Decorated Carbon Ceramic Electrode. *Microchem. J.* **2017**, *133*, 485–489. [CrossRef]
14. Zheng, J.; Sheng, Q.; Li, L.; Shen, Y. Bismuth Hexacyanoferrate-Modified Carbon Ceramic Electrodes Prepared by Electrochemical Deposition and Its Electrocatalytic Activity towards Oxidation of Hydrazine. *J. Electroanal. Chem.* **2007**, *611*, 155–161. [CrossRef]
15. Abbaspour, A.; Norouz-Sarvestani, F. High Electrocatalytic Effect of Au-Pd Alloy Nanoparticles Electrodeposited on Microwave Assisted Sol-Gel-Derived Carbon Ceramic Electrode for Hydrogen Evolution Reaction. *Int. J. Hydrogen Energy* **2013**, *38*, 1883–1891. [CrossRef]
16. Dehgan-Reyhan, S.; Najafi, M. Defective Mesoporous Carbon Ceramic Electrode Modified Graphene Quantum Dots as a Novel Surface-Renewable Electrode: The Application to Determination of Zolpidem. *J. Electroanal. Chem.* **2019**, *832*, 241–246. [CrossRef]
17. Habibi, B.; Jahanbakhshi, M.; Pournaghi-Azar, M.H. Simultaneous Determination of Acetaminophen and Dopamine Using SWCNT Modified Carbon-Ceramic Electrode by Differential Pulse Voltammetry. *Electrochim. Acta* **2011**, *56*, 2888–2894. [CrossRef]
18. Habibi, B.; Pournaghi-Azar, M.H. Simultaneous Determination of Ascorbic Acid, Dopamine and Uric Acid by Use of a MWCNT Modified Carbon-Ceramic Electrode and Differential Pulse Voltammetry. *Electrochim. Acta* **2010**, *55*, 5492–5498. [CrossRef]
19. Habibi, B.; Abazari, M.; Pournaghi-Azar, M.H. Simultaneous Determination of Codeine and Caffeine Using Single-Walled Carbon Nanotubes Modified Carbon-Ceramic Electrode. *Colloids Surfaces B Biointerfaces* **2014**, *114*, 89–95. [CrossRef]
20. Węgiel, J.; Burnat, B.; Skrzypek, S. A Graphene Oxide Modified Carbon Ceramic Electrode for Voltammetric Determination of Gallic Acid. *Diam. Relat. Mater.* **2018**, *88*, 137–143. [CrossRef]
21. Burnat, B.; Brycht, M.; Leniart, A.; Skrzypek, S. Carbon Black-Modified Carbon Ceramic Electrode—Its Fabrication, Characterization, and Electroanalytical Performance. *Diam. Relat. Mater.* **2022**, *130*, 109513. [CrossRef]
22. Robak, J.; Burnat, B.; Kisielewska, A.; Fendrych, K.; Skrzypek, S. Fabrication and Application of Ferrierite–Modified Carbon Ceramic Electrode in Sensitive Determination of Estradiol. *J. Electrochem. Soc.* **2017**, *164*, B574–B580. [CrossRef]
23. Majidi, M.R.; Asadpour-Zeynali, K.; Gholizadeh, S. Nanobiocomposite Modified Carbon-Ceramic Electrode Based on Nano-TiO2-Plant Tissue and Its Application for Electrocatalytic Oxidation of Dopamine. *Electroanalysis* **2010**, *22*, 1772–1780. [CrossRef]
24. Robak, J.; Burnat, B.; Leniart, A.; Kisielewska, A.; Brycht, M.; Skrzypek, S. The Effect of Carbon Material on the Electroanalytical Determination of 4-Chloro-3-Methylphenol Using the Sol-Gel Derived Carbon Ceramic Electrodes. *Sens. Actuators B Chem.* **2016**, *236*, 318–325. [CrossRef]
25. Zhu, L.; Tian, C.; Zhai, J.; Yang, R. Sol-Gel Derived Carbon Nanotubes Ceramic Composite Electrodes for Electrochemical Sensing. *Sens. Actuators B Chem.* **2007**, *125*, 254–261. [CrossRef]
26. Schebeliski, A.H.; Lima, D.; Marchesi, L.F.Q.P.; Calixto, C.M.F.; Pessôa, C.A. Preparation and Characterization of a Carbon Nanotube-Based Ceramic Electrode and Its Potential Application at Detecting Sulfonamide Drugs. *J. Appl. Electrochem.* **2018**, *48*, 471–485. [CrossRef]

27. Habibi, B.; Jahanbakhshi, M. Simultaneous Determination of Ascorbic Acid, Paracetamol and Phenylephrine: Carbon Nanotubes Ceramic Electrode as a Renewable Electrode. *Anal. Bioanal. Electrochem.* **2015**, *7*, 45–58.
28. Robak, J.; Węgiel, K.; Burnat, B.; Skrzypek, S. A Carbon Ceramic Electrode Modified with Bismuth Oxide Nanoparticles for Determination of Syringic Acid by Stripping Voltammetry. *Microchim. Acta* **2017**, *184*, 4579–4586. [CrossRef]
29. Wang, P.; Yuan, Y.; Wang, X.; Zhu, G. Renewable Three-Dimensional Prussian Blue Modified Carbon Ceramic Electrode. *J. Electroanal. Chem.* **2000**, *493*, 130–134. [CrossRef]
30. Teradal, N.L.; Seetharamappa, J. Bulk Modification of Carbon Paste Electrode with Bi2O3 Nanoparticles and Its Application as an Electrochemical Sensor for Selective Sensing of Anti-HIV Drug Nevirapine. *Electroanalysis* **2015**, *27*, 2007–2016. [CrossRef]
31. Li, H.; Guo, Y.; Zeng, S.; Wei, Q.; Sharel, P.E.; Zhu, R.; Cao, J.; Ma, L.; Zhou, K.C.; Meng, L. High-Sensitivity, Selective Determination of Dopamine Using Bimetallic Nanoparticles Modified Boron-Doped Diamond Electrode with Anodic Polarization Treatment. *J. Mater. Sci.* **2021**, *56*, 4700–4715. [CrossRef]
32. Nisanci, F.B.; Yilmaz, B. Electrochemically Grown Bismuth(III) Oxide Nanoparticles on Gold as Sensor for Quantitication of Methimazole. *Rev. Roum. Chim.* **2018**, *63*, 941–952.
33. Anku William, W.; Mamo Messai, A.; Govender Penny, P. Phenolic Compounds in Water: Sources, Reactivity, Toxicity and Treatment Methods. In *Phenolic Compounds—Natural Sources, Importance and Applications*; Soto-Hernández, M., Palma-Tenango, M., García-Mateos, R., Eds.; IntechOpen: Rijeka, Croatia, 2017; pp. 419–443. ISBN 0000957720.
34. Baranowska, I.; Bijak, K. Voltammetric Determination of Disinfectants at Multiwalled Carbon Nanotube Modified Glassy Carbon Electrode. *J. Anal. Chem.* **2013**, *68*, 891–895. [CrossRef]
35. Brycht, M.; Lochyński, P.; Barek, J.; Skrzypek, S.; Kuczewski, K.; Schwarzova-Peckova, K. Electrochemical Study of 4-Chloro-3-Methylphenol on Anodically Pretreated Boron-Doped Diamond Electrode in the Absence and Presence of a Cationic Surfactant. *J. Electroanal. Chem.* **2016**, *771*, 1–9. [CrossRef]
36. Tian, Q.; Xu, J.; Zuo, Y.; Li, Y.; Zhang, J.; Zhou, Y.; Duan, X.; Lu, L.; Jia, H.; Xu, Q.; et al. Three-Dimensional PEDOT Composite Based Electrochemical Sensor for Sensitive Detection of Chlorophenol. *J. Electroanal. Chem.* **2019**, *837*, 1–9. [CrossRef]
37. Baranowska, I.; Bijak, K. Differential Pulse Voltammetry in Analysis of Disinfectants—2-Mercaptobenzothiazole, 4-Chloro-3-Methylphenol, Triclosan, Chloramine-T. *Cent. Eur. J. Chem.* **2010**, *8*, 1266–1272. [CrossRef]
38. Calatayud-Macías, P.; López-Iglesias, D.; Sierra-Padilla, A.; Cubillana-Aguilera, L.; Palacios-Santander, J.M.; García-Guzmán, J.J. Bulk Modification of Sonogel–Carbon with Polyaniline: A Suitable Redox Mediator for Chlorophenols Detection. *Chemosensors* **2023**, *11*, 63. [CrossRef]
39. Cabello-Carramolino, G.; Petit-Dominguez, M.D. Application of New Sol-Gel Electrochemical Sensors to the Determination of Trace Mercury. *Anal. Chim. Acta* **2008**, *614*, 103–111. [CrossRef]
40. Bard, A.J.; Faulkner, L.R. *Electrochemical Methods: Fundamentals and Applications*, 2nd ed.; John Wiley and Sons: New York, NY, USA, 2001; ISBN 978-0-471-04372-0.
41. Swartz, M.E.; Krull, I.S. *Analytical Method Development and Validation*; Marcel Dekker Inc.: New York, NY, USA, 1997; ISBN 0-8247-0115-1.
42. Nady, H.; El-Rabiei, M.M.; El-Hafez, G.M.A. Electrochemical Oxidation Behavior of Some Hazardous Phenolic Compounds in Acidic Solution. *Egypt. J. Pet.* **2017**, *26*, 669–678. [CrossRef]
43. Zuman, P. *The Elucidation of Organic Electrode Processes*; Academic Press: Cambridge, MA, USA, 1969; ISBN 978-1-4832-2725-2.
44. Ozkan, S.A. (Ed.) *Electroanalytical Methods in Pharmaceutical Analysis and Their Validation*; HNB Publishing: New York, NY, USA, 2012.

Disclaimer/Publisher's Note: The statements, opinions and data contained in all publications are solely those of the individual author(s) and contributor(s) and not of MDPI and/or the editor(s). MDPI and/or the editor(s) disclaim responsibility for any injury to people or property resulting from any ideas, methods, instructions or products referred to in the content.

Article

Probing the Use of Homemade Carbon Fiber Microsensor for Quantifying Caffeine in Soft Beverages

Karla Caroline de Freitas Araújo [1], Emily Cintia Tossi de Araújo Costa [1], Danyelle Medeiros de Araújo [1,2], Elisama V. Santos [2,3], Carlos A. Martínez-Huitle [1,2,*] and Pollyana Souza Castro [1,*]

[1] Institute of Chemistry, Federal University of Rio Grande do Norte, Av. Campus Universitário, Av. Salgado Filho 3000, Lagoa Nova, Natal CEP59078-970, RN, Brazil

[2] National Institute for Alternative Technologies of Detection, Toxicological Evaluation and Removal of Micropollutants and Radioactives (INCT-DATREM), Institute of Chemistry, Universidade Estadual Paulista, Araraquara CEP14800-900, SP, Brazil

[3] School of Science and Technology, Federal University of Rio Grande do Norte, Av. Campus Universitário, Av. Salgado Filho 3000, Lagoa Nova, Natal CEP59078-970, RN, Brazil

* Correspondence: carlosmh@quimica.ufrn.br (C.A.M.-H.); pollyana.castro@ufrn.br (P.S.C.)

Citation: de Freitas Araújo, K.C.; de Araújo Costa, E.C.T.; de Araújo, D.M.; Santos, E.V.; Martínez-Huitle, C.A.; Castro, P.S. Probing the Use of Homemade Carbon Fiber Microsensor for Quantifying Caffeine in Soft Beverages. *Materials* 2023, *16*, 1928. https://doi.org/10.3390/ma16051928

Academic Editors: Sławomira Skrzypek, Mariola Brycht, Barbara Burnat and Dimitra Vernardou

Received: 11 November 2022
Revised: 25 January 2023
Accepted: 31 January 2023
Published: 25 February 2023

Copyright: © 2023 by the authors. Licensee MDPI, Basel, Switzerland. This article is an open access article distributed under the terms and conditions of the Creative Commons Attribution (CC BY) license (https://creativecommons.org/licenses/by/4.0/).

Abstract: In the development of electrochemical sensors, carbon micro-structured or micro-materials have been widely used as supports/modifiers to improve the performance of bare electrodes. In the case of carbon fibers (CFs), these carbonaceous materials have received extensive attention and their use has been proposed in a variety of fields. However, to the best of our knowledge, no attempts for electroanalytical determination of caffeine with CF microelectrode (µE) have been reported in the literature. Therefore, a homemade CF-µE was fabricated, characterized, and used to determine caffeine in soft beverage samples. From the electrochemical characterization of the CF-µE in $K_3Fe(CN)_6$ 10 mmol L^{-1} plus KCl 100 mmol L^{-1}, a radius of about 6 µm was estimated, registering a sigmoidal voltammetric profile that distinguishes a µE indicating that the mass-transport conditions were improved. Voltammetric analysis of the electrochemical response of caffeine at the CF-µE clearly showed that no effects were attained due to the mass transport in solution. Differential pulse voltammetric analysis using the CF-µE was able to determine the detection sensitivity, concentration range (0.3 to 4.5 µmol L^{-1}), limit of detection (0.13 µmol L^{-1}) and linear relationship (I (µA) = (11.6 ± 0.09) × 10^{-3} [caffeine, µmol L^{-1}] − (0.37 ± 0.24) × 10^{-3}), aiming at the quantification applicability in concentration quality-control for the beverages industry. When the homemade CF-µE was used to quantify the caffeine concentration in the soft beverage samples, the values obtained were satisfactory in comparison with the concentrations reported in the literature. Additionally, the concentrations were analytically determined by high-performance liquid chromatography (HPLC). These results show that these electrodes may be an alternative to the development of new and portable reliable analytical tools at low cost with high efficiency.

Keywords: caffeine; carbon fiber; microelectrode; cyclic voltammetry; beverages

1. Introduction

Caffeine or 1-3-7-trimethylxantine is a white crystalline xanthine alkaloid that promotes various effects on the body's metabolism when ingested, including stimulating the central nervous system, increasing blood pressure in the short term, and secreting gastric acid. This drug is widely used in different concentrations in coffee, cola nuts, cocoa beans, tea leaves, cola beverages and several pharmaceutical substances used worldwide [1–3]. Therefore, the concentration of caffeine in its various origins should be controlled.

The most common technique used for analyzing/quantifying this bioactive compound is high-performance liquid chromatography (HPLC) [3–7]. However, this instrumental method involves high-cost analysis, is time-consuming, and requires calibration as well as maintenance as compared with other approaches. Among the competitive methods with

HPLC, electrochemical methodologies have received great attention in the last years due to their advantages, allowing significant technical impacts in the form of easy-to-upscale, versatility, effectiveness, cost-effective balance, easy-to-automatize, and the development of small portable devices [8–12]. These technological features are mainly dependent on the sensor materials used because these materials (commonly named electrodes) may be adapted by size and/or different materials characteristics [13]. Therefore, the versatility of electrode materials, in terms of size and nature, has been widely investigated depending on the concentration control area (e.g., food, pharmaceutical and beverages industries) visioning to reach a single-step reagentless analysis with high detection efficiency by using a smaller active electrode area [2,14–19].

Concerning the electroanalytical detection of caffeine, various sensing-materials have been examined and reported in the existing literature; however, these electrode materials are expensive, and their preparation methodologies are dependent on time-consuming procedures and modification steps sensor [20–23]. In this sense, more sensing-elaboration strategies have been investigated for improvements, in sensibility, selectivity, and detection reliability, using traditional materials; for example, the size of the electrochemical sensor [24–27]. Consequently, the miniaturization is a hot-spot investigation topic because it has allowed us to better understand the chemical/electrochemical processes that take place on the micro- and nano- environments as well as provide significant new benefits as real-time monitoring and high sensitivity as a consequence of the high and efficient mass transport [25,28,29].

In the development of electrochemical microsensors, carbon materials have been widely used as supports due to their simplicity and active surface to improve the performances in specific concentration control areas [30,31]. Among the various types of carbon materials, carbon fibers have found extensive use in a variety of fields, including supercapacitors, sensors, biomedical applications, etc., because of their desirable properties. Shape, high chemical stability, high strength, high electrical conductivity, high surface area, outstanding electrocatalytic activity, and compatibility with matrix materials are just a few of its characteristics. Therefore, CF-µE has recently attracted interest due to their remarkable qualities [31,32].

Based on the existing literature, only few reports have described the applicability of carbon-based sensors to determine caffeine in food, beverages, drugs, and medications. However, to the best of our knowledge, no attempts for electroanalytical determination of caffeine with CF-µE have been reported in the literature. Therefore, a homemade CF-µE was fabricated, characterized, and used to determine caffeine, estimating the concentration linear range, calibration function, and determination of caffeine in soft beverage samples.

2. Experimental Methods

2.1. Reagents

All reagents were of analytical grade, and all aqueous solutions were prepared using a high-purity water obtained from a Millipore Milli-Q system with resistivity >18 MΩ at 25 °C. H_2SO_4 and caffeine (purity 98%) were purchased from Quimex and Isofar, respectively. $K_3[Fe(CN)_6]$ and KCl were purchased from Synth (São Paulo, Brazil). Solutions were daily prepared under constant agitation for 30 min before each experiment.

2.2. Homemade CF-µE Fabrication

A cleaned 3 cm length carbon fiber was connected to a Cu wire (⌀ 1 mm diameter and 12 cm length) with silver conductive ink (Joint Metal Comércio LTDA (SP) Brazil). After drying, this set was carefully inserted into a plastic mold, filled with epoxy resin SQ 2119-PT (Avipol (SP), Brazil), and held in an upright position. For curing and demolding, the 48-hour time was respected. Then, the CF-µE was polished using 1200 and 1500 grit sandpaper until obtaining a smooth and flat surface containing a carbon fiber microdisk. Finally, the microelectrode was washed thoroughly with distilled water, air-dried, inspected by using an optical microscope (Olympus BX51M, Thermo Fisher Scientific, Waltham, MA,

USA) to prevent any defects and stored away from dust. A scheme for the CF-µE fabrication is provided in the Figure 1.

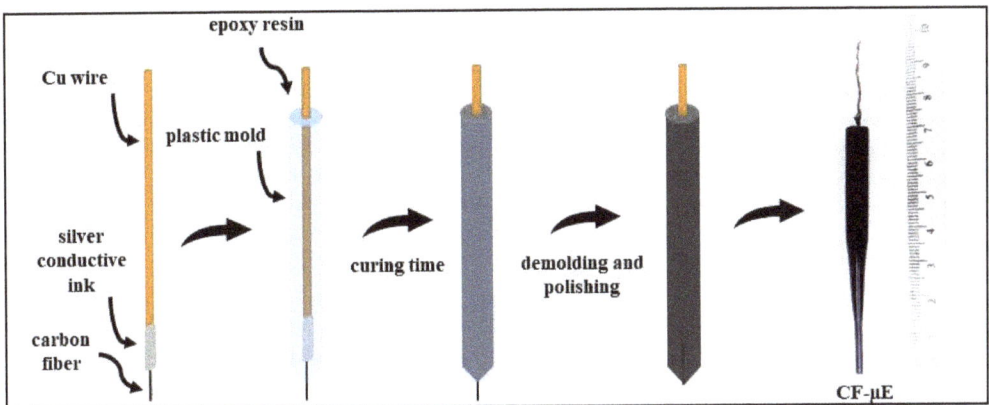

Figure 1. Scheme for the CF-µE homemade fabrication.

2.3. Electrochemical Measurements

Voltammetric analysis (cyclic voltammetry (CV) and differential pulse voltammetry (DPV)) were performed with an Autolab PGSTAT302 by using a three-electrode cell consisting of an Ag/AgCl (3 mol L^{-1} KCl) as a reference electrode, a platinum wire as a counter electrode, and a homemade CF-µE as the working electrode. The CF-µE radius was estimated recording a CV in K$_3$Fe(CN)$_6$ 10 mmol L^{-1} plus KCl 100 mmol L^{-1} solution. The caffeine electroanalytical experiments were carried out at 25 °C and five CVs were recorded for each measurement in the potential range from +0.5 to +1.7 V with scan rate of 50 mV s^{-1} in 0.5 mol L^{-1} H$_2$SO$_4$ as the supporting electrolyte in the presence and absence of caffeine 0.1 mol L^{-1}. The effect of the scan rate was also studied at 20, 40, 60, 80, and 100 mV s^{-1} in the presence of caffeine in 0.5 mol L^{-1} H$_2$SO$_4$. The DPV parameters to quantify caffeine, using 0.5 mol L^{-1} H$_2$SO$_4$ as a supporting electrolyte, were equilibration time = 10 s, initial potential = +0.5 V, final potential = +1.7 V, potential scan rate = 50 mV s^{-1}, pulse amplitude = +0.05 V, and modulation time: 0.04 s. The above-optimized parameters were used for all measurements. Then, the calibrations curves (peak intensity vs. caffeine concentration in the range from 0 to 6 µmol L^{-1}) were studied by least-square linear regression, and the obtained figures (slopes and intercepts) were reported with their confidence interval, p = 95%. Reproducibility and stability parameters were also evaluated. Homemade CF-µE was cleaned recording ten CV cycles from +0.60 V to +1.80 V at 100 mV s^{-1} in 0.5 mol L^{-1} H$_2$SO$_4$ before each one of the measurements. Caffeine determinations with CF-µE were validated by a reverse-phase HPLC (Shimadzu LC-6 Series, Berlin, Germany) equipped with a Nucleosil C18 column, Berlin, Germany (4.6 × 250 mm), and an UV–vis detector set at 273 nm. An acetonitrile/water mixture (25:75 % v/v) was used as the mobile phase at a flow rate of 0.6 mL min^{-1}, injecting 20 µL of each sample. The retention time (t_r) was 6.8 min.

2.4. Electrochemical Determination of Caffeine in Beverages

Evaluating the practical feasibility of the homemade CF-µE, the caffeine concentration in soft beverage samples was determined. Ultrasonication (10 min) was used to eliminate the gas from commercial soft drinks, which were then transferred to the electrochemical cell with the supporting electrolyte to proceed with the caffeine detection (0.5 mL to 4.5 mL of supporting electrolyte). The standard addition method was used to quantify caffeine in the samples to minimize the possible matrix effects due to the presence of other components in the real samples. All experiments were carried out in triplicate, and mean values (standard deviation < 5%) were used for the figures.

3. Results and Discussion

3.1. Fabrication and Characterization of the CF-μE

Prior to evaluating the analytical performance of the CF-μE as a sensor for caffeine, it was necessary to inspect the electrodic surface by using an optical microscope and isolate the area. This procedure is extremely important to ensure the quality of the analytical results by using a flat and defect-free surface containing only a carbon fiber microdisk. Then, the CF-μE was characterized by CV using a $K_3Fe(CN)_6$ 10 mmol L^{-1} plus KCl 100 mmol L^{-1} solution. In Figure 2, a steady-state response owing to the enhanced mass-transport conditions was obtained giving a sigmoidal shape, characteristic of microelectrodes. By using the Equation (1), the radius of the CF-μE was found to be 6 μm.

$$I_L = 4nFDCr \tag{1}$$

where I_L is the limiting current at the steady-state condition (A), n is the number of electrons involved on the electrodic reaction, F is the Faraday constant (96,485 C mol^{-1}), D is the diffusion coefficient ($cm^2\ s^{-1}$), C is the bulk concentration of the electroactive species (mol cm^{-3}), and r is the radius of the microdisk electrode (cm) [33].

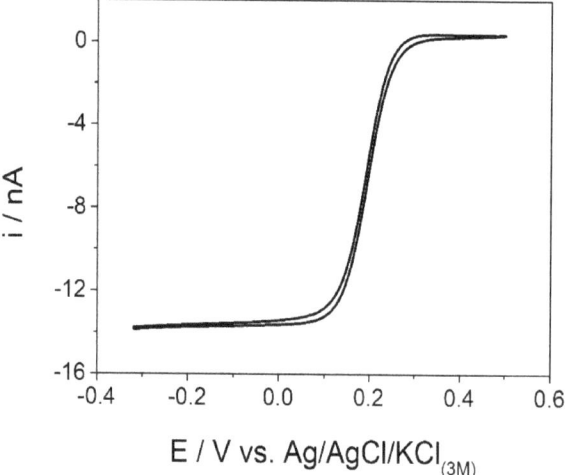

Figure 2. Cyclic voltammogram recorded with the CF-μE using $K_3[Fe(CN)_6]$ 10 mmol L^{-1} in KCl 100 mmol L^{-1} solution; ν = 30 mV s^{-1}; r = 6 μm.

3.2. Cyclic Voltammetry Experiments in Presence of Caffeine

As a preliminary result, the electrochemical response for the electroactive species in solution (supporting electrolyte and caffeine) was investigated by using the CV technique (Figure 3). In the supporting electrolyte (Figure 3a), the CF-μE did not exhibit significant current responses. Conversely, a clear voltammetric oxidation response was registered at +1.5 V in the presence of caffeine in solution (Figure 3a), while no cathodic response was observed on the reverse scan, indicating that the oxidation is an irreversible process [14,27,34,35]. This voltammetric behavior is in agreement with the results reported elsewhere where the caffeine oxidation mechanism is a $4e^-$, $4H^+$ process. Following the mechanism described in the existing literature, oxidation byproducts such as uric acid and its diol-analog are formed and, subsequently, these intermediates are rapidly fragmented [2,36] (see scheme in Figure 4).

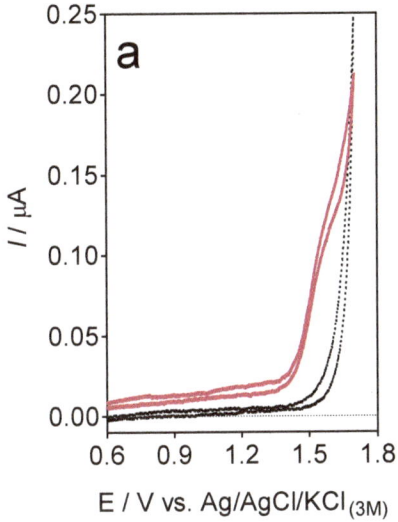
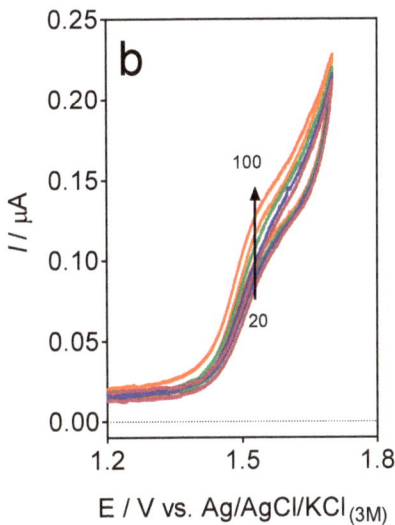
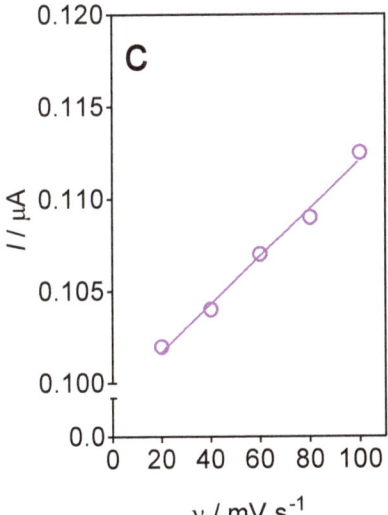
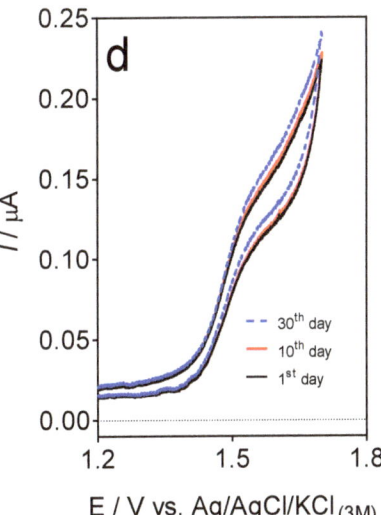

Figure 3. (**a**) CVs recorded at CF-µE in 25 mL of 0.5 mol L^{-1} H$_2$SO$_4$ solution (black curve) and 100 µL of 0.1 mol L^{-1} caffeine in 25 mL of 0.5 mol L^{-1} H$_2$SO$_4$ solution (red curve), scan rate: 10 mV s^{-1}; (**b**) scan rate effect (20 (pink line), 40 (blue line), 60 (green line), 80 (orange line) and 100 mV s^{-1} (red line)) as a function of the electrochemical response of caffeine using 100 µL of 0.1 mol L^{-1} caffeine in 25 mL of 0.5 mol L^{-1} H$_2$SO$_4$; (**c**) oxidation peak current (I_{pa}) values versus the scan rates (v), I_{pa} (µA) = 1.3×10^{-4} (µA mV s^{-1}) + 0.098, r^2 = 0.9928; (**d**) CVs recorded at CF-µE in 100 µL of 0.1 mol L^{-1} caffeine in 25 mL of 0.5 mol L^{-1} H$_2$SO$_4$ solution for the first, tenth, and after 30 days; r = 6 µm.

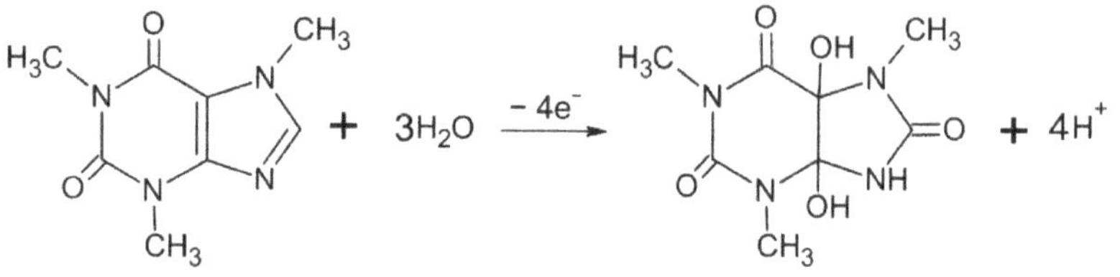

Figure 4. Mechanism of overall oxidation of caffeine.

The voltammetric effect of the scan rate (20–100 mV s^{-1}) as a function of the electrochemical response of caffeine was also investigated using 100 µL of 0.1 mol L^{-1} caffeine in 25 mL of 0.5 mol L^{-1} H$_2$SO$_4$. As observed in Figure 3b, the oxidation peak current (I_{pa}) values at different scan rates showed a negligible difference. In addition, a slight shift on the oxidation potential (E_{pa}) was attained to more positive potential values when the scan rate was increased. These results visibly showed that no effects were attained due to the mass transport in solution because the miniaturization of the electrode surface provides significant profits in the sensitivity.

The oxidation peak current (I_{pa}) values versus the scan rates (v) are shown in Figure 3c, demonstrating a good fitting linear relationship between them (I_{pa} (µA) = 1.3 × 10^{-4} (µA mV s^{-1}) + 0.098), with a coefficient of determination over 0.9928. From these results, it was possible to confirm that the CF-µE has a higher mass transfer rate and reaction rate. This behavior is associated to the dimensions of the CF-µE that reduces its mass transfer rate of the caffeine oxidation, resulting in the amplification of the electrochemical response. In fact, the signal is significantly improved with respect to the result previously achieved when a macro-electrode was used [11], in terms of the voltammetric signal, concentration range, noise, and so on.

The reproducibility of the electrochemical signal of the CF-µE was tested by consecutive 10 cyclic potential scans in the presence of caffeine (100 µL of 0.1 mol L^{-1} caffeine in 25 mL of 0.5 mol L^{-1} H$_2$SO$_4$), at a scan rate of 100 mV s^{-1}. The result clearly showed that no significant changes in the caffeine oxidation-current were observed after five voltammetric cycles, showing that the CF-µE response is stable. After that, the CF-µE was taken out from the solution, washed with deionized water, and exposed to the air for several days. Afterward, a similar voltammetric test was carried out again after 10 and 30 days [14,34]. As can be observed in Figure 3d, all of the CV curves are similar (the voltammetric profiles at 1st, 10th, and 30th day), in terms of the electrochemical current responses, indicating that this homemade microelectrode is exceptionally stable and has significant reproducibility, and consequently, it can be considered as a potential sensing-tool to quantify caffeine in the liquid samples.

3.3. Differential Pulse Voltammetric Experiments

The detection sensitivity of the CF-µE was assessed by the DPV analysis (Figure 5), aiming the quantification applicability in concentration quality-control for the beverages industry. The relevant parameters of this experiment are reported in the Material and Methods section. The DPV signal for caffeine was registered at approximately +1.48 V (Figure 5a), which agrees with the current-signal recorded at the CV study (Figure 3). Then, a linear relationship between the peak current and the caffeine concentration for the CF-µE was obtained via the evaluation of different caffeine concentrations (0–6 µM) in 0.5 mol L^{-1} H$_2$SO$_4$. As can be seen in Figure 5b, an increase in the current-response was attained when the concentration of caffeine in the acidic solution was increased by, at least, twelve analyte concentration additions. This protocol was carried out in triplicate, and after that, the analytical curves were also constructed and compared.

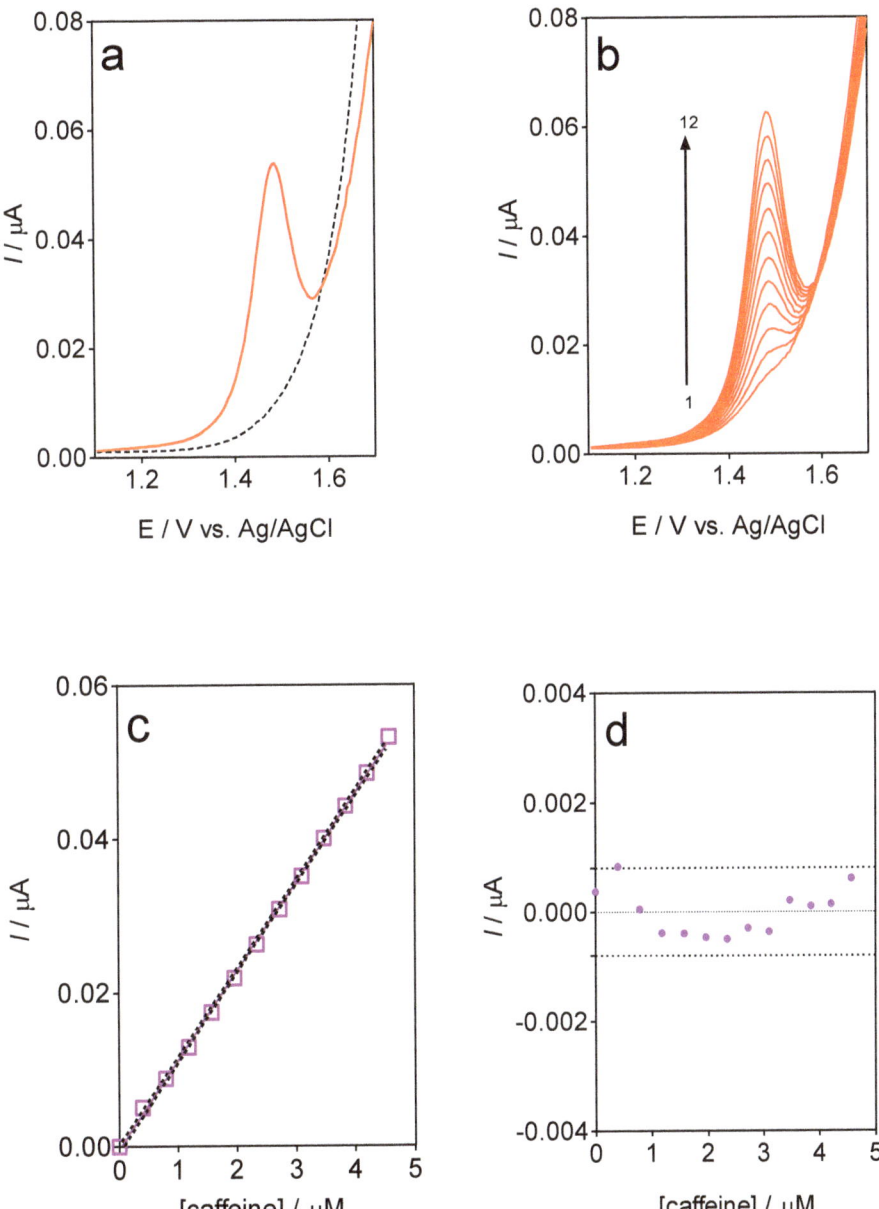

Figure 5. DPV profiles at CF-µE in 0.5 mol L^{-1} H$_2$SO$_4$ as supporting electrolyte (**a**) in absence (dashed line) or presence (orange full line) of caffeine in solution, (**b**) standard additions of caffeine solution (0.1 mol L^{-1}): (1) 0.40, (2) 0.79, (3) 1.19, (4) 1.57, (5) 1.96, (6) 2.34, (7) 2.72, (8) 3.10, (9) 3.47, (10) 3.84, (11) 4.21, and (12) 4.58 µmol L^{-1}. DPV parameters were of initial potential = 0.5 V; final potential = 1.8 V; potential scan rate = 10 mV s^{-1}, pulse amplitude = 50 mV and slow agitation. (**c**) Linear calibration plot of caffeine concentration in solution versus current peak, based on the data collected from (**b**), using CF-µE in acidic (0.1 mol L^{-1} H$_2$SO$_4$) medium; r = 6 µm. (**d**) Graphic displays weighted residuals.

In this sense, a linear relationship between 0.3 to 4.5 µmol L^{-1} (r^2 = 0.9994) (inset in Figure 5c) was found by plotting the peak current intensity as a function of the caffeine concentration, obtaining I (µA) = (11.6 ± 0.09) × 10^{-3} [caffeine, µmol L^{-1}] − (0.37 ± 0.24) × 10^{-3} (slope and intercept were the average of three independent calibrations). The limit of detection (LOD) was estimated from the equation LOD = 3.3 × $S_{y/x}/b$, where $S_{y/x}$ is the residual standard deviation and b is the slope of the calibration plot, which was approximately 0.13 µmol L^{-1}. Additionally, the residuals of the regression were randomly distributed around the zero (inset, Figure 5d), allowing a visual verification of the absence of a significant nonlinearity [27]. These approaches (LOD and residuals) are able to control both false positive and false negative errors (α = β = 0.05), as recommended by IUPAC [37,38] as well as already established by experts in the field [38,39]. It is important to indicate that additional calibration curves were obtained on different days in order to verify the good stability of the CF-µE by no registration of alterations in the statistical values (relative standard deviation (RSD), which were about 1.92% [38]). Then, the results clearly demonstrated that the CF-µE presented good repeatability and reproducibility in the analytical measurements performed, avoiding time-consuming procedures associated with the chemical cleaning or pre-treatment procedures for its surface.

Based on the existing literature, various electrochemical sensors have been constructed and used for quantifying caffeine in beverages. In this context, the proposed CF-µE has been compared with the other results (Table 1); however, even when the figures of merit are similar in some cases, the dimensions of our miniaturized sensor was able to obtain a higher sensitivity and selectivity. For example, the DPV responses achieved with other carbonaceous materials were nearly to 12.5 [1], 6.5 [40], and 3 µA [41] for the concentrations of about 0.4, 28 and 0.8 µM by using areas of approximately 28.3 mm^2, 12.6 mm^2, and 0.72 cm^2, while that, in this work, a DPV signal of 0.038 µA was achieved for 0.79 µM with a reduced surface area of about 113.1 µm^2. These insights evidence the advantages of our CF-µE over other sensors because it can be easily fabricated for sustainable practices in the spirit of green chemistry, avoiding high volumes of reagent wastes and enabling time- and cost-efficient research, achieving significant responses at lower caffeine concentrations. Another feature is that this CF-µE can be inexpensive because it was not mixed or modified with other materials, as illustrated in the examples in Table 1. Generally speaking, our CF-µE can be used in the caffeine quality-control concentration in beverages, but more investigations are needed for evaluating other additional experimental conditions, surface areas, and modifiers to significantly improve the LOD in order to find supplementary applications.

Table 1. Selected examples of quantification of caffeine in soft drink beverages at different electrochemical sensors. Comparison of LOD at different electrodes.

Electrodes	Sample	Method	Electrolyte	LOD/µmol L^{-1}	LOD/ppm	Ref.
[1] CA-ZnFe-modified GCE	Coffee and commercial beverages	DPV	1 mol L^{-1} H$_2$SO$_4$	10.0	0.194	[35]
[2] CuS NPs MCPE	Commercial tea and coffee samples	DPV	Acetate buffer (pH 7.0)	0.018	0.00035	[18]
Pt@ZnCo$_2$O$_4$	Beverage and energy drink	Amperometric	0.1 mol L^{-1} H$_2$SO$_4$	0.05	0.000971	[42]
Nafion/GCE	Cola beverages	DPV	0.1 mol L^{-1} H$_2$SO$_4$	0.798	0.0155	[34]
Nafio-covered lead film electrode	Tea, coffee, soft and energy drink samples, and pharmaceutical formulation	DPV	0.1 mol L^{-1} H$_2$SO$_4$	7.98	0.155	[43]
Nafion/GR/GCE	Soft drinks	DPV	0.01 mol L^{-1} H$_2$SO$_4$	0.12	0.00233	[22]

Table 1. Cont.

Electrodes	Sample	Method	Electrolyte	LOD/μmol L^{-1}	LOD/ppm	Ref.
Nafion/MWCNTs/GCE	Beverage samples	DPV	0.01 mol L^{-1} H$_2$SO$_4$	0.23	0.00447	[44]
Nafion/BDDE	Real cola samples	DPV	0.2 mol L^{-1} H$_2$SO$_4$	0.10	0.00194	[14]
Nafion®/GCE	Energy drinks	DPV	0.1 mol L^{-1} Britton–Robinson buffer (pH 4.5)	18.9	0.367	[45]
Nafion/PST/GCE	Tea	LSD	H$_2$SO$_4$ solution (pH 1.0)	0.10	0.00194	[46]
CTAB/GR/GCE	Soft drink sample	DPV	0.01 mol L^{-1} H$_2$SO$_4$	0.091	0.00177	[47]
PAHNSA/GCE	Coffee extracts	SWV	Acetate buffer solution (pH 5)	0.137	0.00266	[48]
Pt/CNTs/GCE	Chinese tea and Cola beverage	DPV	0.01 mol L^{-1} H$_2$SO$_4$	0.20	0.00388	[20]
×GnP-ZrO$_2$ nanocomposite-modified GCE	Various beverage	DPV	0.1 mol L^{-1} H$_2$SO$_4$	0.0119	0.00231	[49]
Modified MoS$_2$/PANI@g-C$_3$N$_4$ electrode (GCE)	Red Bull energy drink	DPV	1 mmol L^{-1} Phosphate buffer solutions (PBS)	0.062	0.0012	[50]
MIP/CPE	Spiked beverage and tea samples	DPV	Phosphate buffer (pH 7)	0.015	0.000291	[51]
[3] BQMCPE	Coffee	SWV	Phosphate buffer (pH 6)	0.30	0.00583	[52]
SWCNT/CPE	Coffee, tea, and cola nuts	DPV	0.01 mol L^{-1} H$_2$SO$_4$ pH 1.7	0.12	0.00233	[20]
[4] Nitrogen-doped carbon/GCE	Green tea and energy drink sample	DPV	0.01 mol L^{-1} H$_2$SO$_4$–Na$_2$SO$_4$ (pH 1.70)	0.02	0.000388	[53]
Nitrogen-doped grafhene (NGR)	Cookie samples, chocolate and two kinds of milk tea	SWV	0.01 mol L^{-1} H$_2$SO$_4$	0.02	0.000388	[53]
[5] Poly (ARS)	Energy drink	SWV	Acetate buffer	0.06	0.00117	[54]
GrRAC	Soft beverages	DPV	0.1 mol L^{-1} H$_2$SO$_4$	2.94	0.0571	[10]
GrRGC	Soft beverages	DPV	0.1 mol L^{-1} H$_2$SO$_4$	6.05	0.117	[11]
CF-μE	Soft drinks	DPV	0.5 mol L^{-1} H$_2$SO$_4$	0.13	0.00252	This work

[1] Carbon active with ZnFe-modified glass carbon; [2] copper sulfide nanoparticle-modified carbon paste electrode; [3] 1,4-benzoquinone-modified carbon paste electrode; [4] glassy carbon electrode (GCE) modified with nitrogen-doped carbon nanotubes; [5] Simultaneous determination of CAF and VAN.

3.4. Caffeine Determination in Soft Beverage Samples

CF-μE was used to control the caffeine concentration in several soft beverage samples by the standard addition protocol with the DPV analysis (Figure 6). This procedure was implemented in order to diminish the effects of other components in the soft beverage samples (matrix effect). Firstly, as indicated in the Experimental section, all beverage samples were ultrasonicated to remove the gas content. After that, laboratory samples were prepared by diluting 0.2 mL of each one of the samples under investigation with the supporting electrolyte (5 mL) to operate in the linear range of the method. On the one hand, a caffeine-free soft drink was also examined, aiming to test a possible contribution of the matrix itself to the analyte signal. On the other hand, cola beverages were also electrochemically evaluated. As shown in Figure 6a, no voltammetric signal around the caffeine potential was registered. Meanwhile, a current-voltammetric response was achieved at all beverages samples, confirming the presence of caffeine in the composition of these samples (see some examples in Figure 6a).

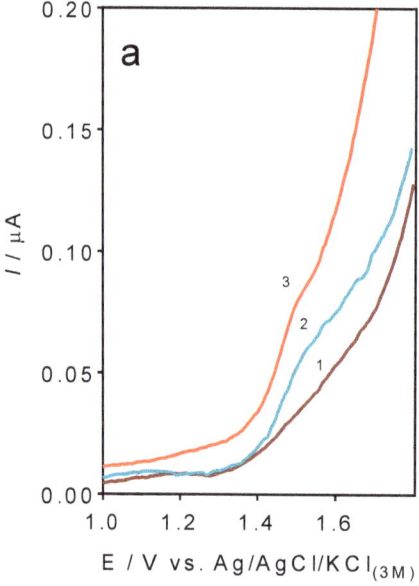 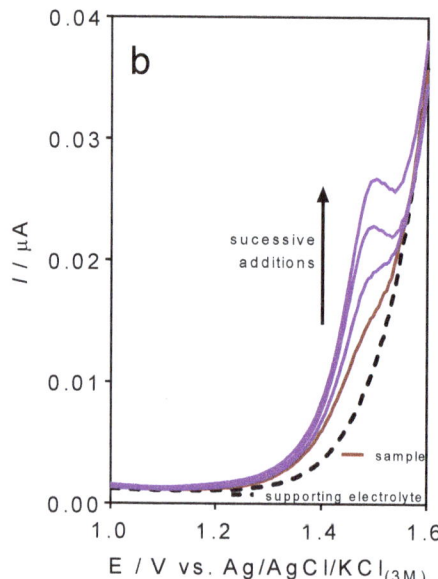

Figure 6. DPV analysis of (**a**) some drink samples ((1) caffeine-free soft drink, (2) cola–soft drink 1 and (3) cola–soft drink 2), and (**b**) the standard addition procedure for the cola–soft drink 2 (plot with the DPV profile of supporting electrolyte), the soft beverage sample as well as the 1° addition, 2° addition, and 3° addition of 0.1 mol L^{-1} caffeine, with CF-µE. DPV parameters were of initial potential = 0.5 V; final potential = 1.8 V; potential scan rate = 10 mV s^{-1}, pulse amplitude = 50 mV.

In the case of soft drinks containing caffeine, as described in the product information, three consecutive standard additions (100 µL of 0.1 M caffeine) were conducted in the electrochemical cell [40,55]. Figure 6b shows that the standard additions promoted an increase in the current-signal, also confirming the presence of the caffeine in the drink. Another important feature is that no significant alterations in the caffeine oxidation peak were achieved (Figure 6b), as observed on other sensors reported in the literature [9,14,40,41,54,55], which is significantly associated with the effects on mass transport that were minimized by the dimensions of the CF-µE. The known amount of caffeine added to the samples was able to estimate the recoveries.

The mean results were obtained for the standard additions protocol at each one of the soft drinks analyzed, recording three measurements with acceptable standard deviations and confidence intervals relating to a probability of 95%. This strategy consents to validate both false positives and false negatives ($\alpha = \beta = 0.05$), as recommended by the IUPAC [37,39]. Subsequently, all samples were also analyzed by HPLC, and the results were compared with the DPV analysis of the CF-µE (Table 2). Analyzing the figures reached, CF-µE can be considered an efficient tool to be employed with good confidence in the caffeine concentration evaluation of soft beverage samples. The caffeine concentrations measured in the soft beverage samples with the CF-µE were like those quantified by HPLC (as an independent method with 95% of confidence [37,39]), and comparable to those reported in the nutritional table of the samples.

Table 2. Caffeine contents, as reported in the soft drinks, free of or containing caffeine, as well as HPLC and microelectrode determinations.

Beverages	Labelled/mg [a]	HPLC/mg	CF-µE/mg	Error [b] (%)	Error [c] (%)
Caffeine-free soft drink	0.00	_ [d]	0.01	5	0
Soft drink 1	32.0	35.1	33.2	−9.68	−3.75
Soft drink 2	32.0	34.5	32.9	−7.81	3.23
Diet Soft drink 3	46.0	45.5	45.8	1.09	0.43
Zero sugar Soft drink 4	68.0	69.2	68.2	−1.76	−0.29
Energy drink	80.0 [e]	78.2	79.5	2.25	0.63

[a] Commercial coke soft drinks contain 32 mg of caffeine per 12-oz. (335 mL) serving. [b] Relative error (%) = [(Labelled value − voltammetric value)/(Labelled value) × 100]. [c] Relative error (%) = [(Labelled value − HPLC value)/(Labelled value) × 100]. [d] Under limit of instrumental detection. [e] Commercial energy drinks contain 80 mg of caffeine per 8.4-oz. (250 mL) serving.

3.5. Interference Studies

As described, CF-µE presented good performance for quantifying caffeine. The selectivity of this sensor was also evaluated by intentionally introducing concentrations of ascorbic acid as interference during caffeine analysis. Ascorbic acid, also known as Vitamin C, is a water-soluble vitamin found in citrus and other fruits and vegetables; also, it is a supplement in the soft beverages. Therefore, this acid was chosen to test the determination of caffeine in the presence of an important interference. The experimental data are reported in Figure 7. The voltammetric signals for the ascorbic acid and caffeine were clearly registered at +0.05 V and + 0.4 V, respectively. The results showed that an excess of the concentration of ascorbic acid into the sample solution did not cause interference on the determination of caffeine, demonstrating the viability of this electrochemical sensor. In fact, no significant modifications were observed on the voltammetric responses of caffeine, in terms of electrical potential and current intensities, when a new analytical curve was obtained in the presence of ascorbic acid.

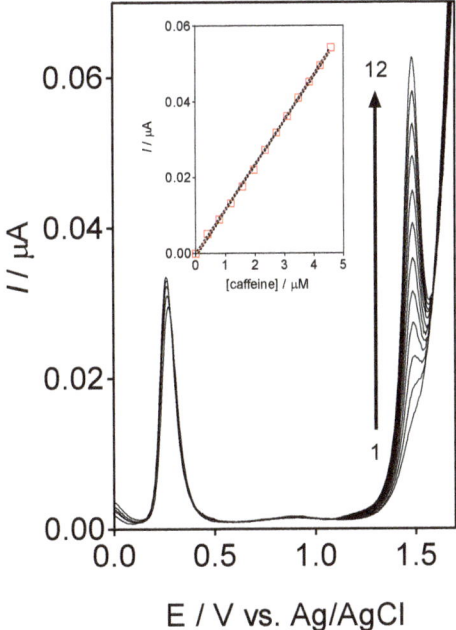

Figure 7. DPV profiles at CF-µE in 0.5 mol L^{-1} H$_2$SO$_4$ as supporting electrolyte for contructing the

analytical curve of caffeine (standard additions of caffeine solution (0.1 mol L^{-1}): (1) 0.40, (2) 0.79, (3) 1.19, (4) 1.57, (5) 1.96, (6) 2.34, (7) 2.72, (8) 3.10, (9) 3.47, (10) 3.84, (11) 4.21, and (12) 4.58 µ mol L^{-1}) in presence of ascorbic acid (50 µmol L^{-1}) in solution. Inset: Linear calibration plot of caffeine concentration in solution versus current peak, based on the data collected from the analytical curve.

4. Conclusions

A homemade CF-µE was fabricated and it was a suitable electrochemical microsensor for caffeine determinations. Under the experimental conditions examined, caffeine displayed irreversible behavior in cyclic voltammetry with the CF-µE. This sensor exhibited a high detection sensitivity, a high mass transfer rate, and an enhanced signal-to-noise ratio; consequently, it is shown to have extraordinary stability and reproducibility, making it more applicable as a sensor in the food control analyses and measurements. For future works of this research field, some modifiers could be investigated in order to increase the surface area of the microsensor to guarantee more stability, sensibility, and selectivity for real-time monitoring.

Author Contributions: Conceptualization, C.A.M.-H. and P.S.C.; methodology, K.C.d.F.A., E.C.T.d.A.C. and D.M.d.A.; validation, K.C.d.F.A., E.C.T.d.A.C. and D.M.d.A.; formal analysis, E.C.T.d.A.C. and D.M.d.A.; investigation, K.C.d.F.A., E.C.T.d.A.C., D.M.d.A. and E.V.S.; data curation, E.V.S. and P.S.C.; writing—original draft preparation, E.V.S. and P.S.C.; writing—review and editing, E.V.S. and P.S.C.; supervision, E.V.S., P.S.C. and C.A.M.-H.; project administration, C.A.M.-H. and E.V.S.; funding acquisition, C.A.M.-H. and E.V.S. All authors have read and agreed to the published version of the manuscript.

Funding: Conselho Nacional de Desenvolvimento Científico e Tecnológico (CNPq, Brazil), 306323/2018-4, 312595/2019-0, 439344/2018-2, 315879/2021-1, 409196/2022-3, 408110/2022-8, and Fundação de Amparo à Pesquisa do Estado de São Paulo (Brazil), FAPESP 2014/50945-4 and 2019/13113-4. The APC was funded by CNPq Process 116925/2022-1 PDJ.

Institutional Review Board Statement: Not applicable.

Informed Consent Statement: Not applicable.

Data Availability Statement: Not applicable.

Conflicts of Interest: The authors declare no conflict of interest.

References

1. Švorc, L.; Tomčík, P.; Svítková, J.; Rievaj, M.; Bustin, D. Voltammetric determination of caffeine in beverage samples on bare boron-doped diamond electrode. *Food Chem.* **2012**, *135*, 1198–1204. [CrossRef] [PubMed]
2. Araujo, D.; Brito, C.; de Oliveira, S.D.; Silva, D.; Martinez-Huitle, C.; Aragao, C. Platinum sensor for quantifying caffeine in drug formulations. *Curr. Pharm. Anal.* **2014**, *10*, 231–238. [CrossRef]
3. Rostagno, M.A.; Manchón, N.; D'Arrigo, M.; Guillamón, E.; Villares, A.; García-Lafuente, A.; Ramos, A.; Martínez, J.A. Fast and simultaneous determination of phenolic compounds and caffeine in teas, mate, instant coffee, soft drink and energetic drink by high-performance liquid chromatography using a fused-core column. *Anal. Chim. Acta* **2011**, *685*, 204–211. [CrossRef] [PubMed]
4. Rajabi Khorrami, A.; Rashidpur, A. Development of a fiber coating based on molecular sol-gel imprinting technology for selective solid-phase micro extraction of caffeine from human serum and determination by gas chromatography/mass spectrometry. *Anal. Chim. Acta* **2012**, *727*, 20–25. [CrossRef] [PubMed]
5. Rahim, A.A.; Nofrizal, S.; Saad, B. Rapid tea catechins and caffeine determination by HPLC using microwave-assisted extraction and silica monolithic column. *Food Chem.* **2014**, *147*, 262–268. [CrossRef] [PubMed]
6. Al-Othman, Z.A.; Aqel, A.; Alharbi, M.K.E.; Yacine Badjah-Hadj-Ahmed, A.; Al-Warthan, A.A. Fast chromatographic determination of caffeine in food using a capillary hexyl methacrylate monolithic column. *Food Chem.* **2012**, *132*, 2217–2223. [CrossRef]
7. Hadad, G.M.; Abdel Salam, R.A.; Soliman, R.M.; Mesbah, M.K. Rapid and simultaneous determination of antioxidant markers and caffeine in commercial teas and dietary supplements by HPLC-DAD. *Talanta* **2012**, *101*, 38–44. [CrossRef] [PubMed]
8. Svorc, L. Determination of caffeine: A comprehensive review on electrochemical methods. *Int. J. Electrochem. Sci.* **2013**, *8*, 5755–5773.
9. Vasilescu, I.; Eremia, S.A.V.; Penu, R.; Albu, C.; Radoi, A.; Litescu, S.C.; Radu, G.L. Disposable dual sensor array for simultaneous determination of chlorogenic acid and caffeine from coffee. *RSC Adv.* **2015**, *5*, 261–268. [CrossRef]

10. Monteiro, M.K.S.; da Silva, D.R.; Quiroz, M.A.; Vilar, V.J.P.; Martínez-Huitle, C.A.; dos Santos, E.V. Applicability of cork as novel modifiers to develop electrochemical sensor for caffeine determination. *Materials* **2021**, *14*, 37. [CrossRef]
11. Monteiro, M.K.S.; Paiva, S.S.M.; da Silva, D.R.; Vilar, V.J.P.; Martínez-Huitle, C.A.; dos Santos, E.V. Novel cork-graphite electrochemical sensor for voltammetric determination of caffeine. *J. Electroanal. Chem.* **2019**, *839*, 283–289. [CrossRef]
12. de Araújo, D.M.; Paiva, S.d.S.S.M.; Henrique, J.M.M.; Martínez-Huitle, C.A.; dos Santos, E.V. Green composite sensor for monitoring hydroxychloroquine in different water matrix. *Materials* **2021**, *14*, 4990. [CrossRef]
13. Ören, T.; Anık, Ü. Voltammetric determination of caffeine by using gold nanoparticle-glassy carbon paste composite electrode. *Measurement* **2017**, *106*, 26–30. [CrossRef]
14. Martínez-Huitle, C.A.; Suely Fernandes, N.; Ferro, S.; de Battisti, A.; Quiroz, M.A. Fabrication and application of Nafion®-modified boron-doped diamond electrode as sensor for detecting caffeine. *Diam. Relat. Mater.* **2010**, *19*, 1188–1193. [CrossRef]
15. Rick Lopes Da Silva, Á.; Medeiros De Araújo, D.; Bernardo Sabino Da Silva, E.; Serradella Vieira, D.; de Kássio Vieira Monteiro, N.; Martínez-Huitle, C.A. Understanding the behavior of caffeine on a boron-doped diamond surface: Voltammetric, DFT, QTAIM and ELF Studies. *New J. Chem.* **2017**, *41*, 7766–7774. [CrossRef]
16. Yardim, Y.; Keskin, E.; Şentürk, Z. Voltammetric determination of mixtures of caffeine and chlorogenic acid in beverage samples using a boron-doped diamond electrode. *Talanta* **2013**, *116*, 1010–1017. [CrossRef] [PubMed]
17. Wong, A.; Santos, A.M.; Silva, T.A.; Fatibello-Filho, O. Simultaneous determination of isoproterenol, acetaminophen, folic acid, propranolol and caffeine using a sensor platform based on carbon black, graphene oxide, copper nanoparticles and PEDOT:PSS. *Talanta* **2018**, *183*, 329–338. [CrossRef]
18. Mahanthappa, M.; Yellappa, S.; Kottam, N.; Srinivasa Rao Vusa, C. Sensitive determination of caffeine by copper sulphide nanoparticles modified carbon paste electrode. *Sens. Actuators A Phys.* **2016**, *248*, 104–113. [CrossRef]
19. Wang, Y.; Ding, Y.; Li, L.; Hu, P. Nitrogen-doped carbon nanotubes decorated Poly (L-Cysteine) as a novel, ultrasensitive electrochemical sensor for simultaneous determination of theophylline and caffeine. *Talanta* **2018**, *178*, 449–457. [CrossRef]
20. Habibi, B.; Abazari, M.; Pournaghi-Azar, M.H. A carbon nanotube modified electrode for determination of caffeine by differential pulse voltammetry. *Chin. J. Catal.* **2012**, *33*, 1783–1790. [CrossRef]
21. AL-Gahouari, T.; Bodkhe, G.; Sayyad, P.; Ingle, N.; Mahadik, M.; Shirsat, S.M.; Deshmukh, M.; Musahwar, N.; Shirsat, M. Electrochemical Sensor: L-Cysteine induced selectivity enhancement of electrochemically reduced graphene oxide–multiwalled carbon nanotubes hybrid for detection of lead (Pb^{2+}) ions. *Front. Mater.* **2020**, *7*, 68. [CrossRef]
22. Sun, J.Y.; Huang, K.J.; Wei, S.Y.; Wu, Z.W.; Ren, F.P. A graphene-based electrochemical sensor for sensitive determination of caffeine. *Colloids Surf. B Biointerfaces* **2011**, *84*, 421–426. [CrossRef] [PubMed]
23. Jeevagan, A.J.; John, S.A. Electrochemical determination of caffeine in the presence of paracetamol using a self-assembled monolayer of non-peripheral amine substituted copper(II) phthalocyanine. *Electrochim. Acta* **2012**, *77*, 137–142. [CrossRef]
24. Ghoreishi, S.M.; Attaran, A.M.; Amin, A.M.; Khoobi, A. Multiwall carbon nanotube-modified electrode as a nanosensor for electrochemical studies and stripping voltammetric determination of an antimalarial drug. *RSC Adv.* **2015**, *5*, 14407–14415. [CrossRef]
25. Rassaei, L.; Marken, F.; Sillanpää, M.; Amiri, M.; Cirtiu, C.M.; Sillanpää, M. Nanoparticles in electrochemical sensors for environmental monitoring. *Trends Anal. Chem.* **2011**, *30*, 1704–1715. [CrossRef]
26. Nebel, C.E.; Yang, N.; Uetsuka, H.; Osawa, E.; Tokuda, N.; Williams, O. Diamond nano-wires, a new approach towards next generation electrochemical gene sensor platforms. *Diam. Relat. Mater.* **2009**, *18*, 910–917. [CrossRef]
27. Carolina Torres, M.; Barsan, M.M.; Brett, C.M.A. Simple electrochemical sensor for caffeine based on carbon and nafion-modified carbon electrodes. *Food. Chem.* **2014**, *149*, 215–220. [CrossRef]
28. Siqueira, G.P.; de Faria, L.V.; Rocha, R.G.; Matias, T.A.; Richter, E.M.; Muñoz, R.A.A.; da Silva, I.S.; Dantas, L.M.F. Nanoporous gold microelectrode arrays using microchips: A highly sensitive and cost-effective platform for electroanalytical applications. *J. Electroanal. Chem.* **2022**, *925*, 116880. [CrossRef]
29. Jose, J.; Subramanian, V.; Shaji, S.; Sreeja, P.B. An electrochemical sensor for nanomolar detection of caffeine based on Nicotinic Acid Hydrazide Anchored on Graphene Oxide (NAHGO). *Sci. Rep.* **2021**, *11*, 1–11. [CrossRef]
30. Sachidananda, T.G.; Chikkanagoudar, R.N.; Pattar, N.; Nandurkar, S. Investigations of the influence of geometrical parameters of carbon nanotube material for sensor and MEMS applications. *Mater. Today Proc.* **2022**, *63*, 745–750. [CrossRef]
31. Mao, Y.; Lu, G.; Chen, J. Nanocarbon-based gas sensors: Progress and challenges. *J. Mater. Chem. A Mater.* **2014**, *2*, 5573–5579. [CrossRef]
32. Wang, B.; Wen, X.; Chiou, P.-Y.; Maidment, N.T. Pt nanoparticle-modified carbon fiber microelectrode for selective electrochemical sensing. *Electroanalysis* **2019**, *31*, 1641–1645. [CrossRef]
33. Aquino de Queiroz, J.L.; Martínez-Huitle, C.A.; Castro, P.S. Real time monitoring of in situ generated hydrogen peroxide in electrochemical advanced oxidation reactors using an integrated Pt microelectrode. *Talanta* **2020**, *218*, 121133. [CrossRef] [PubMed]
34. Brunetti, B.; Desimoni, E.; Casati, P. Determination of caffeine at a nafion-covered glassy carbon electrode. *Electroanalysis* **2007**, *19*, 385–388. [CrossRef]
35. Arroyo-Gómez, J.J.; Villarroel-Rocha, D.; de Freitas-Araújo, K.C.; Martínez-Huitle, C.A.; Sapag, K. Applicability of activated carbon obtained from peach stone as an electrochemical sensor for detecting caffeine. *J. Electroanal. Chem.* **2018**, *822*, 171–176. [CrossRef]

36. Khoo, W.Y.H.; Pumera, M.; Bonanni, A. Graphene platforms for the detection of caffeine in real samples. *Anal. Chim. Acta* **2013**, *804*, 92–97. [CrossRef]
37. Currie, L.A. International Union of Pure and Applied Chemistry Nomenclature in Evaluation of Analytical Methods Including Detection and Quantification Capabilities. *Pure Appl. Chem.* **1995**, *67*, 1699–1723. [CrossRef]
38. Desimoni, E.; Brunetti, B. About estimating the limit of detection of heteroscedastic analytical systems. *Anal. Chim. Acta* **2009**, *655*, 30–37. [CrossRef]
39. Danzer, K.; Currie, L.A. Guideline for calibration in analytical chemistry—Part 1. Fundamentals and single component calibration. *Pure Appl. Chem.* **1998**, *70*, 993–1014. [CrossRef]
40. Redivo, L.; Stredanský, M.; DeAngelis, E.; Navarini, L.; Resmini, M.; Švorc, Ľ. Bare carbon electrodes as simple and efficient sensors for the quantification of caffeine in commercial beverages. *R. Soc. Open Sci.* **2018**, *5*, 172146. [CrossRef]
41. Lourenção, B.C.; Antigo Medeiros, R.; Rocha-Filho, R.C.; Mazo, L.H.; Fatibello-Filho, O. Simultaneous voltammetric determination of paracetamol and caffeine in pharmaceutical formulations using a boron-doped diamond electrode. *Talanta* **2009**, *78*, 748–752. [CrossRef] [PubMed]
42. Jesu Amalraj, A.J.; Umesh, N.; Wang, S.F. Rational design of platinum assimilated 3-D zinc cobalt oxide flowers for the electrochemical detection of caffeine in beverage and energy drink. *J. Ind. Eng. Chem.* **2022**, *106*, 205–213. [CrossRef]
43. Tyszczuk-Rotko, K.; Bęczkowska, I. Nafion covered lead film electrode for the voltammetric determination of caffeine in beverage samples and pharmaceutical formulations. *Food Chem.* **2015**, *172*, 24–29. [CrossRef] [PubMed]
44. Yang, S.; Yang, R.; Li, G.; Qu, L.; Li, J.; Yu, L. Nafion/Multi-wall carbon nanotubes composite film coated glassy carbon electrode for sensitive determination of caffeine. *J. Electroanal. Chem.* **2010**, *639*, 77–82. [CrossRef]
45. Farag, A.S.; Pravcová, K.; Česlová, L.; Vytřas, K.; Sýs, M. Simultaneous determination of caffeine and pyridoxine in energy drinks using differential pulse voltammetry at glassy carbon electrode modified with Nafion®. *Electroanalysis* **2019**, *31*, 1494–1499. [CrossRef]
46. Guo, S.; Zhu, Q.; Yang, B.; Wang, J.; Ye, B. Determination of caffeine content in tea based on Poly(Safranine T) electroactive film modified electrode. *Food Chem.* **2011**, *129*, 1311–1314. [CrossRef]
47. Sun, J.Y.; Huang, K.J.; Wei, S.Y.; Wu, Z.W. Application of cetyltrimethylammonium bromide–graphene modified electrode for sensitive determination of caffeine. *Can. J. Chem.* **2011**, *89*, 697–702. [CrossRef]
48. Amare, M.; Admassie, S. Polymer modified glassy carbon electrode for the electrochemical determination of caffeine in coffee. *Talanta* **2012**, *93*, 122–128. [CrossRef]
49. Okutan, M.; Boran, F.; Alver, E.; Asan, A. One-pot synthesize of graphene-ZrO_2 nanocomposite: A novel modified glassy carbon electrode for the detection of caffeine in beverage samples. *Mater. Chem. Phys.* **2022**, *280*, 125846. [CrossRef]
50. Murugan, E.; Dhamodharan, A. Separate and simultaneous determination of vanillin, theophylline and caffeine using molybdenum disulfide embedded polyaniline/graphitic carbon nitrite nanocomposite modified glassy carbon electrode. *Diam. Relat. Mater.* **2021**, *120*, 108684. [CrossRef]
51. Alizadeh, T.; Ganjali, M.R.; Zare, M.; Norouzi, P. Development of a voltammetric sensor based on a Molecularly Imprinted Polymer (MIP) for caffeine measurement. *Electrochim. Acta* **2010**, *55*, 1568–1574. [CrossRef]
52. Aklilu, M.; Tessema, M.; Redi-Abshiro, M. Indirect voltammetric determination of caffeine content in coffee using 1,4-Benzoquinone modified carbon paste electrode. *Talanta* **2008**, *76*, 742–746. [CrossRef]
53. Jiang, L.; Ding, Y.; Jiang, F.; Li, L.; Mo, F. Electrodeposited nitrogen-doped graphene/carbon nanotubes nanocomposite as enhancer for simultaneous and sensitive voltammetric determination of caffeine and vanillin. *Anal. Chim. Acta* **2014**, *833*, 22–28. [CrossRef] [PubMed]
54. Filik, H.; Avan, A.A.; Mümin, Y. Simultaneous electrochemical determination of caffeine and vanillin by using Poly(Alizarin Red S) modified glassy carbon electrode. *Food Anal. Methods* **2017**, *10*, 31–40. [CrossRef]
55. Tadesse, Y.; Tadese, A.; Saini, R.C.; Pal, R. Cyclic voltammetric investigation of caffeine at anthraquinone modified carbon paste electrode. *Int. J. Electrochem.* **2013**, *2013*, 849327. [CrossRef]

Disclaimer/Publisher's Note: The statements, opinions and data contained in all publications are solely those of the individual author(s) and contributor(s) and not of MDPI and/or the editor(s). MDPI and/or the editor(s) disclaim responsibility for any injury to people or property resulting from any ideas, methods, instructions or products referred to in the content.

Article

The Composite Material of (PEDOT-Polystyrene Sulfonate)/Chitosan-AuNPS-Glutaraldehyde/as the Base to a Sensor with Laccase for the Determination of Polyphenols

Paweł Krzyczmonik [1,*], Marta Klisowska [1], Andrzej Leniart [1], Katarzyna Ranoszek-Soliwoda [2], Jakub Surmacki [3], Karolina Beton-Mysur [3] and Beata Brożek-Płuska [3]

[1] Department of Inorganic and Analytical Chemistry, Faculty of Chemistry, University of Lodz, Tamka 12, 91-403 Lodz, Poland
[2] Department of Materials Technology and Chemistry, Faculty of Chemistry, University of Lodz, Pomorska 163 Street, 90-236 Lodz, Poland; katarzyna.ranoszek.soliwoda@chemia.uni.lodz.pl
[3] Laboratory of Laser Molecular Spectroscopy, Faculty of Chemistry, Institute of Applied Radiation Chemistry, Lodz University of Technology, Wroblewskiego 15, 93-590 Lodz, Poland; jakub.surmacki@p.lodz.pl (J.S.)
* Correspondence: pawel.krzyczmonik@chemia.uni.lodz.pl

Citation: Krzyczmonik, P.; Klisowska, M.; Leniart, A.; Ranoszek-Soliwoda, K.; Surmacki, J.; Beton-Mysur, K.; Brożek-Płuska, B. The Composite Material of (PEDOT-Polystyrene Sulfonate)/Chitosan-AuNPS-Glutaraldehyde/as the Base to a Sensor with Laccase for the Determination of Polyphenols. *Materials* 2023, 16, 5113. https://doi.org/10.3390/ma16145113

Academic Editors: Yuyan Jiang and Abderrahim Yassar

Received: 31 May 2023
Revised: 12 July 2023
Accepted: 17 July 2023
Published: 20 July 2023

Copyright: © 2023 by the authors. Licensee MDPI, Basel, Switzerland. This article is an open access article distributed under the terms and conditions of the Creative Commons Attribution (CC BY) license (https://creativecommons.org/licenses/by/4.0/).

Abstract: The described research aimed to develop the properties of the conductive composite /poly(3,4-ethylenedioxy-thiophene-poly(4-lithium styrenesulfonic acid)/chitosan-AuNPs-glutaraldehyde/ (/PEDOT-PSSLi/chit-AuNPs-GA/) and to develop an electrochemical enzyme sensor based on this composite material and glassy carbon electrodes (GCEs). The composite was created via electrochemical production of an /EDOT-PSSLi/ layer on a glassy carbon electrode (GCE). This layer was covered with a glutaraldehyde cross-linked chitosan and doped with AuNPs. The influence of AuNPs on the increase in the electrical conductivity of the chitosan layers and on facilitating the oxidation of polyphenols in these layers was demonstrated. The enzymatic sensor was obtained via immobilization of the laccase on the surface of the composite, with glutaraldehyde as the linker. The investigation of the surface morphology of the GCE/PEDOT-PSSLi/chit-AuNPs-GA/Laccase sensor was carried out using SEM and AFM microscopy. Using EDS and Raman spectroscopy, AuNPs were detected in the chitosan layer and in the laccase on the surface of the sensor. Polyphenols were determined using differential pulse voltammetry. The biosensor exhibited catalytic activity toward the oxidation of polyphenols. It has been shown that laccase is regenerated through direct electron transfer between the sensor and the enzyme. The results of the DPV tests showed that the developed sensor can be used for the determination of polyphenols. The peak current was linearly proportional to the concentrations of catechol in the range of 2–90 µM, with a limit of detection (LOD) of 1.7 µM; to those of caffeic acid in the range of 2–90 µM, LOD = 1.9 µM; and to those of gallic acid in the range 2–18 µM, LOD = 1.7 µM. Finally, the research conducted in order to determine gallic acid in a natural sample, for which white wine was used, was described.

Keywords: PEDOT; chitosan; Au nanoparticles; laccase; immobilization; electrochemical biosensor; differential pulse voltammetry; SEM; AFM; Raman

1. Introduction

Sensors and biosensors are among the most interesting tools of modern analytical chemistry. The essence of the proper operation of sensors is to design them for the needs of specific analyses. This requires the use of a variety of materials and, sometimes, the development of new materials to play a specific role in the operation of the sensor as a whole. In the described research, a new type of composite, which combines the conductive PEDOT-PSSLi layer with a chitosan layer, was developed.

Chitosan is a substance of natural origin. It is a polysaccharide that is obtained through the partial deacetylation of chitin. It is characterized by a very high level of biocompatibility.

Chitosan is a substance used in medicines and may safely come into direct contact with the body, e.g., some materials made with chitosan are used to stop bleeding and as burn dressings [1,2]. This feature is very desirable in systems that are designed to work with natural samples because they increase the durability of the sensor and, at the same time, have no negative impact on the tested sample. Moreover, chitosan is characterized by its excellent film-forming ability and good mechanical strength. Although chitosan is a non-conductive biomaterial, it can be used together with conductive polymers, such as polyaniline [3,4], polypyrrole [5,6], and PEDOT [7–9], and polymer/chitosan coatings can be used in the construction of sensors with electrochemical detection. Very often, in the developed sensors, solid electrodes such as Au, Pt, or various types of carbon are used, and they are modified with layers of conductive polymers.

In many applications, the properties of PEDOT turned out to be better than those of polypyrrole and polythiophene. Krosa et al. [10] proved in their studies that polypyrrole can only be used as a biosensor component for a very short time, while PEDOT, in the same applications, turned out to be a component that was suitable for continuous use. According to the authors, this was due to the fact that PEDOT has greater electrochemical stability than polypyrrole. In the case of systems containing biochemical objects, it is very important that PEDOT has low toxicity in comparison to, for example, polyaniline, whose degradation products are carcinogenic [11]. PEDOT doped with polystyrene sulfonic acid salts is also very widely used. It is a material in which the anions doping the conductive polymer are permanently bound to the PEDOT layer, which increases the durability of the entire system and does not allow for a complete reduction of the PEDOT. In the described tests, glassy carbon electrodes (GCEs), modified with the PEDOT-PSSLi layer, were used. This type of material was previously used to modify platinum and GCE electrodes [12,13] and as a basis for the construction of enzymatic sensors [14,15].

In order to change its electrical properties, the chitosan was doped with AuNPs, which allowed us to obtain conductivity in the entire layer. In addition, AuNPs were used for the direct oxidation of laccase on the sensor surface. Gold nanoparticles are materials characterized by very good electrical conductivity and a very high degree of biocompatibility; AuNP suspensions are used in various medical therapies. The sizes of gold nanoparticles depend on the method of their synthesis. In the described research, nanoparticles synthesized via reaction with sodium borohydride were used [16]. As a result of this reaction, nanoparticle suspensions with sizes of 3–5 nm were obtained. The disadvantage of chitosan layers is their susceptibility to dissolution in aqueous solutions, which reduces the durability of sensors based on this material. In order to eliminate this drawback, the chitosan layer was modified through cross-linking with glutaraldehyde, which made the layer resistant to dissolution.

It is known from the literature reports that chitosan is very suitable for immobilizing biomolecules on its surface [11–19]. Chitosan molecules contain amino groups and carboxyl groups in their structure. As a result, it is possible to immobilize various biological objects, such as enzymes, to one or the other functional group by forming a covalent bond. It is one of the most frequently used polymers of natural origin in biosensors [17,20,21]. In the described research, the authors decided to immobilize laccase by linking the amino groups of the enzyme and the amino groups of chitosan, using glutaraldehyde as a linker. The use of glutaraldehyde enabled the formation of covalent bonds between laccase and chitosan. Compared to traditional chromatographic methods, electroanalytical techniques based on oxidation–reduction reactions have many advantages, such as simplicity, low cost, high stability and sensitivity, fast response, and excellent repeatability [22,23]. If electrochemical sensors are used for electroanalytical methods, it is possible to obtain analytical techniques for which preliminary sample preparation will not be necessary. This allows us to design systems that will be promising analytical tools for the analysis of real samples.

The purpose of the research was to develop a conductive electrode material based on chitosan, which could be the basis for the production of an electrochemical sensor with an immobilized enzyme. The enzyme chosen to be immobilized was laccase, which was bound to the substrate through the formation of a covalent bond. This method of immobilization is the most effective because it eliminates the problem of leaching the enzyme from the layer. Glutaraldehyde was used to cross-link the structure of the chitosan layer, which increased its durability, and AuNPs were used to increase the electrical conductivity of the chitosan layer. In the first stage of the research, the focus was on obtaining a composite material that would ensure the best electrochemical properties and the possibility of attaching an enzyme molecule. In the second stage of the research, a method for laccase immobilization on the surface of the composite material, using glutaraldehyde, was developed. The electrochemical properties were characterized with cyclic voltammetry in ferricyanides solutions. The electrocatalytic oxidation of polyphenols was tested using a sensor in catechol solutions. The obtained materials were characterized using SEM, EDS, Raman spectroscopy, and cyclic voltammetry. The last stage involved tests for the determination of selected polyphenols in aqueous solutions, in order to investigate the possibility of their use in electroanalysis. The possibility of the determination of polyphenols in a natural sample, for which white wine was used, was also presented.

2. Materials and Methods

All of the chemical reagents were analytically pure and were used without further purification. Laccase, 3,4-ethylenedioxy-thiophene, catechol, caffeic acid, gallic acid, and Rhodanine were supplied by Sigma-Aldrich (St. Louis, MO, USA). Chitosan was obtained from Across Organic (Geel, Belgium) and $HAuCl_4 \cdot 3H_2O$ was obtained from AlfaAesar (Kandel, Germany). The 25% solution of glutaraldehyde, potassium ferrocyanide, sodium chloride, citric acid, trisodium citrate, potassium chloride, acetic acid, disodium hydrogen orthophosphate dodecahydrate ($Na_2HPO_4 \cdot 12H_2O$), potassium dihydrogen orthophosphate, and sodium hydroxide were supplied by POCH Gliwice. All of the solutions were prepared just before use, with water purified using the Millipore (Milli-Q) system. Laccase was stored at 4 °C.

Instrumentation

The measuring equipment comprised a PAR 273A potentiostat (EG&G Princeton Applied Research Company, Princeton, NJ, USA) and a computer with CorrWare 2.9 and CorrView 2.9 software (Scribner Associates, Inc. Southern Pines, NC, USA). All of the electrochemical measurements were carried out in a three-electrode cell. A modified glassy carbon electrode was used as a working electrode, a saturated calomel electrode (SCE) was used as a reference electrode, and a platinum mesh was used as a counter electrode. The morphology of sensors was investigated using an Atomic Force Microscope (DIMENSION ICON ScanAsyst, BRUKER, Billerica, MA, USA). The measurements were taken using the scanning probe TESPA-New 09 in tapping mode. The surface morphology was also investigated with electron microscopy using a High-Resolution Scanning Electron Microscope (HR-SEM, FEI Nova NanoSEM 450, Hillsboro, OR, USA) equipped with a CBS detector for the detection of backscattered electrons. The chemical composition analysis was performed using the energy-dispersive spectrometer (EDS, EDAX/AMETEK, Materials Analysis Division, Model Octane Super, Mahwah, NJ, USA). A WITec alpha 300 RSA+ confocal microscope was used to record Raman spectra. The configuration of the experimental setup was as follows: the diameter of fiber, 50 μm for 532 nm, a monochromator Acton-SP-2300i, and a CCD camera Andor Newton DU970-UVB-353 for 532 nm. The excitation line was focused on the sample through a 40× dry objective (Nikon, objective type CFI Plan Fluor C ELWD DIC-M, numerical aperture (NA) of 0.60 and a 3.6–2.8 mm working distance). The laser excitation power was 10 mW at 532 nm for pure components (laccase and chitosan), and 2.7 mW for samples deposited on the electrodes, with an integration time of 0.5 s and 10 accumulations. Data acquisition and processing were performed using WITec Project Plus software ver. 4.1. The cosmic rays were removed from each Raman spectrum (model:

filter size, 2; dynamic factor, 10), and for the smoothing procedure, the Savitzky–Golay method was also implemented (model: order, 4; derivative, 0). The baseline corrections of Raman spectra were performed using WITec Project Plus and OriginPro.

Synthesis of gold nanoparticles (AuNPs)

The synthesis of gold nanoparticles (AuNPs) was carried out by mixing 30 mL of distilled water, 1.8 mL of 0.0025 M Au (III), and 1.5 mg of sodium borohydride $NaBH_4$ [16]. During preparation and reaction, the solution was continuously stirred on a magnetic stirrer at room temperature. The reaction product was a violet–blue suspension of gold nanoparticles. The SEM image of the obtained gold nanoparticles and the histogram describing the size of the nanoparticles are shown in Figure S1 of the Supplementary Materials.

Preparation of composite layers for GCE/PEDOT-PSSLi/chitosan-AuNPs-GA/

The following procedure was used for the modification of the glassy carbon electrode. The first stage involved cleaning the working electrode surface. The glassy carbon electrode was carefully polished with an aqueous alumina slurry (0.5 μm) on a microcloth pad, and then thoroughly washed with double-distilled water.

(a) The layer of GCE/PEDOT-PSSLi was obtained by means of potentiostatic electrolysis at potential E = 1 V and time t = 20 s. The polymerization solution contained 0.002 mol/dm^3 EDOT and 0.1 mol/dm^3 PSSLi. This procedure was described in previous publications [12,14].

(b) For the layer of GCE/PEDOT-PSSLi/chitosan-AuNPs-GA, a solution containing 0.5 mL of AuNP suspension, 0.25 mL of 1% chitosan solution (in 0.05 M acetic acid), and 5 μL of 2.5% glutaraldehyde solution was prepared. The solution was stirred on a magnetic stirrer, and then 7 μL of the obtained solution was taken and spotted on the surface of the GCE/PEDOT-PSSLi electrode. The electrode was left for 1 h at the room temperature.

Preparation of the GCE/PEDOT-PSSLi/chitosan-AuNPs-GA/laccase sensor

A solution containing 12 mg laccase and 30 μL of 2.5% glutaraldehyde solution in 1 mL phosphate–citrate buffer (pH = 5.0) was prepared. The GCE/PEDOT-PSSLi (chitosan-AuNPs-GA) electrode was immersed in this solution and left for 3 h at 4 °C. Then, the electrode was rinsed with PBS solution and distilled water in order to remove the unbound enzyme.

Due to the durability of the coating, the sensor must not be allowed to dry completely; therefore, the prepared electrode was stored at 4 °C and immersed in a small amount of phosphate–citrate buffer (pH = 5.0). This method of storing the sensors ensured their stability for 30 days, with a signal drop of no more than 10%.

3. Results

Figure 1 shows AFM microscopy images for the GCE/Chit-GA, GCE/Chit-AuNPs-GA, and GCE/Chit-AuNPs-GA/laccase layers. Table 1 shows the roughness parameter values calculated for these surfaces. The differences between the layers of chitosan and those of chitosan doped with AuNPs are slight. The effect of doping with gold nanoparticles on the morphology of the chitosan surface was small: their addition caused a slight smoothing of the chitosan's surface. Small differences in the morphology of both layers may result from the fact that the AuNP doping takes place in the entire volume of the chitosan layer and not only on the surface of this layer. The next image shows clear differences in the surface structure, caused by the immobilization of the enzyme. This is confirmed by a clear increase in the surface roughness value (Table 1). This image also shows that the surface of the sensor is evenly coated with the enzyme.

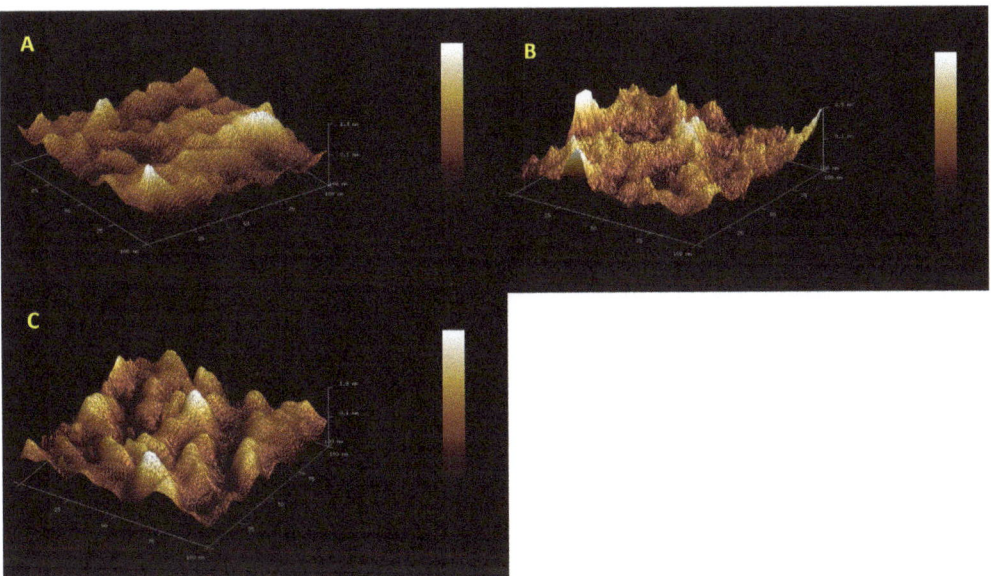

Figure 1. AFM microscopy images for (**A**) GCE/PEDOT-PSSLi/Chit-GA, (**B**) GCE/PEDOT-PSSLi/Chit-AuNPs-GA, and (**C**) GCE/PEDOT-PSSLi/Chit-AuNPs-GA/Laccase.

Table 1. The roughness parameters for the surfaces of the investigated materials.

	GCE/PEDOT-PSSLi/Chit-GA	GCE/PEDOT-PSSLi/Chit-AuNPs-GA	GCE/PEDOT-PSSLi/Chit-AuNPs-GA/Laccase
Image Z Range [nm]	2.22	1.84	2.89
Image Surface Area [nm^2]	11,266	11,357	12,010
Image Projected Surface Area [nm^2]	10,000	10,000	10,000
Surface extension coefficient	1.1266	1.1357	1.2010
Image Surface Area Difference	12.7	13.6	20.1
Image R_q [nm]	0.264	0.188	0.447
Image R_a [nm]	0.202	0.146	0.361
Image R_{max} [nm]	2.22	1.84	2.89

R_q—Quadratic mean, or root mean square average of profile height deviations from the mean line. R_a—Average, or arithmetic average of profile height deviations from the mean line. R max—The Maximum Roughness Depth is the greatest single roughness depth within the evaluation length.

The surface morphology was also investigated using electron microscopy SEM with a CBS detector for the detection of backscattered electrons. The use of the CBS detector allowed us to obtain the "Z-contrast image", which is directly related to the atomic numbers of the elements that are the components of the material. Hence, the CBS detector was very useful for both the surface morphology investigation and the confirmation of the presence of metallic nanoparticles as a dispersed phase in the material matrix. The CBS detects more signals from atoms with higher atomic numbers (AuNPs), and these elements can be seen as brighter spots/areas in the resulting image (Figure 2B,C) compared with the matrix material (Figure 2A), which has a darker color in the image. An analysis of the HR-SEM images revealed the presence of AuNPs homogenously distributed within the material (Figure 2B,C). The EDS analysis confirmed the presence of AuNPs in the sensors as peaks characteristic of Au, at Lα = 9.712 eV and M = 2.120 eV (Figure 2B,C). The small EDS signals from Au were caused by the small size of single AuNPs, and the fact that they are embedded in the matrix. However, the EDS composition analysis and the CBS morphology images confirmed the presence and homogenous distribution of AuNPs in the sensors.

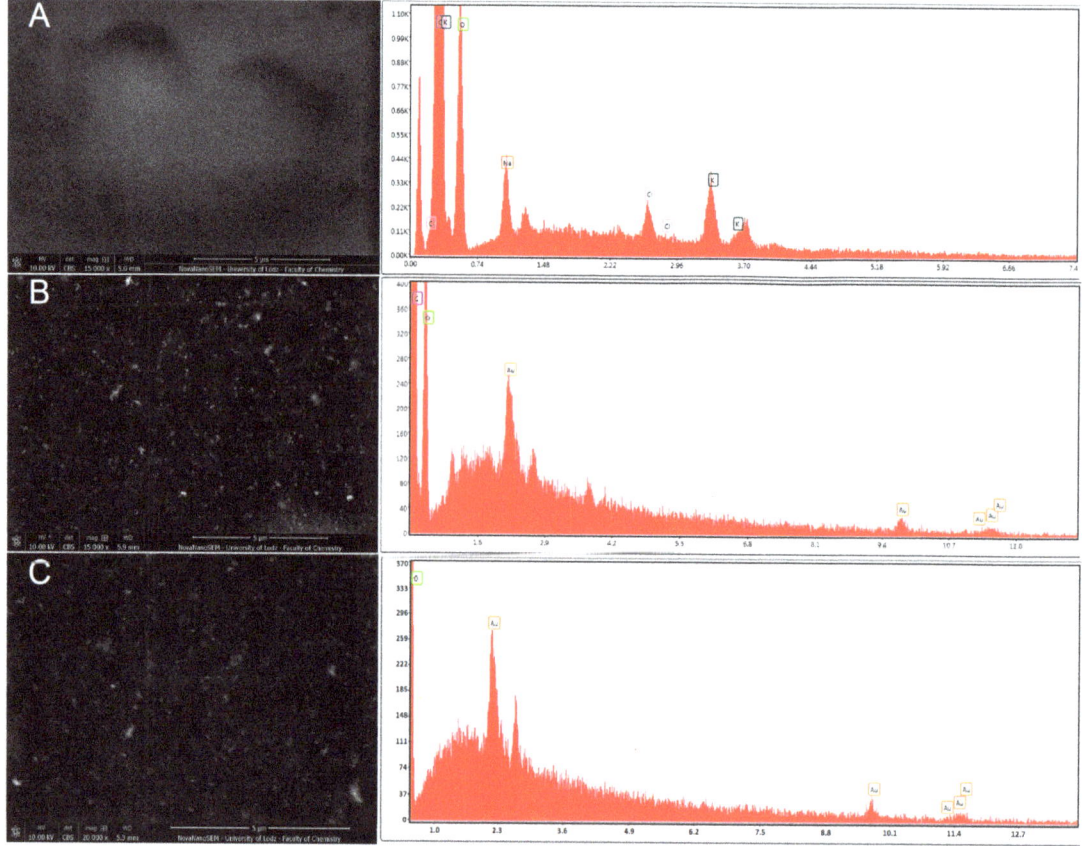

Figure 2. SEM images with EDS spectra of chitosan layers containing AuNPs: (**A**) GCE/PEDOT-PSSLi/Chit-GA, (**B**) GCE/PEDOT-PSSLi/Chit-AuNPs-GA, and (**C**) GCE/PEDOT-PSSLi/Chit-AuNPs-GA/Laccase.

In order to confirm the presence of laccase on the electrode, we performed a Raman spectroscopy analysis. Raman spectroscopy is a non-destructive analytical technique in which inelastically scattered light is used to obtain information about the vibrational energy modes of the analyzed samples. Figure 3 shows the Raman spectra of reference chemical compounds (laccase and chitosan) and chemicals deposited on the electrodes: without (sample A) and with (sample B) laccase. Characteristic Raman bands, at 482, 580, 853, 935, 1121, 1350, 1337, 1386, 1456, and 2906 cm^{-1}, correspond to laccase. A detailed inspection of Figure 3 demonstrates that the most significant Raman bands attributed to laccase are present in the Raman spectrum of the electrode with that enzyme. Broad Raman bands, observed in sample A at 1350 and 1590 cm^{-1}, correspond to amorphous carbon sp^2 (D-band) and amorphous carbon sp^3 (G-band) [24], respectively.

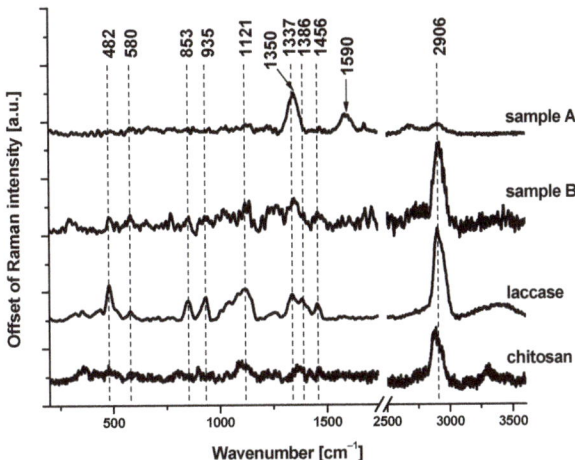

Figure 3. Raman spectra of chitosan, laccase, and two electrodes: without (sample A) and with (sample B) laccase.

Laccase is the enzyme that shows the highest activity at pH = 5 [25,26]. Polyphenols, on the other hand, are compounds whose electrochemical oxidation reactions depend on the pH of the environment. Voltammetric curves of catechol in phosphate–citrate buffer solutions at pH values of 4, 5, 6, 7, and 8 are shown in Figure 4. Analogous curves for gallic acid and caffeic acid are provided in the Supplementary Materials (Figures S2 and S3). Based on these data, a phosphate–citrate buffer environment with pH = 5 was adopted for further research. In order to confirm the validity of this assumption, DPV measurements were performed, using a laccase sensor for solutions of catechol in phosphate–citrate buffers at pH values of 4, 5, 6, 7, and 8. Figure 5 shows the dependence of the peak current on pH. As can be seen, the highest peak currents were observed in the solutions when pH = 5.

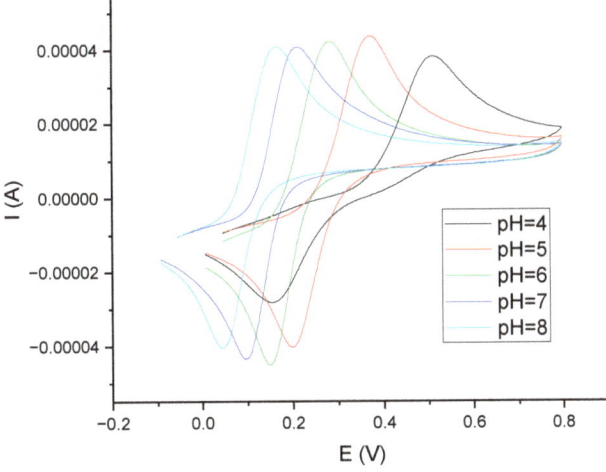

Figure 4. Voltammetric curves obtained on GCE in a solution of catechol C = 0.001 M in phosphate–citrate buffer for different pH values (v = 200 mV/s). pH = 4.0 (black), pH = 5.0 (red), pH = 6.0 (green), pH = 7.0 (blue), and pH = 8.0 (cyan).

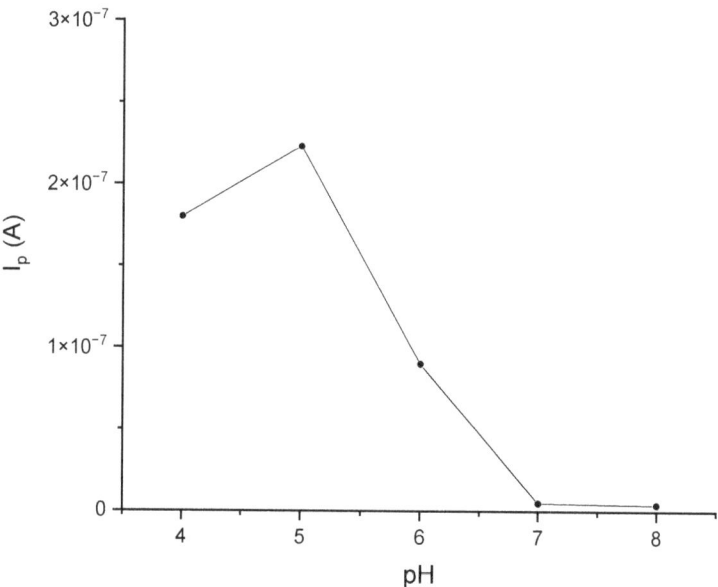

Figure 5. The effect of pH on the response of the GCE/PEDOT-PSSLi/Chit-AuNPs-GA/laccase sensor in catechol solutions 0.001 M in phosphate–citrate buffer. Dependences of the current peak on pH.

In addition to the pH value, other parameters of the sensor manufacturing process were also optimized. The thickness of the produced chitosan layer was optimized by determining the volume of the chitosan solution applied to the electrode. Too-thick layers peeled off after drying and fell off of the substrate. Moreover, thick layers of non-conductive chitosan deteriorated the electrochemical properties of the electrode. The application of chitosan solution drops in different volumes on the surface of the electrode was tested. Droplets of 4, 5, 6, 7 and 8 µL were applied. The layers prepared by applying 7 µL of chitosan solution to the electrode (electrode diameter 3 mm) produced the best results and were used for further studies. Droplets of less than 7 µL did not cover the entire electrode. On the other hand, droplets of more than 7 µL formed thicker layers of lower conductivity, with a tendency to exfoliate.

In the next stage, the influence of gold nanoparticles on the electrochemical properties of the chitosan layers was investigated. The amount of the AuNP mixture added to the chitosan solution was optimized. The addition of AuNPs was intended to increase the electrical conductivity of the layer and improve its electrochemical properties. Too much AuNP suspension added to the chitosan solution diluted it, and, as a result, a too-thin chitosan layer formed. On the other hand, the layer obtained from a solution with a smaller amount of AuNPs demonstrated worse electrical conductivity. The assumed optimal composition contained 0.5 mL of AuNP suspension, 0.25 mL of 1% chitosan solution, and 5 µL of 2.5% glutaraldehyde solution. The effect caused by the addition of nanoparticles was assessed using the measurements from cyclic voltammetry of the tested electrodes in solutions of ferricyanides. For this purpose, voltammetric measurements were performed for the following three sensors: GCE/PEDOT-PSSLi, GCE/PEDOT-PSSLi/Chit-GA, and GCE/PEDOT-PSSLi/Chit-AuNPs-GA.

All measurements were carried out in solutions of ferrocyanide, at a concentration of 0.002 M in 2 M KCl (Figure 6). Both peak currents and differences in peak potential were evaluated. In the case of the first, the GCE/PEDOT-PSSLi sensor, the peak potential difference was 76 mV, and the peaks were symmetrical and well-formed. In the case of the GCE/PEDOT-PSSLi/Chit-GA sensor, the chitosan layer without the addition of nanoparticles worsened the properties of the electrode. The currents were very low, and the difference in peak potential was 119 mV. This is understandable because the PEDOT-PSSLi layer provides better conductivity and facilitates redox processes, while chitosan cross-linked via glutaraldehyde is an insulator and will deteriorate the properties of the electrode.

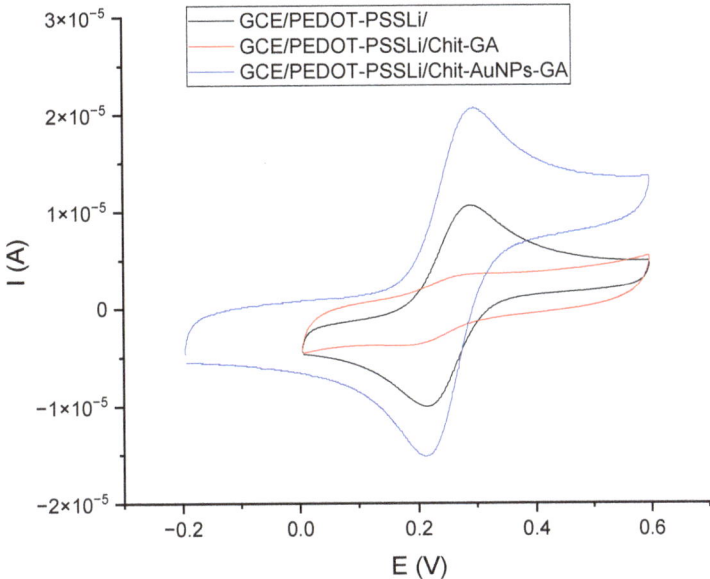

Figure 6. Voltammetric curves for sensors obtained in solutions of 0.002 M $K_2Fe[(CN)_6]_4$ in 2 M KCl (v = 200 mV/s). GCE/PEDOT-PSSLi/Chit-AuNPs-GA (black), GCE/PEDOT-PSSLi/Chit-GA (red), and GCE/Chit-AuNPs-GA (blue).

For the GCE/PEDOT-PSSLi/Chit-AuNPs-GA sensor, the highest current values were obtained, the difference in peak potential was 83 mV, and the peaks were symmetrical and well-shaped. This proves that the addition of gold nanoparticles increased the conductivity of the chitosan layer.

Thus, it can be concluded that the doping of the chitosan layer with gold nanoparticles significantly increased the conductivity of the layer and, at the same time, the presence of AuNPs allowed for a rapid charge exchange through the interface of the solution/chitosan doped with AuNPs.

The volume of glutaraldehyde solution added to the mixture of AuNPS and chitosan was optimized. Glutaraldehyde cross-links the chitosan layer by binding amino groups, but using too much aldehyde will block all amino groups, which will prevent the immobilization of the enzyme at the next stage of work. The level of saturation of the chitosan layer with glutaraldehyde was assessed using the immobilization of laccase on the prepared layers. The research began with the addition of 240 µL of 2.5% solution of glutaraldehyde to the mixture of AuNPs with chitosan. This amount completely saturated all amino groups in chitosan and prevented enzyme immobilization. In the next tests, the amount of glutaraldehyde was reduced successively to 120, 60, and 30 µL, until a layer with immobilized laccase was obtained. For the volumes of 60 and 30 µL, the oxidation current of polyphenols was obtained. This means that, after using such volumes of glutaraldehyde, free amino groups, which are capable of binding to the enzyme, remained in the chitosan structure. We decided to use 30 µL of 2.5% glutaraldehyde solution for the cross-linking of chitosan layers.

In the last stage of sensor development, the amount of immobilized enzyme was optimized. The results were evaluated on the basis of measurements of cyclic voltammetry of the sensor in the catechol solution. The enzyme was immobilized from solutions containing 3.0, 6.0, 9.0, and 12 mg laccase/1 mL solution. The peak current grew with the increasing amount of immobilized enzyme, up to a layer obtained from a solution of 12 mg laccase/ 1 mL phosphate–citrate buffer (pH = 5.0). The layers obtained from solutions with a higher amount of laccase yielded a peak current of a similar value, but they dissolved more easily.

Another characteristic is the comparison of the voltammetric curves of the developed sensor at the subsequent stages of its production. These studies were performed in catechol solutions (c = 0.001 M) in phosphate–citrate buffer at pH = 5. The results of these measurements are shown in Figure 7. Measurements were made on four sensors. Comparison showed the influence of individual sensor components on the catechol oxidation reaction. The first sensor was the GCE/PEDOT-PSSLi one, for which the oxidation peak potential is 0.214 V, and the peaks are symmetrical. The second sensor was the GCE/PEDOT-PSSLi/Chit one. In this case, the oxidation reaction was inhibited. The peak current was much lower, and the peak potential was shifted to about 0.680 V. This is due to the fact that the chitosan layer effectively insulates the electrode surface. The potential of the oxidation peak shifted to 0.675 V. The third sensor was the GCE/PEDOT-PSSLi/Chit-AuNPs one, which differs from the previous sensor only in the addition of AuNPs to the chitosan layer. As can be seen, this greatly improved the properties of the electrode. The oxidation peak potential, in this case, was 0.390 V. The fourth sensor was the complete GCE/PEDOT-PSSLi/Chit-AuNPs/Laccase system, for which the peak potential became 0.343 V, and the peak current had the highest value.

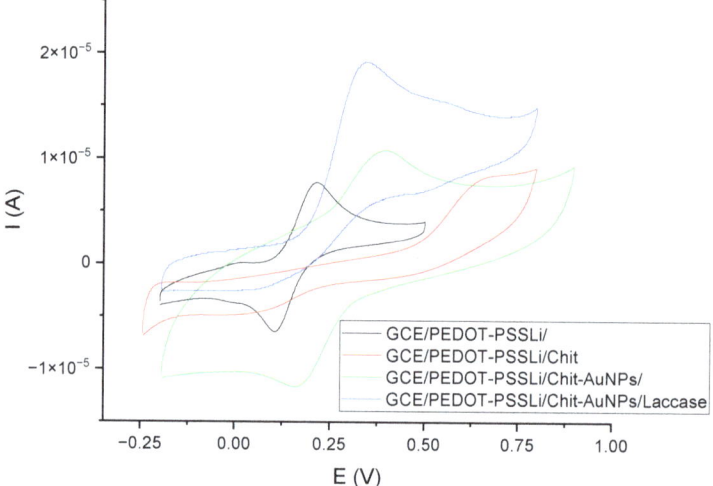

Figure 7. Voltammetric curves obtained in a solution of catechol C = 0.001 M (in phosphate–citrate buffer pH = 5.0) for sensors: GCE/PEDOT-PSSLi (black), GCE/PEDOT-PSSLi/Chit (red), GCE/PEDOT-PSSLi/Chit-AuNPs-GA (green), GCE/PEDOT-PSSLi/Chit-AuNPs-GA/laccase (blue).

The next step in the design of the sensor was to study the operation of the GCE/PEDOT-PSSLi/Chit-AuNPs-GA system in combination with laccase as a mediator in solution. Figure 8 shows the voltammetric curves made for the GCE/PEDOT-PSSLi/Chit-AuNPs-GA sensor in a catechol solution, with a concentration of $C = 9.09 \times 10^{-5}$ mol/dm^3 in a phosphate–citrate buffer at pH = 5.0, with the addition of 3 mg laccase/1 mL (black curve) and 6 mg/1 mL (red curve). In this case, at a constant concentration of catechol, the value of the peak current depends on the concentration of laccase. This proves that laccase is involved in the oxidation of catechol as a mediator.

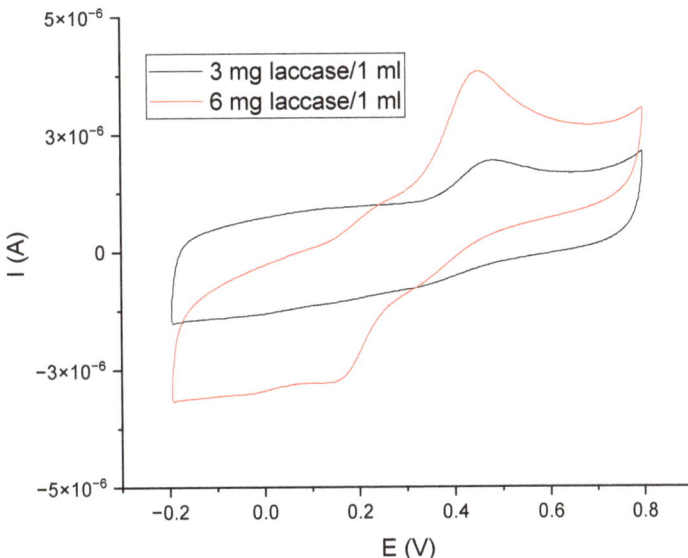

Figure 8. Voltammetric curves for GCE/PEDOT-PSSLi/Chit-AuNPs-GA sensors, obtained in a solution of catechol C = 9.09 E−05 M (in phosphate–citrate buffer pH = 5.0) and laccase in concentrations of 3 mg/1 mL (black) and 6 mg/1 mL (red). v = 200 mV/s.

The last step was to test the performance of the sensor with immobilized laccase (GCE/PEDOT-PSSLi/Chit-AuNPs-GA/laccase) with the GCE/PEDOT-PSSLi/Chit-AuNPs-GA sensor immersed in a solution containing laccase. Immobilization of the enzyme was carried out in a solution containing 12 mg laccase and 30 μL of 2.5% glutaraldehyde solution in 1 mL phosphate–citrate buffer (pH = 5.0). The voltammetric curve obtained in this measurement was compared with the curve obtained for the GCE/PEDOT-PSSLi/chit-AuNPs-GA sensor in a catechol solution with a concentration of C = 9.09 × 10^{-5} mol/dm^3 (in phosphate–citrate buffer pH = 5.0), with the addition of 3 mg/1 mL (the red curve from Figure 9). As can be seen for the sensor with laccase immobilized on the surface, the obtained peak current was about six times higher than that without. Such a large catalytic effect of immobilized laccase makes the tested sensor a promising analytical tool for the determination of polyphenols.

Using the developed sensor (GCE/PEDOT-PSSLi/chit-AuNPs-GA/Laccase), catechol was determined using the amperometric method. The measurements were carried out in a phosphate–citrate buffer solution (pH = 5.0) at a potential of 0.6 V for 30 s while the solution was agitated. The amperometric curves of the measurements performed are shown in Figure S4 of the Supplementary Materials. The measurements were made five times, on a newly prepared sensor each time. The average values of currents from all five measurements were taken for the analysis. The measurements were performed for the concentration range from 1.96 × 10^{-5} to 9.09 × 10^{-5} mol/dm^3. For the analysis of the results, the values of the currents read for the time t = 25 s were used. The R^2 value was 0.9958, LOD = 9.5 μmol/dm^3, and the linear range was 19–90 μmol/dm^3.

The next step was to determine the polyphenols catechol, gallic acid, and caffeic acid using the DPV method. The measurements of DPV were carried out for the following parameters: potential range from 0.0 V to 0.8 V, potential jump 2 mV, duration of the jump 0.4 s, pulse amplitude 50 mV. The DPV voltammetry curves and the caffeic acid standard line are shown in Figure 10. The other polyphenol curves and standard lines are provided in the Supplementary Materials (Figures S5 and S6). The measurements were performed for the concentration range from 9.99 × 10^{-7} to 9.09 × 10^{-5} mol/dm^3.

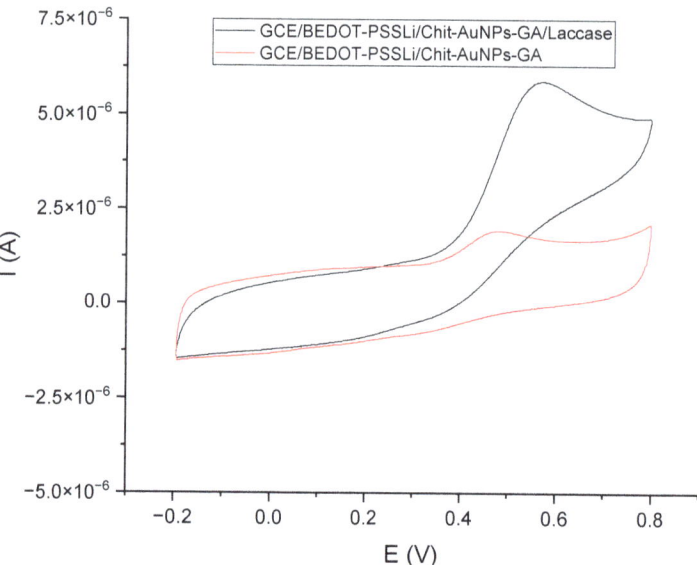

Figure 9. Voltammetric curves: black—GCE/PEDOT-PSSLi/Chit-AuNPs-GA/laccase sensor in a solution of catechol C = 9.09 E−05 M (in buffer pH = 5.0); red—GCE/PEDOT-PSSLi/Chit-AuNPs-GA sensor in a solution of catechol C = 9.09 E−05 M (in phosphate–citrate buffer pH = 5.0) and laccase in concentrations of 3 mg/1 mL.

For each of the polyphenols, the measurements were performed five times, on a newly prepared sensor each time. Standard lines were determined from the average values of peak currents. Standard deviations and error values were calculated. The R^2 coefficient, sensitivity, precision, accuracy, RSD, and recovery were calculated for the assays performed. The lower limit of detection (LOD) was calculated from the relationship between the slope and the standard deviation of the intercept (LOD = s_b/a). The calculated values are listed in Table 2. The presented parameters show that, in all cases, the GCE/PEDOT-PSSLi/chit-AuNPs-GA/Laccase sensor worked adequately and was suitable for determining polyphenols. The amperometric method showed the highest sensitivity and allowed for the determination of catechol in higher concentration ranges, from 19 to 90 µM. DPV assays for all three polyphenols allowed for the detection of much lower concentrations. In the case of gallic acid, it was possible to determine it in the range from 2 to 18 µmol/dm³; for higher concentrations, the dependence on the concentration was no longer linear. The best results were obtained for the determination of catechol and caffeic acid, for which the linear range was from 2 to 90 µmol/dm³, and the sensitivity of these determinations was higher than that for gallic acid. The DPV voltammetry method allowed for the determination of polyphenols in a wider concentration range, with lower LOD values than the amperometric method. Moreover, the amperometric method is troublesome in practical application. The measurements are less reproducible, and the operation of the sensor for a long time, mixed with the solution, often causes much faster wear of the sensor and, thus, the need to repeat the entire procedure anew. Very often, the layer peels off from the substrate. When using DPV as a measurement method, the determination is performed without mixing it with the solution and takes much less time than the amperometric measurement. Sensors in this case are much more durable, which allows for longer use. Cases of detachment of the layer from the substrate occurred sporadically. For these reasons, it was decided to present the results of DPV measurements for all tested polyphenols, and the amperometric method was used only for catechol.

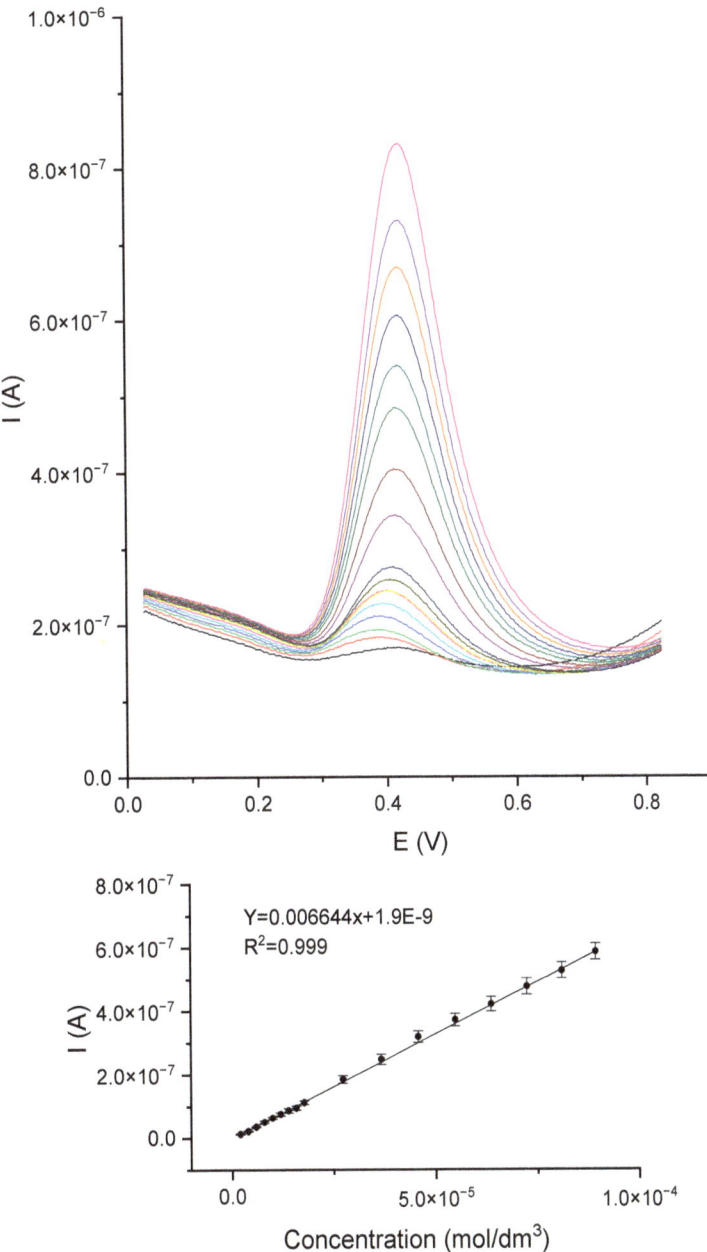

Figure 10. DPV voltammetry curves for the GCE/PEDOT-PSSLi/Chit-AuNPs-GA/laccase sensor in solutions of caffeic acid in phosphate–citrate buffer pH = 5.0.

Table 2. Parameters for the determination of polyphenols.

Lp	Polyphenol—Method	R^2	Sensitivity [A/mol/dm^3]	CV [%]	Recovery [%]	RSD [%]	LOD [µM]	Linear Range [µM]
1	Catechol—Amperometric	0.996	0.0248	0.082	99.1	15.64	9.5	19–90
2	Catechol—DPV	0.998	0.00508	0.002	100.1	22.83	1.7	2–90
3	Caffeic acid—DPV	0.999	0.0066	0.020	108.7	5.77	1.9	2–90
4	Gallic acid—DPV	0.993	0.0039	0.111	114.7	5.80	1.7	2–18

Table 3 shows examples of electrochemical sensors with immobilized laccase. Sensors with different types of detection are presented, both amperometric and voltammetric, as well as sensors with DPV and SWV detection. Compared to these electrodes, the sensors developed by us are characterized by a wide linear range of concentrations. The widest concentration ranges were achieved for the determination of catechol and caffeic acid. For this reason, the developed sensor may be an interesting proposal as a convenient tool for the determination of polyphenols.

The last stage of the research was an attempt to determine gallic acid in a natural sample, for which white wine was used. The analysis was performed using the GCE/PEDOT-PSSLi/Chit-AuNPs-GA/laccase sensor in phosphate–citrate buffer, pH = 5. This analysis was carried out in order to estimate the accuracy of the presented biosensor. DPV measurements were carried out using the standard addition method. There are DPV curves for this measurement in the Supplementary Materials (Figure S7). Figure 11 shows the dependence of the peak current value on the increasing amount of added gallic acid in the standard addition method for the sensor in a white wine sample. The results for the determination of gallic acid on the GCE/PEDOT-PSSLi/Chit-AuNPs-GA/laccase electrode were as follows. The linear regression equation was expressed as y (A) = $5.71 \times 10^{-8} \times$ (µmol/dm^3) + 3.24×10^{-7} (R^2 = 0.997). The relative standard deviation was 0.96%, and the estimated concentration in the sample was 5.66 µmol/dm^3. The peak current measurements for each standard addition were performed three times. The standard line was determined from the average values of peak currents, and the results, with a confidence interval for a probability (p) of 95%, were analyzed using linear least-square regression. After the calculations, the content of gallic acid in the determined wine was 4.8 mg/dm^3. After careful examination of the presented results, it can be concluded that the presented biosensors may be suitable tools for measuring the concentration of gallic acid in real samples, such as white wine. As a reference method in selected natural samples, the spectrophotometric method for determining the concentration of gallic acid, as described by Inoue and Hagerman [27] was used. The idea behind this method is the reaction between rhodamine and the detected polyphenol. The color change in the sample is assessed by measuring the absorbance at l = 520 nm. The content of gallic acid in the determined wine was 4.8 mg/dm^3. The content of gallic acid in the wine as obtained with the reference method was 4.63 mg/dm^3. The calculated recovery for the assay performed was 103.67%.

Table 3. Examples of different electrochemical sensors with immobilized laccase for the determination of polyphenols.

Phenolic Compound	Sensor	Method	LOD [µM]	Linear Range [µM]	Ref.
Caffeic acid			0.14	1–50	
Rosmarinic acid	graphite/ePDA-Lac	amperometry	0.09	1–20	[28]
Gallic acid			0.29	1–150	
Gallic acid	TvL-MWCNTs-SPEd	amperometry	0.6	0.6–99.9	[29]
Catechol	GCE /FYSSns-2-Lac	DPV	1.6	12.5–450	[30]

Table 3. Cont.

Phenolic Compound	Sensor	Method	LOD [μM]	Linear Range [μM]	Ref.
Caffeic acid Rosmarinic acid Gallic acid	AuSPE/laccase/Nafion	amperometry	2.5 2.4 1.55	3–15 3–15 2–7	[31]
Caffeic acid Rosmarinic acid	Au/Lacc-CS-MWCNT	amperometry	0.151 0.233	0.735–10.5 0.91–112.1	[32]
Catechol	(CNTs–CS)	voltammetry	0.66	1.2–30	[33]
Caffeic acid	Graphite/Lac	amperometry	0.56	1–10	[34]
Mixtures of catechin and caffeic acid	Polyethersulfone/Lac	voltammetry	0.18	12–14	[35]
Catechol	PEI-AuNP-Lac	SWV	0.03	0.36–11	[36]
Caffeic acid Gallic acid	Carbon-Sonogel/Nafion-Lac	amperomtery	0.06 0.41	0.04–2.2 0.01–22	[37]
Catechol Catechol Gallic acid Caffeic acid	GCE/PEDOT-PSSLi/Chit-AuNPs-GA/laccase	amperomtery DPV DPV DPV	9.5 1.7 1.7 1.9	19–90 2–90 2–18 2–90	This work

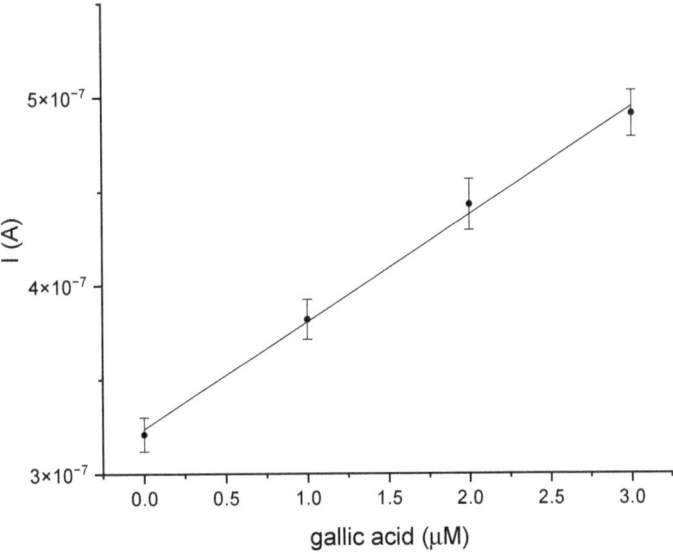

Figure 11. The dependence of the current value on the concentration of gallic acid, obtained using the standard addition method for the GCE/PEDOT-PSSLi/Chit-AuNPs-GA/laccase sensor in white wine samples.

4. Conclusions

The developed sensor is a skillful combination of materials including PEDOT, chitosan, AuNPs, and laccase. PEDOT is a modern conductive polymer ensuring fast charge exchange. Chitosan, as a product of natural origin, is characterized by high biocompatibility, which is of great importance when contact with natural samples is necessary. At the same time, chitosan enables the covalent attachment of the enzyme. On the one hand, AuNPs are able to ensure a sufficiently high conductivity of the chitosan layers; on the other hand, they are also characterized by high biocompatibility. Laccase is an enzyme with a large spectrum of specificity that works as a mediator in the reaction with the analyte. This article described the research leading to the development of the sensor /PEDOT-polystyrene sulfonate)/chitosan-AuNPs-glutaraldehyde/Laccase. This sensor was designed for the electrochemical determination of polyphenols. The method of manufacturing the sensor was optimized. The effect of the addition of gold nanoparticles on the performance of the sensor, including the possibility of direct oxidation of the immobilized laccase, was described. The influence of AuNPs on the increase in the electrical conductivity of the chitosan layers was presented. A catalytic effect from laccase was demonstrated in the oxidation reaction of polyphenols, using a sensor with an immobilized enzyme. The developed sensor, for the electrochemical determination of catechol using the amperometric method, was used. That sensor was used for the electrochemical determination of polyphenols such as catechol, gallic acid, and caffeic acid with the DPV method. The results of our investigations showed that the sensor can be applied for the determination of these polyphenols. Linear proportional relationships of peak currents to the concentrations of polyphenols were obtained. The determination of gallic acid in wine samples was also conducted, demonstrating the possibility for the practical application of the developed sensor. The sensor had a shelf life of 30 days when stored at 4 °C. It can therefore be concluded that a good analytical tool for the determination of polyphenols has been obtained.

Supplementary Materials: The following supporting information can be downloaded at https://www.mdpi.com/article/10.3390/ma16145113/s1. Figure S1. (A) SEM image of gold nanoparticles obtained in the synthesis with sodium borohydride. (B) Size distribution histogram of the obtained AuNPs. Figure S2. Voltammetric curves obtained on GCE in a solution of gallic acid C = 0.001 M in phosphate-citrate buffer different pH (v = 200mV/s). Figure S3. Voltammetric curves obtained on GCE in a solution of caffeic acid C = 0.001 M in phosphate-citrate buffer different pH (v = 200mV/s). Figure S4. (A) Amperometric curves for sensor GCE/PEDOT-PSSLi/Chit-AuNPs-GA/laccase in solutions of catechol in phosphate-citrate buffer pH = 5.0. (B) Standard line for catechol. Figure S5. (A) DPV voltammetry curves for catechol solutions in phosphate-citrate buffer pH = 5.0. (B) Standard line for catechol. Figure S6. (A) DPV voltammetry curves for gallic acid solutions in phosphate-citrate buffer pH = 5.0. (B) Standard line for gallic acid. Figure S7. DPV voltammetry curves of gallic acid were obtained using the standard addition method for the GCE/PEDOT-PSSLi/Chit-AuNPs-GA/laccase sensor in white wine samples.

Author Contributions: Conceptualization, P.K.; methodology, P.K., A.L., K.R.-S. and J.S.; validation, P.K.; investigation, P.K., M.K., A.L., K.R.-S., J.S., K.B.-M. and B.B.-P.; visualization, P.K., K.R.-S. and J.S.; writing—original draft preparation, P.K., J.S. and K.R.-S.; writing—review and editing, P.K. All authors have read and agreed to the published version of the manuscript.

Funding: This research received no external funding.

Institutional Review Board Statement: Not applicable.

Informed Consent Statement: Not applicable.

Data Availability Statement: All of the involved data are available upon request.

Conflicts of Interest: The writers claim that no conflict of interest exist.

References

1. Barbosa, M.A.; Pego, A.P.; Amaral, I.F. Chitosan. In *Comprehensive Biomaterials*; Ducheyne, P., Healy, K., Hutmacher, D.E., Grainger, D.W., Kirkpatrick, C.J., Eds.; Elsevier: Amsterdam, The Netherlands, 2011; ISBN 9780080552941.
2. Baldrick, P. The safety of chitosan as a pharmaceutical excipient. *Regul. Toxicol. Pharmacol.* **2010**, *56*, 290–299. [CrossRef] [PubMed]
3. Yavuz, A.G.; Uygun, A.; Can, H.K. The effect of synthesis media on the properties of substituted polyaniline/chitosan composites. *Carbohydr. Res.* **2011**, *346*, 2063–2069. [CrossRef] [PubMed]
4. Tiwari, A.; Gong, S. Electrochemical detection of a breast cancer susceptible gene using cDNA immobilized chitosan-co-polyaniline electrode. *Talanta* **2009**, *77*, 1217–1222. [CrossRef] [PubMed]
5. Abdi, M.M.; Abdullah, L.C.; Sadrolhosseini, A.R.; Yunus, W.M.M.; Moksin, M.M.; Tahir, P.M. Surface plasmon resonance sensing detection of mercury and lead ions based on conducting polymer composite. *PLoS ONE* **2011**, *6*, e24578. [CrossRef]

6. Fang, Y.; Ni, Y.; Zhang, G.; Mao, C.; Huang, X.; Shen, J. Biocompatibility of CS-PPy nanocomposites and their application to glucose biosensor. *Bioelectrochemistry* **2012**, *88*, 1–7. [CrossRef]
7. Sui, L.; Zhang, B.; Wang, J.; Cai, A. Polymerization of PEDOT/PSS/Chitosan-Coated Electrodes for Electrochemical Bio-Sensing. *Coatings* **2017**, *7*, 96. [CrossRef]
8. Kuralay, F.; Demirci, S.; Kiristi, M.; Oksuz, L.; Oksuz, A.U. Poly(3,4-ethylenedioxy-thiophene) coated chitosan modified disposable electrodes for DNA and DNA–drug interaction sensing. *Colloids Surf. B Biointerfaces* **2014**, *123*, 825–830. [CrossRef]
9. Kiristi, M.; Oksuz, A.U.; Oksuz, L.; Seyhan, U. Electrospun chitosan/PEDOT nanofibers. *Mater. Sci. Eng. C* **2013**, *33*, 3845–3850. [CrossRef]
10. Krosa, A.; Sommerdijkc, N.A.J.M.; Nolteb, R.J.M. Poly(pyrrole) versus poly(3,4-ethylenedioxy-thiophene): Implications for biosensor applications. *Sens. Actuators B* **2005**, *106*, 289–295. [CrossRef]
11. Vineis, P.; Pirastu, R. Aromatic amines and cancer. *Cancer Causes Control* **1997**, *8*, 346–355. [CrossRef]
12. Krzyczmonik, P.; Socha, E.; Andrijewski, G. Determination of Ascorbic Acid by a Composite-Modified Platinum Electrode. *Anal. Lett.* **2017**, *50*, 806–818. [CrossRef]
13. Krzyczmonik, P.; Socha, E.; Skrzypek, S.; Soliwoda, K.; Celichowski, G.; Grobelny, J. Honeycomb-structured porous poly(3,4-ethylenedioxy-thiophene) composite layers on a gold electrode. *Thin Solid. Film.* **2014**, *565*, 54–61. [CrossRef]
14. Krzyczmonik, P.; Socha, E.; Skrzypek, S. Immobilization of glucose oxidase on modified electrodes with composite layers based on poly(3,4-ethylenedioxy-thiophene). *Bioelectrochemistry* **2015**, *101*, 8–13. [CrossRef]
15. Krzyczmonik, P.; Socha, E.; Skrzypek, S. Electrochemical detection of glucose in beverages samples using poly(3,4-ethylenedioxy-thiophene) modified electrodes with immobilized glucose oxidase. *Electrocatalysis* **2018**, *9*, 380–387. [CrossRef]
16. Deraedt, C.; Salmon, L.; Gatard, S.; Ciganda, R.; Hernandez, R.; Ruiz, J.; Astruc, D. Sodium borohydride stabilizes very active gold nanoparticle catalysts. *Chem. Commun.* **2014**, *50*, 14194–14196. [CrossRef]
17. Suginta, W.; Khunkaewla, P.; Schulte, A. Electrochemical Biosensor Applications of Polysaccharides Chitin and Chitosan. *Chem. Rev.* **2013**, *113*, 5458–5479. [CrossRef]
18. Bo, Y.; Wang, W.; Qi, J.; Huang, S. A DNA biosensor based on graphene paste electrode modified with Prussian blue and chitosan. *Analyst* **2011**, *136*, 1946–1951. [CrossRef]
19. Junbo, B.; Cunxian, S. Chitosan chip and application to evaluate DNA loading on the surface of the metal. *Biomed. Mater.* **2009**, *4*, 011002. [CrossRef]
20. Pavinatto, F.J.; Caseli, L.; Oliveira, O.N. Chitosan in nanostructured thin films. *Biomacromolecules* **2010**, *11*, 1897–1908. [CrossRef]
21. Jayakumar, R.; Prabaharan, M.; Nair, S.V.; Tamura, H. Novel chitin and chitosan nanofibers in biomedical applications. *Biotechnol. Adv.* **2010**, *28*, 142–150. [CrossRef]
22. Krzyczmonik, P.; Bozal-Palabiyik, B.; Skrzypek, S.; Uslu, B. Quantum dots-based sensors using solid electrodes. In *Electroanalytical Applications of Quantum Dot-Based Biosensors*; Uslu, B., Ed.; Elsevier: Amsterdam, The Netherlands, 2021; Chapter 3, pp. 81–120. [CrossRef]
23. Skrzypek, S.; Krzyczmonik, P. Solid electrodes in pharmaceutical and biomedical analysis. In *Recent Advances in Analytical Techniques*; Bentham Science: Sharjah, United Arab Emirates, 2018; Volume 2, pp. 208–248, ISBN 978-1-68108-574-6.
24. Dychalska, A.; Popielarski, P.; Franków, W.; Fabisiak, K.; Paprocki, K.; Szybowicz, M. Study of CVD diamond layers with amorphous carbon admixture by Raman scattering spectroscopy. *Mater. Sci.-Pol.* **2015**, *33*, 799–805. [CrossRef]
25. Kumar, R.; Kaur, J.; Jain, V.; Kumar, A. Optimization of laccase production from Aspergillus flavus by design of experiment technique: Partial purification and characterization. *J. Genet. Eng. Biotechnol.* **2016**, *14*, 125–131. [CrossRef] [PubMed]
26. Selvia, K.; James, A.N. Characterization of Trametes versicolor laccase for the transformation of aqueous phenol. *Bioresour. Technol.* **2008**, *99*, 7825–7834. [CrossRef]
27. Inoue, K.H.; Hagerman, A.E. Determination of Gallotannin with Rhodanine. *Anal. Biochem.* **1988**, *169*, 363–369. [CrossRef]
28. Almeida, L.C.; Correia, R.D.; Squillaci, G.; Morana, A.; La Cara, F.; Correia, J.P.; Viana, A.S. Electrochemical deposition of bio-inspired laccase-polydopamine films for phenolic sensors. *Electrochim. Acta* **2019**, *319*, 462–471. [CrossRef]
29. Di Fusco, M.; Tortolini, C.; Deriu, D.; Mazzei, F. Laccase-based biosensor for the determination of polyphenol index in wine. *Talanta* **2010**, *81*, 235–240. [CrossRef]
30. Zheng, Y.J.; Wang, D.D.; Li, Z.K.; Sun, X.F.; Gao, T.T.; Zhou, G.W. Laccase biosensor fabricated on flower-shaped yolk-shell SiO_2 nanospheres for catechol detection. *Colloid. Surf.* **2018**, *538*, 202–209. [CrossRef]
31. Litescu, S.C.; Eremia, S.A.V.; Bertoli, A.; Pistelli, L.; Radu, G.-L. Laccase-nafion based biosensor for the determination of polyphenolic secondary metabolites. *Anal. Lett.* **2010**, *43*, 1089–1099. [CrossRef]
32. Diaconu, M.; Litescu, S.C.; Radu, G.L. Laccase–MWCNT–chitosan biosensor—A new tool for total polyphenolic content evaluation from in vitro cultivated plants. *Sens. Actuators B Chem.* **2010**, *145*, 800–806. [CrossRef]
33. Liu, Y.; Qu, X.; Guo, H.; Chen, H.; Liu, B.; Dong, S. Facile preparation of amperometric laccase biosensor with multifunction based on the matrix of carbon nanotubes-chitosan composite. *Biosens. Bioelectron.* **2006**, *21*, 2195–2201. [CrossRef]
34. Jarosz-Wilkołazka, A.; Ruzgas, T.; Gorton, L. Amperometric detection of mono and diphenols at Cerrena unicolor laccase-modified graphite electrode: Correlation between sensitivity and substrate structure. *Talanta* **2005**, *66*, 1219–1224. [CrossRef] [PubMed]
35. Gomes, S.A.S.S.; Nogueira, J.M.F.; Rebelo, M.J.F. An amperometric biosensor for polyphenolic compounds in red wine. *Biosens. Bioelectron.* **2004**, *20*, 1211–1216. [CrossRef]

36. Brondani, D.; de Souza, B.; Souza, B.S.; Neves, A.; Vieira, I.C. PEI-coated gold nanoparticles decorated with laccase: A new platform for direct electrochemistry of enzymes and biosensing applications. *Biosens. Bioelectron.* **2013**, *42*, 242–247. [CrossRef] [PubMed]
37. ElKaoutit, M.; Naranjo-Rodriguez, I.; Temsamani, K.R.; Hernández-Artiga, M.P.; Bellido-Milla, D.; Cisneros, J.L.H.-H.d. A comparison of three amperometric phenoloxidase-Sonogel-Carbon based biosensors for determination of polyphenols in beers. *Food Chem.* **2008**, *110*, 1019–1024. [CrossRef] [PubMed]

Disclaimer/Publisher's Note: The statements, opinions and data contained in all publications are solely those of the individual author(s) and contributor(s) and not of MDPI and/or the editor(s). MDPI and/or the editor(s) disclaim responsibility for any injury to people or property resulting from any ideas, methods, instructions or products referred to in the content.

Article

Good Choice of Electrode Material as the Key to Creating Electrochemical Sensors—Characteristics of Carbon Materials and Transparent Conductive Oxides (TCO)

Anna Cirocka *, Dorota Zarzeczańska and Anna Wcisło *

Department of Analytical Chemistry, Faculty of Chemistry, University of Gdansk, ul. Wita Stwosza 63, 80-308 Gdansk, Poland; dorota.zarzeczanska@ug.edu.pl
* Correspondence: anna.cirocka@ug.edu.pl (A.C.); anna.wcislo@ug.edu.pl (A.W.); Tel.: +48-58523-5106 (A.C.); +48-58523-5157 (A.W.)

Abstract: The search for new electrode materials has become one of the goals of modern electrochemistry. Obtaining electrodes with optimal properties gives a product with a wide application potential, both in analytics and various industries. The aim of this study was to select, from among the presented electrode materials (carbon and oxide), the one whose parameters will be optimal in the context of using them to create sensors. Electrochemical impedance spectroscopy and cyclic voltammetry techniques were used to determine the electrochemical properties of the materials. On the other hand, properties such as hydrophilicity/hydrophobicity and their topological structure were determined using contact angle measurements and confocal microscopy, respectively. Based on the research carried out on a wide group of electrode materials, it was found that transparent conductive oxides of the FTO (fluorine doped tin oxide) type exhibit optimal electrochemical parameters and offer great modification possibilities. These electrodes are characterized by a wide range of work and high chemical stability. In addition, the presence of a transparent oxide layer allows for the preservation of valuable optoelectronic properties. An important feature is also the high sensitivity of these electrodes compared to other tested materials. The combination of these properties made FTO electrodes selected for further research.

Keywords: contact angle; conductive materials; electrochemical measurements; FTO electrodes; carbon electrodes

1. Introduction

An important aspect determining the application potential of new electrodes is to know their electrochemical and surface properties. The new electrode material is expected to have high electrical conductivity, fast electron transfer for a wide range of redox systems, as well as structural and electrochemical stability over a wide range of potentials. An additional advantage of an ideal electrode is its simplicity and low production cost. Meeting these expectations is a guarantee of using a given electrode material as a modification platform in further electroanalytical tests.

The need to quantify various chemical compounds in environmental, medical, or industrial analysis motivates scientists to search for new electrode materials with well-defined electrochemical properties. Embedding molecules with specific properties on the surface of electrode materials allows them to be used as molecular recognition systems in many analytical aspects. As the base material, semiconductors (silicon) or dielectrics (glass, ceramics) are usually used, which, as a result of the formation of conductive or semiconductor structures on their surface, obtain interesting properties from the point of view of electronic technology and physicochemical issues. Most often, we apply chemically defined layers of inorganic oxides and carbon materials with various electrical properties to the surface of the starting material.

The most popular electrode material in electroanalytical measurements is glassy carbon (GC) [1]. It is a carbon material with a very complex structure, consisting of intertwined graphite fibers that do not show long-range order. The presence of the sp^2 hybridized carbon atom in the structure is associated with the presence of various types of oxygen groups located at the end of the fibers, which may interact with the analyte. Unlike other structures, it is characterized by a glass fracture and the possibility of polishing its surface to a mirror effect. This non-graphite carbon material combines the properties of glass and ceramics with those of graphite. It is impermeable to gases and liquids, has a compact isotropic microstructure, is characterized by high hardness, and has high thermal and chemical resistance, which is superior to other forms of carbon structure used as electrodes. Vitreous carbon is susceptible to machining, which allows for the production of various electrode structures, in the form of rods, plates, or disc electrodes [2,3]. Glassy carbon electrodes are characterized by high electrochemical stability over a wide range of potentials, while showing low electrical resistance. These properties make the glassy carbon electrode an ideal substrate for many electroanalytical tests. Despite the passage of time, the GC electrode is still one of the most used electrode materials in electrochemistry. This is proven by scientific articles describing new application possibilities of this material [4–6].

The inclusion of a boron dopant in the diamond crystal structure of sp^3 hybridization creates very interesting p-type semiconductor materials known as boron doped diamond (BDD) electrodes. BDD electrodes show excellent electrochemical properties [7–9]. Compared to other materials used in electroanalytical measurements (Au, Pt, or GC), they are characterized by a wide range of potentials in aqueous and non-aqueous environments, low capacitive current, and microstructural stability at extreme cathode and anodic potentials. High stability, combined with biocompatibility, chemical inertness, and mechanical resistance to contamination make it an interesting material for the construction of electrochemical sensors [10–16]. Moreover, the BDD electrodes, due to the aliphatic nature of the surface composed of sp^3 hybridized carbon atoms, show poor adsorption properties. Nevertheless, the lack of chemically reactive functional groups in the structure of this material precludes direct attachment of organic compounds to its surface [17]. The major part of research on BDD electrodes concerns surface modification, enabling covalent coupling of organic modifiers or metal nanoparticles [18,19]. BDD electrodes are used in classical electroanalytical measurements concerning the determination of redoxactive compounds, but also in the construction of biosensors [20–23].

Thin carbon films made of boron doped nanocrystalline diamond (B-NCD) is another interesting material. It is usually synthesized on silicon substrates, but it can also be successfully obtained on the surface of an amorphous dielectric—quartz glass [24,25]. It has optical parameters similar to quartz and, at the same time, due to the electrical conductivity of the diamond layer, it enables electrochemical measurements [26,27]. The undoubted advantage of optically transparent diamond electrodes is the transmission in a wide range of optical radiation, from ultraviolet to far infrared [28]. However, an important aspect is also the high refractive index of the B-NCD film, which allows a clear optical contrast between the diamond electrode and the glass substrate to be obtained ($n_{diamCVD}$ = 2.4: n_{quarc} = 1.45). Boron-doped diamond films are commonly used as an electrode material for the construction of biosensors by functionalization of the B-NCD surface with DNA [29,30]. Moreover, these nanocrystalline structures have found application as optically transparent electrodes (OTE) for spectroelectrochemical measurements [31].

Intensive research on carbon nanotubes contributed to the discovery of the so-called carbon nanowalls (CNWs), which are a system of graphite walls set vertically to the substrate [32]. Carbon nanowalls are variously branched networks with a morphological structure resembling a labyrinth [33–36]. The basic properties of carbon nanowalls, which are of fundamental importance for their potential applications, are primarily the interesting structure of the material, i.e., sharp edges or a high surface-to-volume ratio, which makes it an ideal functional support for synthesizing a new composite material with a large area. CNWs have also been used in energy storage, as electrodes for fuel cells, catalyst carriers, lithium-ion batteries or field emission devices [37–42]. In addition, the unique properties of CNWs make these electrodes an interesting starting material for the creation of electrochemical sensors for environmental, medical, and industrial analysis [21,43–45].

One of the most frequently and intensively studied oxide compositions, due to high conductivity and transparency compared to SnO_2 or ZnO, is indium tin oxide (ITO) (In_2O_3: Sn). Nevertheless, due to the low availability of indium, and thus the increasing costs associated with the production of this electrode, new materials were sought with comparable optoelectronic properties. The FTO (fluorine doped tin oxide) electrode turned out to be an alternative to the ITO electrode. It is a glass covered with a thin layer of conductive inorganic material, fluorine doped tin oxide (SnO_2: F) [46,47].

FTO electrodes were found to be a very promising material due to their greater weather stability and resistance to high temperatures, compared to ITO electrodes. Moreover, this material is chemically inert, mechanically resistant, and has a high resistance to physical abrasion [48,49]. The key feature of these electrodes is the combination of optical and electrochemical properties. Both FTO and ITO electrodes find practical application in a wide range of devices; among others, they are used to create transparent conductive coatings, in touch panels, flat screens, aircraft cockpit windows, and plasma monitors. Thin oxide layers are also used in the production of organic light emitting diodes (OLED) and in solar cells [50,51]. Moreover, thanks to their unique properties, these materials are a good substrate for modifying their surface [48,52–56]. They are also used in classical electroanalytical measurements as working electrodes in the determination of a wide range of electroactive compounds [57].

The aim of this study was to select, from among the presented electrode materials (carbon and oxide), the one whose parameters will be optimal in the context of using them to create sensors. Based on the research carried out on a wide group of electrode materials, it was found that transparent conductive oxides of the FTO type exhibit optimal electrochemical parameters and offer great modification possibilities. These electrodes are characterized by a wide potential range and high chemical stability. In addition, the presence of a transparent oxide layer allows for the preservation of valuable optoelectronic properties. An important feature is also the high sensitivity of these electrodes compared to other tested materials. The combination of these properties resulted in them being selected for further research. Electrochemical impedance spectroscopy and cyclic voltammetry techniques were used to determine the electrochemical properties of the materials. Properties such as hydrophilicity/hydrophobicity and their topological structure were determined using contact angle measurements and confocal microscopy, respectively.

2. Materials and Methods
2.1. Electrochemical Measurements

Cyclic voltammetry measurements were carried out in an aqueous solution of Na_2SO_4 (0.5 M) containing a reference redox system. Appropriate redox systems, i.e., potassium ferri/ferrocyanide (($Fe(CN)_6)^{3-/4-}$), hydroquinone/quinone (H_2Q/Q) at 5 mM concentration were selected for the study. The prepared solutions were deoxidized with a stream of inert gas (argon). Electrochemical measurements were made in a standard three-electrode system consisting of a working electrode (GC, Si/CNW, Si/B-NCD, glass/B-NCD, ITO, and FTO), a reference electrode (silver wire coated with a layer of silver chloride (Ag/AgCl), immersed in a saturated solution of potassium chloride (0.1 M KCl)) and anti-electrodes (platinum). The surface of the working electrode exposed to the electrolyte was about 0.50 cm^2 or 0.13 cm^2. Measurements were recorded at appropriate potential scan rates, i.e., 10, 50, and 100 mV/s. Electrochemical experiments were carried out using the Autolab PGSTAT30 potentiostat/galvanostat (Metrohm Autolab B.V., Utrecht, The Netherlands) and Nova software (1.11, 2005–2013, Metrohm Autolab B.V, Utrecht, The Netherlands).

2.2. Contact Angle Measurements

The wettability of the electrode materials surface was determined by measuring the contact angle at room temperature using standard liquids based on the static method of a sessile drop. Water, formamide, diiodomethane and glycerol were chosen as measuring liquids with known surface tension. The contact angle values given are mean values measured at various positions on the electrode surface. Drop shape analysis (with a volume of 2 µL or 4 µL) was carried out using the circle method and the Young-Laplace method [58–60]. Measurements were made using a KRÜSS Drop Shape Analyzer—DSA100 (Hamburg, Germany). Free surface energy and its components were determined based on the results of direct contact angle measurements calculated by means of the OWRK method [61–67].

2.3. Electrode Materials

The subjects of the study were electrode materials on silicon or glass substrates. Based on the differences resulting from the structure of the conductive layer, the materials were divided into two groups. The first group were FTO glass electrodes. In this case, the conductive layer was transparent fluorine doped tin oxide (SnO_2/F). These electrodes were purchased from Sigma Aldrich (St. Louis, MO, USA). For FTO electrodes, the layer resistance was ~7 and ~13 Ω/sq, respectively, with a transparency of 80–82% and 82–84.5%. In addition to transparent conductive coatings, the subject of the study were electrodes made of conductive carbon layers deposited on silicon or glass substrates. This group of electrodes includes carbon nanowalls (CNW) and nanocrystalline boron doped diamond (B-NCD). The BDD and TCO electrodes required preliminary surface preparation. These materials were treated to obtain a hydrogenated BDD surface and an oxidized surface of the TCO electrodes [12,18,61,68]. For comparison purposes, popular carbon electrodes, boron doped diamond (BDD), and glassy carbon (GC) disk electrode were selected. As part of the study of the same electrode materials, they were differentiated in terms of chemical composition. Introducing different boron content into the structure of carbon layers, they were given new electrochemical properties. Therefore, the B-NCD electrodes name includes the ratio of boron to carbon atoms (B)/(C) determining the degree of doping. For example, an electrode described as Si/B-NCD-2k has a (B)/(C) ratio of 2000 ppm. The synthesis of carbon layers (BDD, CNW, and B-NCD) on the silicon/quartz glass surface, along with the selection of deposition parameters, was carried out in accordance with the procedures described in earlier works [18,25,31,33,69] by the team of Prof. Robert Bogdanowicz from the Department of Metrology and Optoelectronics from the Faculty of Electronics, Telecommunications and Informatics of the Gdańsk University of Technology.

3. Results and Discussion

Searching for new materials that could form the platform for future sensors, a number of conductive materials that differed in both chemical and surface structure were investigated. Among the tested group of electrodes were three types of carbon electrodes (Si/CNW, Si/B-NCD, and glass/B-NCD) and one TCO electrode (FTO). Due to the fact that the subjects of the study were electrodes with different chemical composition and surface structure, differences in properties were expected. The operation of the sensor based on the modified electrode is represented by an electrochemical signal (or a change in the impedance spectrum). Therefore, these materials are tested using standard redox systems to assess their properties and choose the one with the best parameters. The model system $(Fe(CN)_6)^{3-/4-}$ is one of the most frequently chosen analytes to assess the electrocatalytic performance of electrode materials. As an outer sphere system, sensitive to changes in the construction of the electrode layer, it is a perfect model to characterize a wide range of materials. Nevertheless, the conducted research showed that on carbon materials, especially those characterized by different hybridization of the carbon atom, this system behaves completely differently.

Due to the fact that specific interactions play an important role in the reactions of the outer sphere, we decided to use the measurements of wettability and free surface energy parameters to assess these phenomena.

3.1. Electrochemical Properties

Determining the characteristics of electrode materials usually begins by determining the potential range in which the work of the electrode is possible. The wider it is, the larger the group of analytes that differ significantly in oxidation and reduction potential which can be studied. Each of the tested electrodes is characterized by a so-called wide "electrochemical window" (Figure 1).

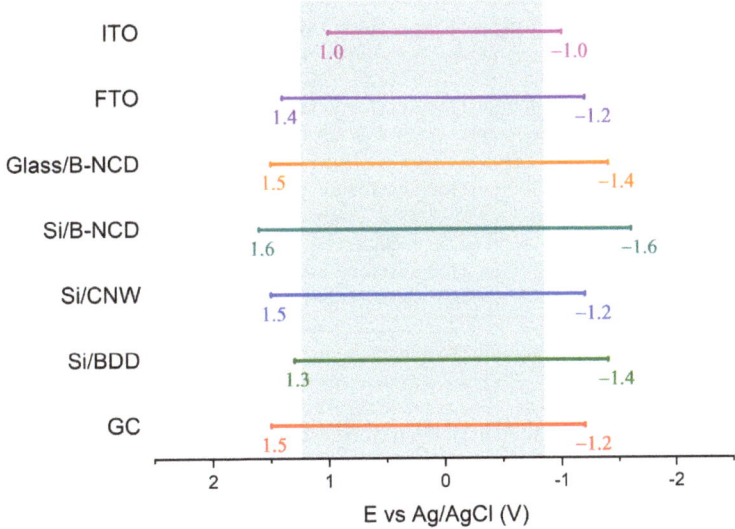

Figure 1. Schematic representation of the range of electrochemical stability of the tested group of electrodes in 0.5 M Na_2SO_4 together with the designated area in which water electrolysis takes place (gray field).

The electrochemical stability range of classic GC and Si/BDD electrodes is 2.7 V, which is comparable with other carbon materials under investigation. The Si/B-NCD electrode deserves special mention, as it is the only one with a potential range above 3 V. The wider stability range results from differences in the structural properties of diamond electrodes. It matters whether they are microcrystalline (BDD) or nanocrystalline boron doped diamonds (B-NCD). According to the available literature, ultra-nanocrystalline (UNCD) diamonds with crystals smaller than B-NCD are even more stable [70]. Undoubtedly, despite the wide range of Si/B-NCD electrode potentials, the disadvantage is the presence in the anode range of a peak at a potential of 1.2 V. This peak is associated with a carbon oxidation reaction with sp^2 hybridization occurring at the grain boundary, where electrochemically active carbon–oxygen forms can be formed [31,70]. Comparing B-NCD electrodes with each other, it seems that the type of substrate affects the range of the electrode. However, these materials could differ in the rate of carbon layer growth during the deposition process, which in turn resulted in electrodes of different thicknesses. The growth time was 1 h for silicon substrates and 3 h for glass substrate. In the first case nanocrystalline layers less than 100 nm thick were obtained, in the second case about 250 nm thick [25,31]. It should be noted that this is a factor that significantly affects the range of potentials. Therefore, a direct comparison of electrochemical stability exhibited by a very thin and thick layer of boron-doped diamond is not reliable [70,71]. The ITO electrode (2 V) has the narrowest potential range in 0.5 M Na_2SO_4. The narrow operating range of the electrode from the group of transparent conductive oxides is related not only to the decomposition of the supporting electrolyte, but also to the gradual disappearance of the optical and electrical properties of the oxide layer in far cathode ranges. Descriptions in the literature studies indicate that during anodic polarization the substrate undergoes slow degradation by oxidation of the ITO oxide layer. The nature of the ions present in the supporting electrolyte and the exposure time of the electrode in the solution strongly affects the rate of electro-dissolution in the anodic and cathodic range [72,73]. On the other hand, the FTO electrode is characterized by electrochemical stability similar to that of carbon materials. The potential range is wider by 0.6 V compared to the ITO electrode, which proves a better chemical stability of this material in 0.5 M Na_2SO_4. The reactions related to the electrolysis of the supporting electrolyte at the FTO electrode are similar to those of ITO, but the signal intensity is much lower. It is worth noting that the literature does not mention any information about the degradation of the conductive layer in the anode range, as is the case with the indium-tin electrode, which is an additional advantage of this electrode [73]. The wide operating range of the FTO electrode while maintaining valuable optoelectronic properties makes this material the most suitable material for further electroanalytical research.

In sensory applications, an important feature is the sensitivity of the system to a given analyte. Therefore, another important feature is the high level of depolarizer signal in relation to the background, which is associated with a low capacitive current that allows us to reduce the level of interference and, consequently, to lower the detection limit of the analyte. The presence of electroactive labeled substances in the solution, $(Fe(CN)_6)^{3-/4-}$ or H_2Q/Q (Figure 2), generates the appearance of electrochemical signals on voltammograms. The differences in the intensity of the analytical response illustrate the different behavior of selected redox systems in relation to the tested electrode materials. This is influenced by many factors originating both from the electrode itself and from the redox system, which was analyzed in the presented studies.

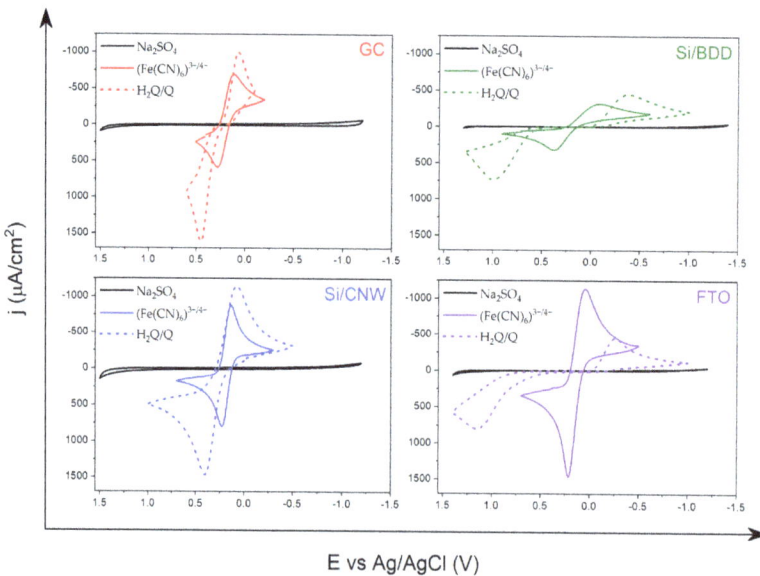

Figure 2. Comparison of cyclic voltammograms of the reference redox systems $(Fe(CN)_6)^{3-/4-}$ and H_2Q/Q in an aqueous solution of Na_2SO_4 (0.5 M) registered on the following electrodes: glassy carbon (GC), boron doped diamond on silicone (Si/BDD), carbon nanowalls on silicone (Si/CNW) and fluorine doped tin oxide (FTO). Scan rate: 0.1 Vs^{-1}.

The recorded voltammograms for the model $(Fe(CN)_6)^{3-/4-}$ in Na_2SO_4 (Figure 2) confirm that the rate of this redox reaction is strongly influenced by the nature of the surface exposed to the solution. Electrochemical reversibility for this redox process is much better on sp^2 carbon electrodes than on a classic diamond electrode. The very slow electron transfer kinetics on the Si/BDD electrode is demonstrated by the high value of ΔE, as well as low values of current densities, j_a and j_k (Table 1). The best reversibility for the redox process $(Fe(CN)_6)^{3-/4-}$ was observed on the Si/CNW electrode, where the ΔE value is up to five times lower compared to the Si/BDD electrode, and is only 0.092 V. Moreover, with this electrode material the current density is more than twice as high for the anodic and cathode responses as compared to the diamond electrode. For comparison, the separation of the oxidation and reduction peaks on a classic GC electrode is 0.171 V. For the B-NCD-10k electrode on a silicon substrate, the shift of the oxidation peak $(Fe(CN)_6)^{3-}$ to $(Fe(CN)_6)^{4-}$ is observed—towards more positive potential values (0.328 V) and the reduction peak towards negative values (0.022 V), which results in increased peak separation of 0.306 V. This type of behavior was observed on the Si/BDD electrode. In turn, the electrochemical parameters of the B-NCD-10k (Table 1) electrode on a glass substrate are oriented at values close to the Si/CNW electrode, as ΔE is 0.126 V, and the current response signals have similar densities. Of all the electrode materials tested, the FTO electrode deserves special mention. The electrochemical parameters recorded for the oxidation and reductions reaction of the $(Fe(CN)_6)^{3-/4-}$ system are extremely interesting. Despite the similar reversibility to the GC electrode, oscillating around 0.165 V, it shows much higher current signals. This proves that the detection limit of the FTO electrodes is significantly lower for this system compared to other electrodes. The additional benefit of this electrode is that the conductive layer is not a carbon material, but a transparent inorganic oxide (SnO_2: F).

Table 1. Electrochemical parameters of the reference redox systems $(Fe(CN)_6)^{3-/4-}$, and H_2Q/Q in an aqueous solution of Na_2SO_4 (0.5 M) registered on the following electrodes: GC, Si/BDD, Si/CNW, and FTO.

Electrode	Redox System	E_a (V)	E_k (V)	ΔE (V)	j_a (µA/cm^2)	j_k (µA/cm^2)
GC	$(Fe(CN)_6)^{3-/4-}$	0.283	0.112	0.171	585.5	−692.9
	H_2Q/Q	0.450	0.063	0.387	1617.4	−993.7
Si/BDD	$(Fe(CN)_6)^{3-/4-}$	0.376	−0.092	0.468	318.7	−319.3
	H_2Q/Q	0.990	−0.387	1.377	730.2	−450.7
Si/CNW	$(Fe(CN)_6)^{3-/4-}$	0.229	0.137	0.092	794.0	−877.1
	H_2Q/Q	0.403	0.079	0.324	1465.0	−1143.9
FTO	$(Fe(CN)_6)^{3-/4-}$	0.214	0.049	0.165	1460.2	−1139.6
	H_2Q/Q	1.151	−0.260	1.411	798.9	−451.9

$\Delta E = E_a - E_k$.

Figure 2 also shows the electrochemical behavior of the hydroquinone/quinone model system. According to the literature, the redox reaction is not affected by the presence of oxidized groups on the surface of the electrode material, and electron transfer is preceded by the adsorption step of the depolarizer particles [74].

For the hydroquinone/quinone pair, a typical oxidation/reduction reaction of single peaks was observed in Figure 2, which is the electrode response for the transfer of two electrons and protons on a GC substrate. More significant effects were observed in the case of carbon nanowalls. With respect to the classic GC electrode, the electrochemical reversibility for this process on the surface of the Si/CNW electrode increased by 63 mV. Similar properties were observed when modifying the GC electrode with single-walled carbon nanotubes (SWCNT). SWCNT as cylindrical rolled graphene layers contributed to an increase in the reversibility of the redox process due to adsorption of hydroquinone on the walls of carbon nanotubes. This phenomenon was confirmed by narrow, symmetrical oxidation and reduction peaks on the CV curve of the H_2Q/Q system for a modified

GC surface [74]. As carbon nanowalls belong to the same group of carbon materials as nanotubes, similar effects were expected from the H$_2$Q/Q system. It is worth noting that the voltammograms obtained for the Si/BDD and FTO electrodes are completely different (Figure 2). The analyzed redox system shows different behavior on a carbon surface rich in carbon atoms with sp^3 hybridization and on transparent oxide coatings. Oxidation and reduction peaks are characterized by increased separation and lower current density. ΔE takes values up to 1 V with a twice lower current response signal compared to sp^2 carbon electrodes. This demonstrates the lower affinity of this redox system to the surface of Si/BDD and FTO electrodes, which results in inhibition of the electron transfer process. A significant improvement in electron transfer kinetics on the Si/CNW surface was observed compared to GC or BDD electrodes in the case of redox pair having delocalized π electrons in their structure. This manifests itself in a different behavior of the H$_2$Q/Q system relative to (Fe(CN)$_6$)$^{3-/4-}$. This may indicate the interaction of π–π between organic molecules and carbon walls on the surface of the Si/CNW electrode.

As the level of boron doping increases, the electrochemical properties of the electrodes change (Figure 3). On the basis of obtained cyclic voltammograms of B-NCD electrodes, it can be observed that the redox reaction rate increases with an increasing (B)/(C) ratio, which is due to the increased density of electronic states. The highest value of ΔE was recorded for the Glass/B-NCD-10k electrode (0.126 V), which indicates greater efficiency of electron transfer compared to other electrodes, and above all for the Glass/B-NCD-2k electrode (0.657 V) (Table 2). The largest changes in the reversibility of the redox process were noted for Si/B-NCD electrodes. The difference in ΔE between the electrode with the largest doping (10k) and the smallest (2k) is 585 mV. Although it is the same conductive material, it exhibits different behavior relative to the electron transfer process for the (Fe(CN)$_6$)$^{3-/4-}$ system. This is due to the different growth time of the carbon layer, and consequently the thickness of the conductive material, and not the substrate on which the layer was deposited, which was noted when determining the electrochemical stability of these electrodes [25,31].

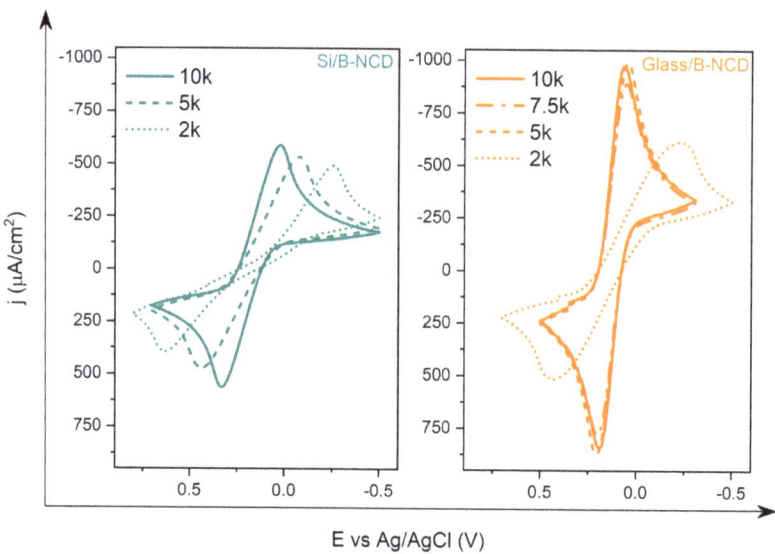

Figure 3. Comparison of cyclic voltammograms of the model redox system (Fe(CN)$_6$)$^{3-/4-}$ (5 mM) in an aqueous solution of Na$_2$SO$_4$ (0.5 M) registered on B-NCD electrodes on silicon and glass substrate with different (B)/(C) ratios.

Table 2. Electrochemical parameters of the reference redox systems $(Fe(CN)_6)^{3-/4-}$ in an aqueous solution of Na_2SO_4 (0.5 M) registered on B-NCD electrodes on silicon and glass substrate with different (B)/(C) ratios.

Electrode	(B)/(C)	E_a (V)	E_k (V)	ΔE (V)	j_a ($\mu A/cm^2$)	j_k ($\mu A/cm^2$)
Si/B-NCD	10k	0.328	0.022	0.306	562.0	−588.9
	5k	0.436	−0.077	0.513	473.1	−533.4
	2k	0.634	−0.257	0.891	395.8	−493.3
Glass/B-NCD	10k	0.188	0.062	0.126	844.9	−972.8
	7.5k	0.197	0.053	0.144	779.8	−889.1
	5k	0.206	0.044	0.162	875.0	−991.3
	2k	0.436	−0.221	0.657	515.9	−612.8

$\Delta E = E_a - E_k$.

3.2. Wettability

The tested electrode materials are characterized by different wettability. The most hydrophobic surface has an Si/CNW electrode (112.79°) (Figure 4a), and a hydrophilic nanodiamond doped with boron on a glass substrate with 2000 ppm doping (31.20°) (Figure 4b). The value of the contact angle for the FTO electrode surface (55.93°) indicates that it is also a material with hydrophilic properties. This is undoubtedly due to their chemical nature, but also to differences in the surface roughness.

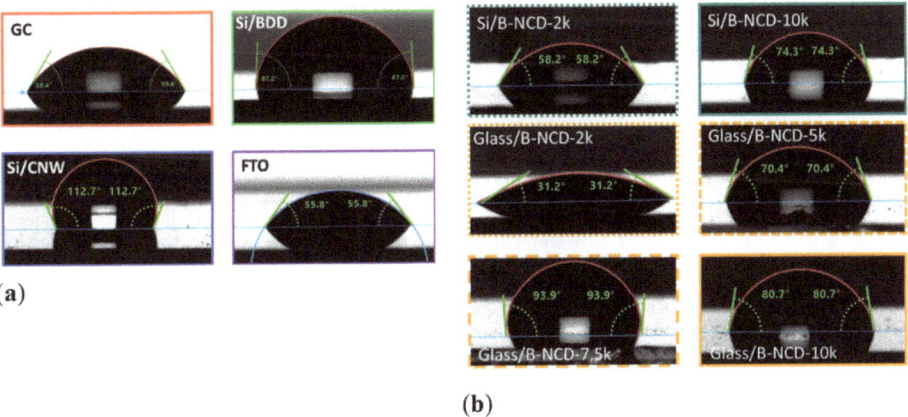

Figure 4. Pictures of contact angle measurement (WCA) of electrode materials (**a**) GC, Si/BDD, Si/CNW, and FTO; (**b**) boron-doped nanodiamonds (B-NCD) on glass and silicon substrates with different (B)/(C) ratios.

The research shows that, regardless of the substrate (glass, silicon) used, electrode materials with a higher degree of doping are characterized by higher hydrophobicity, which is visible at higher contact angle (>70°) (Figure 4b).

3.3. Surface Energy

Surface Free Energy (SFE) is the effect of intermolecular interactions at the liquid–solid interface, and allows us to describe these interactions. It can be divided into two components—(1) disperse (γ^D), attributed to van der Waals interactions and other non-specific interactions, and (2) polar (γ^P) resulting from dipole–dipole, dipole-induced dipole, hydrogen bonds, and other specific interactions at the liquid–solid interface [75].

All tested conductive materials have a high disperse component value and a low polar component value, but they differ significantly (Table 3). As the polar component increases, the surface hydrophilicity increases. The largest share of the polar part (approx.

8–9 mN/m) has GC and FTO electrodes, which confirms the contact angle (approx. 55–59°) (Figure 5a). In the case of carbon materials, a greater degree of boron doping results in a reduced proportion of the SFE polar part, and therefore most surface interactions will be dispersive (Figure 5b). In addition, the total free surface energy is lower for materials with a larger (B)/(C) ratio.

Table 3. Contact angle parameters and surface free energy for selected conductive materials.

Measured Value	Electrode	GC	Si/BDD	Si/CNW	FTO
Contact angle θ (°)	Water	59.03 (±1.17)	87.13 (±0.30)	112.79 (±1.23)	55.93 (±0.59)
Surface Free Energy (mN/m)	γ_S	48.54 (±5.26)	42.46 (±2.49)	50.82 (±0.46)	51.77 (±4.02)
	γ_D	39.32 (±1.76)	41.32 (±1.77)	50.72 (±0.44)	43.22 (±1.82)
	γ_P	9.22 (±3.50)	1.14 (±0.72)	0.11 (±0.02)	8.55 (±2.20)

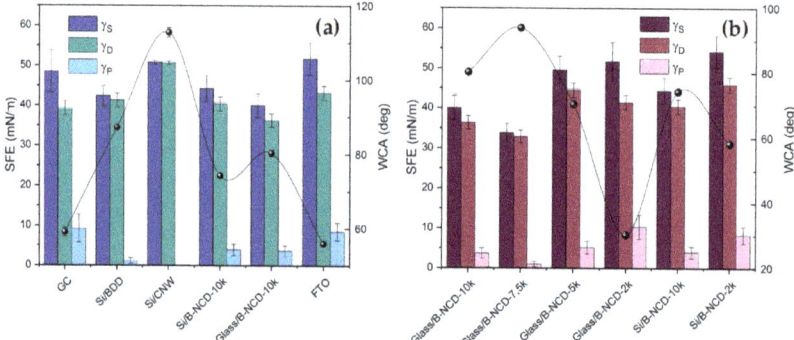

Figure 5. Free surface energy (SFE) graph and contact angles of tested conductive materials: (a) GC, Si/BDD, Si/CNW, Si/B-NCD-10k, Glass/B-NCD-10k, and FTO; (b) boron doped nanodiamonds (B-NCD) on glass and silicon substrates with different (B)/(C) ratios; γ^S-SFE, γ^D: disperse part, γ^P: polar part, and WCA: water contact angle.

In the case of different materials, the correlation between the contact angle and ΔE is much more complex, which probably results directly from the morphology of the material. The most hydrophobic nature, and at the same time the best reversibility of the redox process, was observed for Si/CNW—these are materials with high roughness (Figure 6a). Such structure of the material may cause the reaction of, in the wetting process, it acting on a drop of water similar to a lotus leaf, by reflecting the drop off the surface of the material [76]. In turn, GC and FTO electrodes are characterized by the highest uniformity of surface structure (Figure 7). Compared to others, both materials are hydrophilic and have good reversibility of the redox process (Figure 6a, Tables 1 and 3).

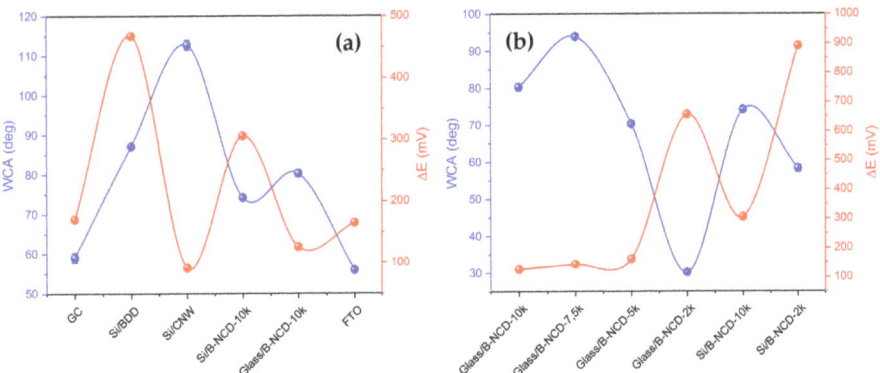

Figure 6. Graph of the water contact angle (WCA) and reversibility of the redox process for $(Fe(CN)_6)^{3-/4-}$ couple (ΔE) on the type of electrode: (**a**) GC, Si/BDD, Si/CNW, Si/B-NCD-10k, Glass/B-NCD-10k, and FTO; (**b**) from different levels of doping B-NCD electrodes on a glass and silicon substrate.

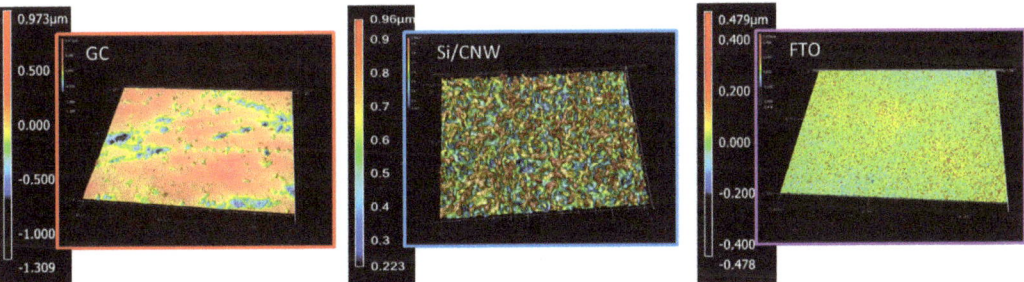

Figure 7. Keyence VK-X1000 confocal laser microscope photos of the surface of electrode materials: GC, Si/CNW, and FTO.

In the case of doped B-NCD electrodes, the reversibility of the redox process is opposite to the contact angle—for highly doped electrodes, we observe better reversibility, but weaker wettability. Additionally, for 2k electrodes, weaker reversibility at higher hydrophilicity (Figure 6b).

4. Conclusions

In recent years, materials chemistry has been one of the most intensively developing fields of science. It mainly includes the synthesis of new electrode materials or modification of their surface to give them specific properties, increasing their applicability. Among the wide range of electrode materials, there are classic metallic electrodes, such as gold and platinum, as well as electrode materials based on carbon and oxide layers.

Carbon and its various structural forms (graphite, graphene, diamond, nanotubes, and other structural forms with sp^2 and sp^3 hybridization) make it possible to use it in a wide range of scientific and technological aspects, due to the great diversity in structure, and thus also in physical and chemical properties.

Despite the availability of many metallic and carbon electrodes on the market, materials with transparent conductive oxides are also very popular. Their important feature is optical transparency, which, combined with high electrical conductivity, allows them to be used in electrochemistry and optoelectronics. The best application properties are demonstrated by electrodes based on zinc, indium, and tin oxide.

The subjects of the presented research were electrode materials on silicon or glass substrates. Based on the differences resulting from the structure of the conductive layer, the materials were divided into two groups. The first group were glass electrodes of the FTO and ITO types. In this case, the conductive layer consisted of transparent inorganic oxides, tin oxide with an admixture of fluorine, and indium-tin oxide. In addition to transparent conductive coatings, the subject of the research were electrodes made of conductive carbon layers deposited on silicon or glass substrates. This group of electrodes includes, e.g., carbon nanowalls (CNW) (being a system of graphite walls set vertically to the substrate) and nanocrystalline boron-doped diamond (B-NCD). For comparison purposes, the popular carbon electrodes, boron doped diamond (BDD) and glassy carbon disc electrode (GC), were selected.

In order to assess how the new materials compare with other commercially available electrodes, their characteristics were made. On the basis of the voltametric measurements recorded in the supporting electrolyte, we determined the potential limits reflecting the operating range of each electrode. Another important aspect that determines the application potential of the electrodes is the high level of the signal of the electroactive substance in relation to the capacitive current of the electrode. To assess the electrochemical reactivity of the analyzed electrode materials, we chose two redox systems: potassium hexacyanoferrate (III)/(II) and the hydroquinone/quinone system.

The conducted research allowed us to choose, from a wide range of tested electrode materials, that with optimal parameters. The FTO electrode is characterized by its distinctive features—electrochemical stability similar to carbon materials and higher compared to the ITO electrode. Moreover, the FTO electrode is characterized by a similar reversibility of the redox process of the model potassium ferrocyanide (III)/(II) system to the GC electrode (165 mV). It shows significantly higher current response signals. This proves a higher level of detection of this electrode for this system compared to other electrodes.

In addition, when examining the wettability, the most hydrophobic surface has the electrode of the carbon nanowalls type (113°), and the hydrophilic FTO electrode (56°). This is undoubtedly due to their chemical nature, but also to differences due to surface roughness. The FTO electrode has the smoothest surface.

The wide operating range of the FTO electrode while maintaining the valuable optoelectronic properties, as well as the high sensitivity of these electrodes compared to other materials tested, meant that we chose the FTO electrode for further electroanalytical tests. In the following works, we will present in detail the modification process of the FTO electrode, the influence of various conditions on the obtained layers, as well as the potential application of the obtained devices.

Author Contributions: Conceptualization, A.C. and A.W.; methodology, A.C.; investigation, A.C. and A.W.; writing—original draft preparation, A.C. and A.W.; writing—review and editing, A.C., D.Z. and A.W.; visualization, A.C. and A.W.; and supervision, D.Z. and A.W. All authors have read and agreed to the published version of the manuscript.

Funding: This research received no external funding.

Institutional Review Board Statement: Not applicable.

Informed Consent Statement: Not applicable.

Data Availability Statement: All the data is available within the manuscript.

Acknowledgments: The authors would like to thank Robert Bogdanowicz and Michał Sobaszek for providing carbon materials for research.

Conflicts of Interest: The authors declare no conflict of interest.

References

1. Van der Linden, W.; Dieker, J. Glassy carbon as electrode material in electro- analytical chemistry. *Anal. Chim. Acta* **1980**, *119*, 1–24. [CrossRef]
2. Pesin, L.A. Review Structure and properties of glass-like carbon. *J. Mater. Sci.* **2002**, *37*, 1–28. [CrossRef]
3. McCreery, R.L. Advanced Carbon Electrode Materials for Molecular Electrochemistry. *Chem. Rev.* **2008**, *108*, 2646–2687. [CrossRef]

4. Tajik, S.; Beitollahi, H. A Sensitive Chlorpromazine Voltammetric Sensor Based on Graphene Oxide Modified Glassy Carbon Electrode. *Anal. Bioanal. Chem. Res.* 2019, *6*, 171–182. [CrossRef]
5. Masikini, M.; Ghica, M.E.; Baker, P.; Iwuoha, E.; Brett, C.M.A. Electrochemical Sensor Based on Multi-walled Carbon Nanotube/Gold Nanoparticle Modified Glassy Carbon Electrode for Detection of Estradiol in Environmental Samples. *Electroanalysis* 2019, *31*, 1925–1933. [CrossRef]
6. Sohouli, E.; Keihan, A.H.; Shahdost-Fard, F.; Naghian, E.; Plonska-Brzezinska, M.E.; Rahimi-Nasrabadi, M.; Ahmadi, F. A glassy carbon electrode modified with carbon nanoonions for electrochemical determination of fentanyl. *Mater. Sci. Eng. C* 2020, *110*, 110684. [CrossRef] [PubMed]
7. Ferreira, N.G.; Silva, L.; Corat, E.J.; Trava-Airoldi, V. Kinetics study of diamond electrodes at different levels of boron doping as quasi-reversible systems. *Diam. Relat. Mater.* 2002, *11*, 1523–1531. [CrossRef]
8. Fujishima, A.; Einaga, Y.; Rao, T.N.; Tryk, D.A. *Diamond Electrochemistry*, 1st ed.; Elsevier: Amsterdam, The Netherlands, 2005.
9. Yang, N. Diamond Electrochemical Devices. In *Novel Aspects of Diamond: From Growth to Applications*; (Topics in Applied Physics); Yang, N., Ed.; Springer International Publishing: Cham, Switzerland, 2019; pp. 223–256, ISBN 978-3-030-12469-4.
10. Compton, R.G.; Foord, J.S.; Marken, F. Electroanalysis at Diamond-Like and Doped-Diamond Electrodes. *Electroanalysis* 2003, *15*, 1349–1363. [CrossRef]
11. Qureshi, A.; Kang, W.P.; Davidson, J.L.; Gurbuz, Y. Review on carbon-derived, solid-state, micro and nano sensors for electrochemical sensing applications. *Diam. Relat. Mater.* 2009, *18*, 1401–1420. [CrossRef]
12. Niedzialkowski, P.; Ossowski, T.; Zięba, P.; Cirocka, A.; Rochowski, P.; Pogorzelski, S.; Ryl, J.; Sobaszek, M.; Bogdanowicz, R. Poly-l-lysine-modified boron-doped diamond electrodes for the amperometric detection of nucleic acid bases. *J. Electroanal. Chem.* 2015, *756*, 84–93. [CrossRef]
13. Zhou, Y.L.; Tian, R.H.; Zhi, J.F. Amperometric biosensor based on tyrosinase immobilized on a boron-doped diamond electrode. *Biosens. Bioelectron.* 2007, *22*, 822–828. [CrossRef]
14. Geng, R.; Zhao, G.; Liu, M.; Li, M. A sandwich structured SiO_2/cytochrome c/SiO_2 on a boron-doped diamond film electrode as an electrochemical nitrite biosensor. *Biomaterials* 2008, *29*, 2794–2801. [CrossRef]
15. Costa, D.J.; Santos, J.C.; Sanches-Brandão, F.A.; Ribeiro, W.F.; Banda, G.; Araujo, M. Boron-doped diamond electrode acting as a voltammetric sensor for the detection of methomyl pesticide. *J. Electroanal. Chem.* 2017, *789*, 100–107. [CrossRef]
16. Komkova, M.A.; Pasquarelli, A.; Andreev, E.; Galushin, A.A.; Karyakin, A.A. Prussian Blue modified boron-doped diamond interfaces for advanced H_2O_2 electrochemical sensors. *Electrochim. Acta* 2020, *339*, 135924. [CrossRef]
17. Szunerits, S.; Boukherroub, R. Different strategies for functionalization of diamond surfaces. *J. Solid State Electrochem.* 2007, *12*, 1205–1218. [CrossRef]
18. Bogdanowicz, R.; Sawczak, M.; Niedzialkowski, P.; Zieba, P.; Finke, B.; Ryl, J.; Ossowski, T. Direct amination of boron-doped diamond by plasma polymerized allylamine film. *Phys. Status Solidi (A)* 2014, *211*, 2319–2327. [CrossRef]
19. Toghill, K.E.; Compton, R.G. Metal Nanoparticle Modified Boron Doped Diamond Electrodes for Use in Electroanalysis. *Electroanalysis* 2010, *22*, 1947–1956. [CrossRef]
20. Pecková, K.; Musilová, J.; Barek, J. Boron-Doped Diamond Film Electrodes—New Tool for Voltammetric Determination of Organic Substances. *Crit. Rev. Anal. Chem.* 2009, *39*, 148–172. [CrossRef]
21. Niedziałkowski, P.; Cebula, Z.; Malinowska, N.; Białobrzeska, W.; Sobaszek, M.; Ficek, M.; Bogdanowicz, R.; Anand, J.S.; Ossowski, T. Comparison of the paracetamol electrochemical determination using boron-doped diamond electrode and boron-doped carbon nanowalls. *Biosens. Bioelectron.* 2019, *126*, 308–314. [CrossRef]
22. Martin, K.K.; Ouattara, L. Electroanalytical Investigation on Paracetamol on Boron-Doped Diamond Electrode by Voltammetry. *Am. J. Anal. Chem.* 2019, *10*, 562–578. [CrossRef]
23. Kowalcze, M.; Jakubowska, M. Voltammetric determination of nicotine in electronic cigarette liquids using a boron-doped diamond electrode (BDDE). *Diam. Relat. Mater.* 2020, *103*, 107710. [CrossRef]
24. Stotter, J.; Zak, J.; Behler, Z.; Show, Y.; Swain, G. Optical and Electrochemical Properties of Optically Transparent, Boron-Doped Diamond Thin Films Deposited on Quartz. *Anal. Chem.* 2002, *74*, 5924–5930. [CrossRef]
25. Sobaszek, M.; Siuzdak, K.; Skowroński, Ł.; Bogdanowicz, R.; Pluciński, J. Optically transparent boron-doped nanocrystalline diamond films for spectroelectrochemical measurements on different substrates. *IOP Conf. Ser. Mater. Sci. Eng.* 2016, *104*, 012024. [CrossRef]
26. Azevedo, A.; Souza, F.; Matsushima, J.; Baldan, M.R.; Ferreira, N.G. Detection of phenol at boron-doped nanocrystalline diamond electrodes. *J. Electroanal. Chem.* 2011, *658*, 38–45. [CrossRef]
27. Dincer, C.; Ktaich, R.; Laubender, E.; Hees, J.J.; Kieninger, J.; Nebel, C.E.; Heinze, J.; Urban, G.A. Nanocrystalline boron-doped diamond nanoelectrode arrays for ultrasensitive dopamine detection. *Electrochim. Acta* 2015, *185*, 101–106. [CrossRef]
28. Zak, J.K.; Butler, J.E.; Swain, G.M. Diamond Optically Transparent Electrodes: Demonstration of Concept with Ferri/Ferrocyanide and Methyl Viologen. *Anal. Chem.* 2001, *73*, 908–914. [CrossRef] [PubMed]
29. Yang, W.; Auciello, O.; Butler, J.; Cai, W.; Carlisle, J.A.; Gerbi, J.E.; Gruen, D.M.; Knickerbocker, T.; Lasseter, T.L.; Russell, J.N., Jr.; et al. DNA-modified nanocrystalline diamond thin-films as stable, biologically active substrates. *Nat. Mater.* 2002, *1*, 253–257. [CrossRef] [PubMed]
30. Bajaj, P.; Akin, D.; Gupta, A.; Sherman, D.; Shi, B.; Auciello, O.; Bashir, R. Ultrananocrystalline diamond film as an optimal cell interface for biomedical applications. *Biomed. Microdevices* 2007, *9*, 787–794. [CrossRef] [PubMed]

31. Sobaszek, M.; Skowroński, Ł.; Bogdanowicz, R.; Siuzdak, K.; Cirocka, A.; Zięba, P.; Gnyba, M.; Naparty, M.; Gołuński, Ł.; Płotka, P. Optical and electrical properties of ultrathin transparent nanocrystalline boron-doped diamond electrodes. *Opt. Mater.* **2015**, *42*, 24–34. [CrossRef]
32. Wu, Y.; Qiao, P.; Chong, T.; Shen, Z. Carbon Nanowalls Grown by Microwave Plasma Enhanced Chemical Vapor Deposition. *Adv. Mater.* **2002**, *14*, 64–67. [CrossRef]
33. Siuzdak, K.; Ficek, M.; Sobaszek, M.; Ryl, J.; Gnyba, M.; Niedzialkowski, P.; Malinowska, N.; Karczewski, J.; Bogdanowicz, R. Boron-Enhanced Growth of Micron-Scale Carbon-Based Nanowalls: A Route toward High Rates of Electrochemical Biosensing. *ACS Appl. Mater. Interfaces* **2017**, *9*, 12982–12992. [CrossRef]
34. Kobayashi, K.; Tanimura, M.; Nakai, H.; Yoshimura, A.; Yoshimura, H.; Kojima, K.; Tachibana, M. Nanographite domains in carbon nanowalls. *J. Appl. Phys.* **2007**, *101*, 94306. [CrossRef]
35. Kondo, S.; Kawai, S.; Takeuchi, W.; Yamakawa, K.; Den, S.; Kano, H.; Hiramatsu, M.; Hori, M. Initial growth process of carbon nanowalls synthesized by radical injection plasma-enhanced chemical vapor deposition. *J. Appl. Phys.* **2009**, *106*, 094302. [CrossRef]
36. Teii, K.; Shimada, S.; Nakashima, M.; Chuang, A.T.H. Synthesis and electrical characterization of n-type carbon nanowalls. *J. Appl. Phys.* **2009**, *106*, 84303. [CrossRef]
37. Choi, H.; Kwon, S.; Kang, H.; Kim, J.H.; Choi, W. Adhesion-Increased Carbon Nanowalls for the Electrodes of Energy Storage Systems. *Energies* **2019**, *12*, 4759. [CrossRef]
38. Giorgi, L.; Makris, T.; Giorgi, R.; Lisi, N.; Salernitano, E. Electrochemical properties of carbon nanowalls synthesized by HF-CVD. *Sens. Actuators B Chem.* **2007**, *126*, 144–152. [CrossRef]
39. Wang, H.; Quan, X.; Yu, H.; Chen, S. Fabrication of a TiO2/carbon nanowall heterojunction and its photocatalytic ability. *Carbon* **2008**, *46*, 1126–1132. [CrossRef]
40. Krivchenko, V.A.; Itkis, D.; Evlashin, S.; Semenenko, D.A.; Goodilin, E.A.; Rakhimov, A.T.; Stepanov, A.S.; Suetin, N.V.; Pilevsky, A.A.; Voronin, P.V. Carbon nanowalls decorated with silicon for lithium-ion batteries. *Carbon* **2012**, *50*, 1438–1442. [CrossRef]
41. Li, B.; Li, S.; Liu, J.; Wang, B.; Yang, S. Vertically Aligned Sulfur–Graphene Nanowalls on Substrates for Ultrafast Lithium–Sulfur Batteries. *Nano Lett.* **2015**, *15*, 3073–3079. [CrossRef]
42. Shin, S.C.; Yoshimura, A.; Matsuo, T.; Mori, M.; Tanimura, M.; Ishihara, A.; Ota, K.-I.; Tachibana, M. Carbon nanowalls as platinum support for fuel cells. *J. Appl. Phys.* **2011**, *110*, 104308. [CrossRef]
43. Yang, J.; Wei, D.; Tang, L.; Song, X.; Luo, W.; Chu, J.; Gao, T.; Shi, H.; Du, C. Wearable temperature sensor based on graphene nanowalls. *RSC Adv.* **2015**, *5*, 25609–25615. [CrossRef]
44. Slobodian, P.; Cvelbar, U.; Riha, P.; Olejnik, R.; Matyas, J.; Filipič, G.; Watanabe, H.; Tajima, S.; Kondo, H.; Sekine, M.; et al. High sensitivity of a carbon nanowall-based sensor for detection of organic vapours. *RSC Adv.* **2015**, *5*, 90515–90520. [CrossRef]
45. Choi, H.; Kwon, S.H.; Kang, H.; Kim, J.H.; Choi, W. Zinc-oxide-deposited Carbon Nanowalls for Acetone Sensing. *Thin Solid Films* **2020**, *700*, 137887. [CrossRef]
46. Andersson, A.; Johansson, N.; Bröms, P.; Yu, N.; Lupo, D.; Salaneck, W.R. Fluorine Tin Oxide as an Alternative to Indium Tin Oxide in Polymer LEDs. *Adv. Mater.* **1998**, *10*, 859–863. [CrossRef]
47. Patni, N.; Sharma, P.; Pillai, S. Newer approach of using alternatives to (Indium doped) metal electrodes, dyes and electrolytes in dye sensitized solar cell. *Mater. Res. Express* **2018**, *5*, 045509. [CrossRef]
48. Lee, K.-T.; Liu, D.-M.; Liang, Y.-Y.; Matsushita, N.; Ikoma, T.; Lu, S.-Y. Porous fluorine-doped tin oxide as a promising substrate for electrochemical biosensors-demonstration in hydrogen peroxide sensing. *J. Mater. Chem. B* **2014**, *2*, 7779–7784. [CrossRef]
49. Banyamin, Z.Y.; Kelly, P.J.; West, G.; Boardman, J. Electrical and Optical Properties of Fluorine Doped Tin Oxide Thin Films Prepared by Magnetron Sputtering. *Coatings* **2014**, *4*, 732–746. [CrossRef]
50. Bierwagen, O. Indium oxide—A transparent, wide-band gap semiconductor for (opto)electronic applications. *Semicond. Sci. Technol.* **2015**, *30*, 024001. [CrossRef]
51. Ouerfelli, J.; Djobo, S.O.; Bernède, J.; Cattin, L.; Morsli, M.; Berredjem, Y. Organic light emitting diodes using fluorine doped tin oxide thin films, deposited by chemical spray pyrolysis, as anode. *Mater. Chem. Phys.* **2008**, *112*, 198–201. [CrossRef]
52. Jasiecki, S.; Czupryniak, J.; Ossowski, T.; Schroeder, G. FTO Coated Glass Electrode Functionalization with Transition Metal Cations Receptors via Electrostatic Self-Assembly. *Int. J. Electrochem. Sci.* **2013**, *8*, 12543–12556.
53. Ahuja, T.; Rajesh; Kumar, D.; Tanwar, V.K.; Sharma, V.; Singh, N.; Biradar, A.M. An amperometric uric acid biosensor based on Bis[sulfosuccinimidyl] suberate crosslinker/3-aminopropyltriethoxysilane surface modified ITO glass electrode. *Thin Solid Films* **2010**, *519*, 1128–1134. [CrossRef]
54. Kim, C.O.; Hong, S.-Y.; Kim, M.; Park, S.-M.; Park, J.W. Modification of indium–tin oxide (ITO) glass with aziridine provides a surface of high amine density. *J. Colloid Interface Sci.* **2004**, *277*, 499–504. [CrossRef]
55. Pruna, R.; Palacio, F.; Martínez, M.; Blázquez, O.; Hernández, S.; Garrido, B.; de Miguel, M.L. Organosilane-functionalization of nanostructured indium tin oxide films. *Interface Focus* **2016**, *6*, 20160056. [CrossRef]
56. Muthurasu, A.; Ganesh, V. Electrochemical characterization of Self-assembled Monolayers (SAMs) of silanes on indium tin oxide (ITO) electrodes – Tuning electron transfer behaviour across electrode–electrolyte interface. *J. Colloid Interface Sci.* **2012**, *374*, 241–249. [CrossRef]
57. Göbel, G.; Talke, A.; Lisdat, F. FTO—An Electrode Material for the Stable Electrochemical Determination of Dopamine. *Electroanalysis* **2018**, *30*, 225–229. [CrossRef]

58. Kwok, D.Y.; Gietzelt, T.; Grundke, K.; Jacobasch, A.H.-J.; Neumann, A.W. Contact Angle Measurements and Contact Angle Interpretation. 1. Contact Angle Measurements by Axisymmetric Drop Shape Analysis and a Goniometer Sessile Drop Technique. *Langmuir* **1997**, *13*, 2880–2894. [CrossRef]
59. Kwok, D.; Neumann, A. Contact angle measurement and contact angle interpretation. *Adv. Colloid Interface Sci.* **1999**, *81*, 167–249. [CrossRef]
60. Kwok, D.Y.; Neumann, A.W. Contact angle measurements and interpretation: Wetting behavior and solid surface tensions for poly(alkyl methacrylate) polymers. *J. Adhes. Sci. Technol.* **2000**, *14*, 719–743. [CrossRef]
61. Cirocka, A.; Zarzeczańska, D.; Wcisło, A.; Ryl, J.; Bogdanowicz, R.; Finke, B.; Ossowski, T. Tuning of the electrochemical properties of transparent fluorine-doped tin oxide electrodes by microwave pulsed-plasma polymerized allylamine. *Electrochim. Acta* **2019**, *313*, 432–440. [CrossRef]
62. Niedziałkowski, P.; Bojko, M.; Ryl, J.; Wcisło, A.; Spodzieja, M.; Magiera-Mularz, K.; Guzik, K.; Dubin, G.; Holak, T.A.; Ossowski, T.; et al. Ultrasensitive electrochemical determination of the cancer biomarker protein sPD-L1 based on a BMS-8-modified gold electrode. *Bioelectrochemistry* **2021**, *139*, 107742. [CrossRef]
63. Szczepańska, E.; Synak, A.; Bojarski, P.; Niedziałkowski, P.; Wcisło, A.; Ossowski, T.; Grobelna, B. Dansyl-Labelled Ag@SiO$_2$ Core-Shell Nanostructures—Synthesis, Characterization, and Metal-Enhanced Fluorescence. *Materials* **2020**, *13*, 5168. [CrossRef] [PubMed]
64. Dąbrowa, T.; Wcisło, A.; Majstrzyk, W.; Niedziałkowski, P.; Ossowski, T.; Więckiewicz, W.; Gotszalk, T. Adhesion as a component of retention force of overdenture prostheses-study on selected Au based dental materials used for telescopic crowns using atomic force microscopy and contact angle techniques. *J. Mech. Behav. Biomed. Mater.* **2021**, *121*, 104648. [CrossRef]
65. Jańczuk, B.; Białłopiotrowicz, T. Surface free-energy components of liquids and low energy solids and contact angles. *J. Colloid Interface Sci.* **1989**, *127*, 189–204. [CrossRef]
66. Hołysz, L.; Szcześ, A. Determination of surface free energy components of organic liquids by the thin layer wicking method. *Ann. Univ. Mariae Curie-Sklodowska Sect. AA—Chem.* **2017**, *71*, 11. [CrossRef]
67. Swebocki, T.; Niedziałkowski, P.; Cirocka, A.; Szczepańska, E.; Ossowski, T.; Wcisło, A. In pursuit of key features for constructing electrochemical biosensors—Electrochemical and acid-base characteristic of self-assembled monolayers on gold. *Supramol. Chem.* **2020**, *32*, 256–266. [CrossRef]
68. Niedzialkowski, P.; Bogdanowicz, R.; Zięba, P.; Wysocka, J.; Ryl, J.; Sobaszek, M.; Ossowski, T. Melamine-modified Boron-doped Diamond towards Enhanced Detection of Adenine, Guanine and Caffeine. *Electroanalysis* **2015**, *28*, 211–221. [CrossRef]
69. Bogdanowicz, R.; Sawczak, M.; Niedzialkowski, P.; Zieba, P.; Finke, B.; Ryl, J.; Karczewski, J.; Ossowski, T. Novel Functionalization of Boron-Doped Diamond by Microwave Pulsed-Plasma Polymerized Allylamine Film. *J. Phys. Chem. C* **2014**, *118*, 8014–8025. [CrossRef]
70. Wang, S.; Swope, V.M.; Butler, J.E.; Feygelson, T.; Swain, G.M. The structural and electrochemical properties of boron-doped nanocrystalline diamond thin-film electrodes grown from Ar-rich and H2-rich source gases. *Diam. Relat. Mater.* **2009**, *18*, 669–677. [CrossRef]
71. Hupert, M.; Muck, A.; Wang, J.; Stotter, J.; Cvackova, Z.; Haymond, S.; Show, Y.; Swain, G.M. Conductive diamond thin-films in electrochemistry. *Diam. Relat. Mater.* **2003**, *12*, 1940–1949. [CrossRef]
72. Matveeva, E. Electrochemistry of the Indium-Tin Oxide Electrode in 1 M NaOH Electrolyte. *J. Electrochem. Soc.* **2005**, *152*, H138–H145. [CrossRef]
73. Benck, J.D.; Pinaud, B.A.; Gorlin, Y.; Jaramillo, T.F. Substrate Selection for Fundamental Studies of Electrocatalysts and Photoelectrodes: Inert Potential Windows in Acidic, Neutral, and Basic Electrolyte. *PLoS ONE* **2014**, *9*, e107942. [CrossRef] [PubMed]
74. Salinas-Torres, D.; Huerta, F.; Montilla, F.; Morallon, E. Study on electroactive and electrocatalytic surfaces of single walled carbon nanotube-modified electrodes. *Electrochim. Acta* **2011**, *56*, 2464–2470. [CrossRef]
75. Owens, D.K.; Wendt, R.C. Estimation of the surface free energy of polymers. *J. Appl. Polym. Sci.* **1969**, *13*, 1741–1747. [CrossRef]
76. Barthlott, W.; Neinhuis, C. Purity of the sacred lotus, or escape from contamination in biological surfaces. *Planta* **1997**, *202*, 1–8. [CrossRef]

Article

Controlled Silanization of Transparent Conductive Oxides as a Precursor of Molecular Recognition Systems

Anna Domaros *, Dorota Zarzeczańska, Tadeusz Ossowski and Anna Wcisło *

Department of Analytical Chemistry, Faculty of Chemistry, University of Gdansk, ul. Wita Stwosza 63, 80-308 Gdansk, Poland
* Correspondence: anna.domaros@ug.edu.pl (A.D.); anna.wcislo@ug.edu.pl (A.W.); Tel.: +48-58523-5106 (A.D.); +48-58523-5457 (A.W.)

Abstract: The search for new molecular recognition systems has become the goal of modern electrochemistry. Creating a matrix in which properties can be controlled to obtain a desired analytical signal is an essential part of creating such tools. The aim of this work was to modify the surface of electrodes based on transparent conductive oxides with the use of selected alkoxysilanes (3-aminopropyltrimethoxysilane, trimethoxy(propyl)silane, and trimethoxy(octyl)silane). Electrochemical impedance spectroscopy and cyclic voltammetry techniques, as well as contact angle measurements, were used to determine the properties of the obtained layers. Here, we prove that not only was the structure of alkoxysilanes taken into account but also the conditions of the modification process—reaction conditions (time and temperature), double alkoxysilane modification, and mono- and binary component modification. Our results enabled the identification of the parameters that are important to ensure the effectiveness of the modification process. Moreover, we confirmed that the selection of the correct alkoxysilane allows the surface properties of the electrode material to be controlled and, consequently, the charge transfer process at the electrode/solution interface, hence enabling the creation of selective molecular recognition systems.

Keywords: conductive materials; FTO electrodes; electrochemical measurements; silanization; electrode modification

Citation: Domaros, A.; Zarzeczańska, D.; Ossowski, T.; Wcisło, A. Controlled Silanization of Transparent Conductive Oxides as a Precursor of Molecular Recognition Systems. *Materials* **2023**, *16*, 309. https://doi.org/10.3390/ma16010309

Academic Editor: Mario Culebras Rubio

Received: 7 November 2022
Revised: 18 December 2022
Accepted: 26 December 2022
Published: 29 December 2022

Copyright: © 2022 by the authors. Licensee MDPI, Basel, Switzerland. This article is an open access article distributed under the terms and conditions of the Creative Commons Attribution (CC BY) license (https://creativecommons.org/licenses/by/4.0/).

1. Introduction

In recent years, materials chemistry has become one of the most rapidly developing fields of science. It mainly comprises the synthesis of new electrode materials or the modification of their surfaces to obtain specific properties and increase their applicability [1–8].

Currently, there is growing interest in scientific research focusing on semiconductor electrode materials based on oxide and carbon layers, which are characterized by various electrical properties [9–13]. However, the combination of two different properties characteristic of transparent conductive oxides, i.e., optical transparency and high electrical conductivity, makes this material extremely interesting for further research [14,15]. Fluorine-doped tin oxide (FTO) electrodes are considered a very promising material due to their greater stability in atmospheric conditions and higher temperature resistance than indium tin oxide (ITO) electrodes. Moreover, this material is chemically inert, mechanically resistant, and has high resistance to physical abrasion [16,17]. Both FTO and ITO electrodes find practical applications in a wide range of devices, among others, to provide transparent conductive coatings for use in touch panels, flat screens, aircraft cockpit windows, or plasma monitors. Thin oxide films are also used in the production of organic light-emitting diodes (OLEDs) and in solar cells [18,19]. In addition, due to their unique features, these materials provide a good basis for the modification of surface properties [16,20–24]. Moreover, they are also used in classical electroanalytical measurements as working electrodes, in the determination of a wide range of electroactive compounds [25].

A specific molecular recognition system can be synthesized by depositing systems on their surface with specific reducing and oxidizing properties (e.g., quinone derivatives), compounds showing the ability to form hydrogen bonds (e.g., edetic acid derivatives), coordination and donor systems (e.g., bipyridyl, tetraxetane), or biomolecules (e.g., proteins, enzymes). Thanks to these properties, FTO electrodes are used in biosensors, as well as in classical electroanalytical measurements concerning the determination of redox-active compounds [26–29]. Organosilicon monolayers bind most effectively to a surface with a silicon atom in its structure [30–32]. In the literature, one can find plenty of reports on the modification of ITO- and FTO-type glass electrodes with silane compounds. Thus, the electrode material also shows excellent silane-binding efficiency, due to the presence of surface inorganic oxides deposited in the easily oxidized form of a thin layer on the glass [20,21,23,33].

The covalent attachment of specific biomolecules to the surface of a conducting material is of great importance for the development of sensors based on molecular recognition. The functionalization of a surface by inclusion in the structure of a compound with specific properties requires the presence of an intermediate organic layer. In the case of oxide surfaces, such as an FTO electrode, organosilanes are the most common compounds and mediate in the anchoring of molecules to the substrate [9,34–38]. The silanization process is the first step toward the construction of a biosensor based on the FTO electrode. Reports available in the literature show a wide variety of conditions used at this stage, starting from the choice of the solvent and silane concentration to the temperature and time of electrode incubation in the silane solution [23,24,30,39]. However, the articles do not provide details as to how these specific conditions affect the properties of the obtained layers, and, thus, what parameters to choose to achieve the desired effect. Therefore, we decided to optimize the deposition conditions of the selected group of alkoxysilanes and their mixtures, paying particular attention to the settling time and the reaction temperature [40]. Focusing on the selection of appropriate conditions, our aim was to obtain the best possible repeatability of the silane deposition process for the FTO electrode surface. In addition, we verified the stability of the modified electrodes to determine their working time and, consequently, their susceptibility to further modification stages.

The aim of the presented work was the synthesis of organic films on conductive electrode materials and the characterization of their properties, which are important for electrochemical applications in the context of creating new molecular recognition systems.

Bearing in mind the set goal, we decided to compare the properties of the currently manufactured electrodes, with a particular emphasis on the electrochemical parameters and their modification potential. Based on previous research, we were able to select such a material—FTO—which met the expected requirements [40]. Subsequently, we modified the materials with alkoxysilanes to provide them with new properties. Modifying the electrode surfaces with chemicals can be a simple but efficient way to obtain a material with the specific properties desired for a given application. Thanks to the appropriate selection of modifiers and the skillful handling of the parameters of their combination, the scope of application of a given electrode is significantly extended. The possibilities are endless as one does not have to set the limits to just one modification stage. More and more often, extensive molecular recognition systems are being prepared in which specific functional groups that can act as an electrochemical probe are attached to linker compounds. The issue that should be taken into account, however, is the appropriate and reliable characterization of the obtained surface, so that it can serve as an innovative solution when researching problems in the future.

2. Materials and Methods

Preparation of the electrode surface: Cleaning the surface of the FTO electrodes began with sonication of the plates in three successive solvents: in acetone and ethyl alcohol for 5 min and in distilled water for 10 min. The electrodes were then dried at 70 °C for 60 min and immersed in an "alkaline piranha" solution for about two hours. The temperature was maintained at 60 °C during the process. The aim of this treatment was to activate the electrode surface by oxidizing the groups on its surface to silanol groups. Subsequently, the electrodes were rinsed with distilled water and placed back in the oven at 70 °C for about 30 min [41,42].

2.1. Modification Procedure

Silanization with alkoxysilanes: FTO electrodes with cleaned and activated surfaces were immersed in 2.5% (m/m) alkoxysilane solutions, which were prepared immediately before this step in 96% ethyl alcohol. Table 1, below, shows the names and formulas of the silanes used for modification.

Table 1. Silanes used in the surface modification process.

Silane	Acronym	Molecular Structure
3-aminopropyltrimethoxysilane	APTMS	(structure)
Trimethoxy(propyl)silane	PTMS	(structure)
trimethoxy(octyl)silane	OTMS	(structure)

The silanizing solutions consisted of one or a mixture of two silanes. The modifications were carried out under various process conditions. The times of electrode incubation in solutions and the temperature at which the silanization reaction was carried out varied. The experimental conditions are shown in Table 2. After the specified time had elapsed (16 or 72 h), the electrodes were rinsed several times with ethanol to remove the excess silanes. The modified electrodes were then heated in an oven at 120 °C for 2 h.

Table 2. Conditions for the silanization process on the surface of the oxidized FTO electrode.

Silane	Reaction Conditions (Temperature, Time)	Electrode	Second Reaction Conditions (Temperature, Time)	Electrode
APTMS	23 °C; 16 h	FTO/APTMS (16 h)	23 °C; 16 h	FTO/APTMS/OTMS (16 h)
	23 °C; 72 h	FTO/APTMS (72 h)	23 °C; 72 h	FTO/APTMS/OTMS (72 h)
	50 °C; 0.5 h	FTO/APTMS (0.5 h)		
PTMS	50 °C; 0.5 h	FTO/PTMS (0.5 h)		
OTMS	23 °C; 16 h	FTO/OTMS (16 h)	23 °C; 16 h	FTO/OTMS/APTMS (16 h)
	23 °C; 72 h	FTO/OTMS (72 h)	23 °C; 72 h	FTO/OTMS/APTMS (72 h)
	50 °C; 0.5 h	FTO/OTMS (0.5 h)		
APTMS and OTMS	23 °C; 16 h	FTO/APTMS_OTMS (16 h)		
	23 °C; 72 h	FTO/APTMS_OTMS (72 h)		

The modified FTO electrodes listed in Table 2 were further silanized (in a second reaction) according to the same procedure, to yield four new electrodes.

Three alkoxysilanes were used to functionalize the FTO substrates: 3-aminopropyltrimethoxysilane (APTMS), trimethoxypropylsilane (PTMS), and trimethoxyoctylsilane (OTMS), differing in the structure of the side chain (Table 1, Figure 1). The electrochemical properties of the modified FTO/APTMS, FTO/OTMS, and FTO/APTMS_OTMS electrodes were compared to those of the unmodified FTO electrode. Additionally, we assessed how the presence of the amino group in the side chain of the alkoxysilane (FTO/APTMS) changed the surface properties of the electrode, compared to its alkyl derivative (FTO/PTMS). We verified whether the long alkyl chain in the trimethoxysilane structure (FTO/OTMS) hinders the transfer of electrons at the electrode/solution interface and whether the deposition of a mixture of silanes (FTO/APTMS_OTMS) on the surface of the FTO electrode would show better conductive properties than their one-component counterparts (FTO/APTMS and FTO/OTMS). In addition, we examined whether the order of silane deposition is important in the case of a two-step silanization of the FTO electrode surface (e.g., FTO/APTMS/OTMS and FTO/OTMS/APTMS). We also assessed the changes in the surface wettability on the FTO/APTMS, FTO/PTMS, and FTO/OTMS electrodes.

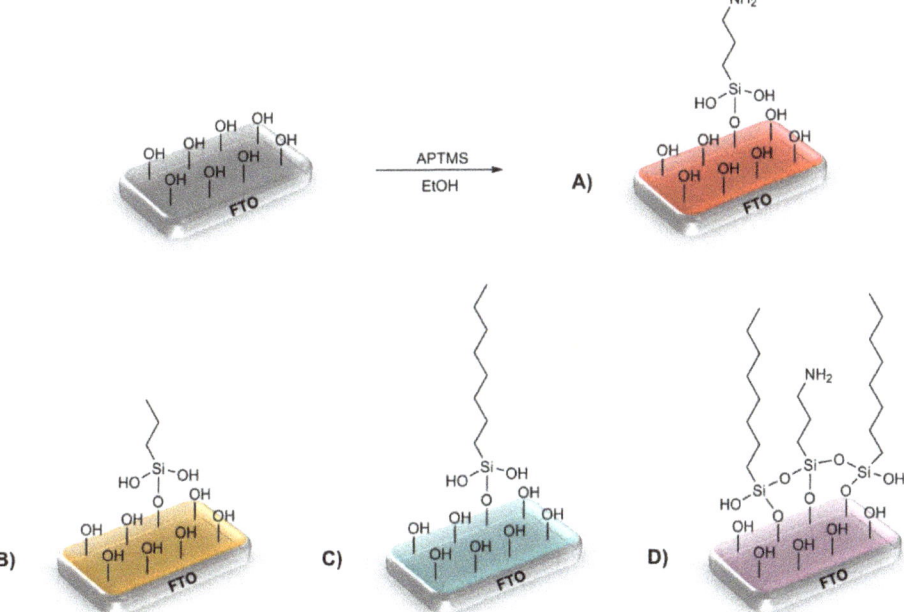

Figure 1. Schematic diagram of the silanization process, with an example of the surface structure of the modified electrodes—FTO/APTMS (**A**), FTO/PTMS (**B**), FTO/OTMS (**C**), and FTO/ATMS_OTMS (**D**).

2.2. Electrochemical Measurements

For the electrochemical evaluation of the modified FTO electrode surfaces, we employed impedance spectroscopy and cyclic voltammetry, which are suitable for the study and monitoring of the changes taking place on the electrode surfaces as a result of the performed silanization reactions. Using the electrochemical impedance spectroscopy (EIS) technique enabled the determination of the charge transfer resistance (Rct), which is a measure of the resistance of self-organizing silane layers to electron transfer, from the solution to the electrode surface. Recorded impedance spectra are displayed in the form of Nyquist diagrams.

The electrochemical measurements were carried out using an apparatus consisting of a Multi Autolab/M204 potentiostat, equipped with an FRA module (Metrohm, Barendrecht, The Netherlands) and a glass measuring cell. The measurements were performed in a Faraday cage, in order to isolate the system from the influence of external factors in a three-electrode measurement system, consisting of a working electrode (unmodified and modified FTO, dimensions 1.5 × 3.5 cm) (Sigma Aldrich, Schnelldorf, Germany), silver chloride reference electrode (Ag/AgCl, 0.1 M KCl) (Mineral, Warsaw, Poland), and platinum wire counter-electrodes (Mennica-Metale Sp. z.o.o., Radzymin, Poland).

Nova 2.11 software was used to operate the potentiostat, as well as to analyze and simulate the electrochemical impedance spectroscopy (EIS) spectra (Metrohm AG, Herisau, Switzerland). OriginPro 2020 software (OriginLab, Northampton, MA, USA) was used for data analysis.

Cyclic voltammetry measurements were performed for a scanning speed of V = 100 mV/s. The range of measurements (initial and final measurements, as well as peak potentials) was adjusted for the best visualization of the processes taking place on the electrode surface. Measurements of EIS were performed in the frequency range from 10 kHz to 100 mHz, with a wave amplitude of 0.01 V and a sinusoidal excitation signal.

Two types of basic electrolytes were used in the electrochemical measurements—0.5 M potassium chloride and 0.5 M sodium sulfate solutions. The model redox system used in the research was potassium hexacyanoferrate (III/II). All solutions were prepared from weights by dissolving them in distilled water in such amounts as to ultimately obtain solutions with a concentration of 0.01 M. The prepared solutions were stored in a refrigerator; however, before the tests, they were brought back to ambient temperature [42].

2.3. Contact Angle Measurements

The wettability of the surface of the electrode materials was determined by measuring the contact angle at room temperature, using standard liquids based on the sitting drop static method. The reported contact angle values are average values, measured at different positions on the electrode surface. Drop shape analysis (with a volume of 2 µL or 4 µL) was performed using the circle method and the Young–Laplace method [1,5,40,42–45]. Measurements were made using the KRÜSS Drop Shape Analyzer—DSA100 apparatus.

3. Results and Discussion

3.1. Alkoxysilane Modification

The impedance spectrum for the unmodified FTO electrode is characterized by a semicircle in the high-frequency range, while in the low-frequency range, it is characterized by a straight line inclined at an angle of 45° (Figure 2A) [46]. This is a classic example of a spectrum, showing that the electron transfer process for the redox $[Fe(CN)_6]^{3-/4-}$ pair is quasi-reversible and diffusion-controlled [47]. The measurement of the charge transfer resistance, Rct, is the diameter of the semicircle obtained in the Nyquist plots. The larger the diameter, the greater the charge transfer resistance. As the value of Rct changes, the capacity of the double layer (Q) also changes. Low Q values indicate an increase in material thickness, due to the surface modification. The n parameter, which defines the degree of heterogeneity in the newly formed structure, is inherent in the capacity of the double layer. In the case of an electrode with a perfectly smooth surface, this equals 1. The Rct value, calculated on the basis of the equivalent circuit used, R(Q(RW)), for the FTO electrode is approximately 58 Ω (Table 3). Anchoring on the surface of a self-assembled layer with an amine group (FTO/APTMS (16 h)) reduces the Rct value by 10 Ω (Figure 2A). As expected, increasing the APTMS deposition time from 16 to 72 h still lowers the surface resistance to 30 Ω. As a result, after 72 h of APTMS deposition, the Rct value was almost doubled (from 58 Ω to 30 Ω) compared to the Rct value for the unmodified FTO electrode. Moreover, we observe the changes in the shape of the Nyquist plot. The diameter of the semicircle decreased, and the linear section lengthened in the low-frequency range. This means that the electron transfer process is faster than the diffusion of the redox system, close to the electrode surface. This effect appears to be due to the increased amount of amine groups on the electrode material, as well as the greater order of the layer. Moreover, as the deposition time lengthened, the capacity of the double layer increased, while the value of the n parameter decreased, which proves the heterogeneity of the FTO/APTMS surface.

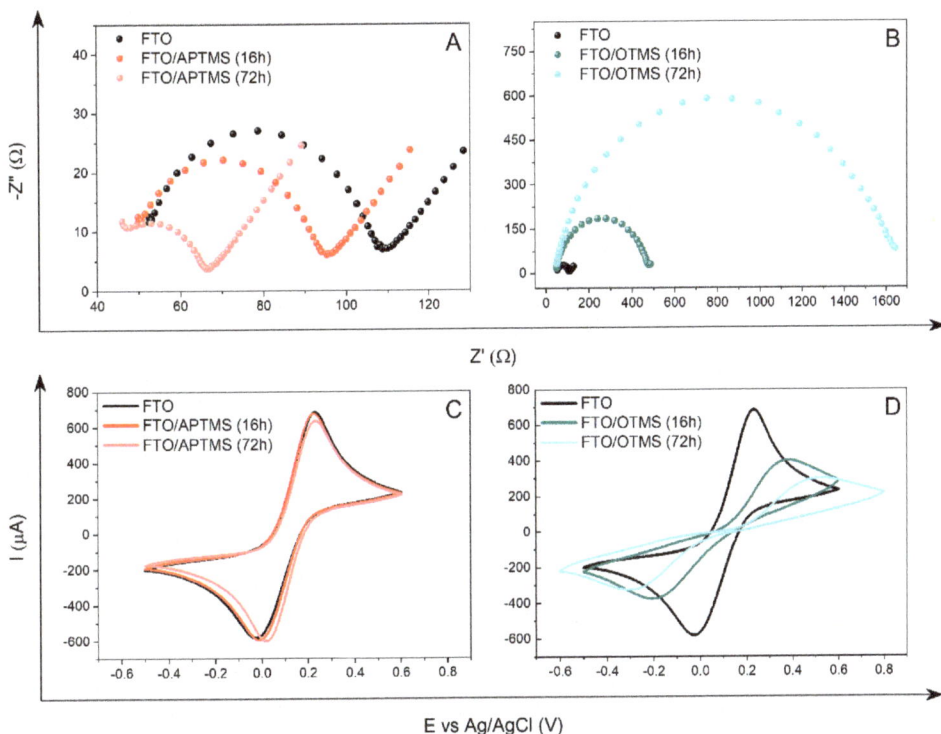

Figure 2. Nyquist plots (**A**,**B**) and voltammograms (**C**,**D**) showing the modification of the FTO electrode with APTMS (**A**,**C**) and OTMS (**B**,**D**) after different silanization times (16 h, 72 h), recorded in 0.5 M KCl (EIS) and Na$_2$SO$_4$ (CV) containing a 5 mM redox system [Fe(CN)$_6$]$^{3-/4-}$.

Table 3. Electrochemical parameters of FTO electrodes after modification with alkoxysilanes, obtained on the basis of CV and EIS measurements by fitting the experimental data with the equivalent R(Q(RW)) or R(QR) circuits.

Electrode	R_{ct} [Ω]	Q [μF]	n	W [μSs^{½}]	Chi^2	E_a [V]	E_k [V]	ΔE [V]	I_a [μA]	I_k [μA]
FTO	57.87	6.65	0.91	0.012	3.27 × 10^{-4}	0.233	−0.026	0.259	683.3	−583.2
FTO/APTMS (16 h) *	48.46	8.28	0.88	0.012	6.19 × 10^{-4}	0.226	−0.012	0.238	676.3	−594.2
FTO/APTMS (72 h) *	30.74	11.23	0.79	0.012	2.69 × 10^{-4}	0.233	0.016	0.217	632.3	−600.9
FTO/OTMS (16 h) *	426.40	5.69	0.89	0.015	7.31 × 10^{-4}	0.384	−0.209	0.593	399.2	−376.9
FTO/OTMS (72 h) *	1545.00	3.52	0.84	0.004	3.47 × 10^{-4}	0.521	−0.311	0.832	298.5	−326.8
FTO/APTMS_OTMS (16 h) *	55.05	8.03	0.88	0.012	5.85 × 10^{-4}	0.244	−0.016	0.260	609.1	−591.4
FTO/APTMS_OTMS (72 h) **	22,450.00	0.25	0.86	-	8.04 × 10^{-4}	0.676	−0.514	1.190	180.3	−195.1
FTO/APTMS/OTMS (16 h) **	336.20	6.35	0.87	-	1.05 × 10^{-3}	0.346	−0.167	0.513	478.2	−429.3
FTO/APTMS/OTMS (72 h) *	1664.00	1.36	0.89	0.007	2.27 × 10^{-4}	0.546	−0.423	0.969	262.6	−288.1
FTO/OTMS/APTMS (16 h) *	439.30	4.39	0.89	0.009	4.91 × 10^{-4}	0.388	−0.219	0.607	403.4	−406.6
FTO/OTMS/APTMS (72 h) *	98.19	6.86	0.88	0.010	8.54 × 10^{-4}	0.293	−0.047	0.340	506.6	−433.5

* R(Q(RW)); ** R(QR); ΔE = $E_a - E_k$.

In turn, the electrodes covered with the OTMS layer show completely different properties (Figure 2B). For the FTO/OTMS modification (16 h), we observed a curve in the Nyquist plot consisting of a large semicircle and a small linear range at low frequencies. This spectral image illustrates a process limited by the rate of electron transfer through the double layer. As a result, diffusion is more efficient than electron transport. This is because the OTMS layer blocks the access of the $[Fe(CN)_6]^{3-/4-}$ system from the solution to the electrode surface. However, the shape of the spectrum suggests the presence of defects in the layer that is formed, which is reflected in an increase in the Rct value. The blocking degree is higher with the FTO/OTMS electrode (72 h). Due to the extension of the deposition time, for this modification, the impedance spectrum consists only of a semicircle in the tested frequency range. This demonstrates an even slower charge transfer reaction, due to the formation of a long aliphatic chain layer. The presence of very large semicircles confirms the excellent electrochemical blocking ability of the OTMS layers. The Rct values for the FTO/OTMS electrodes are much higher than for the FTO/APTMS electrodes. After the first modification (FTO/OTMS (16 h)), the electron transfer resistance increased from 58 Ω to approximately 426 Ω. Conversely, after 72 h of deposition (FTO/OTMS (72 h)), the resistance increased up to 1.5 kΩ. A higher Rct value implies a lower double-layer capacity, as evidenced by the data in Table 3.

It is worth emphasizing that the impedance results are compatible with the measurements obtained for cyclic voltammetry. In CV measurements, similarly to EIS measurements, different electrochemical properties of the modified FTO/APTMS and FTO/OTMS electrodes are observed (Figure 2C,D). In the case of the FTO/APTMS modification, the cathode peak shifts slightly towards the positive values of the potential, thus increasing the reversibility of the redox process of the $[Fe(CN)_6]^{3-/4-}$ system. In total, the ΔE value decreased by 40 mV (Table 3), which confirms that the presence of amino groups on the FTO electrode surface improves the electron transfer process.

Conversely, the deposition of the OTMS layer shows a blocking effect for the same process. After 16 h, the ΔE value increased from 259 to 593 mV, indicating the reduced reversibility of the test system. At the same time, a significant decrease in the intensity of the oxidation current and reduction peaks was noted, by 284 µA and 206 µA, respectively, compared to the unmodified FTO electrode. The further process of surface silanization with OTMS increases the peak separation up to 832 mV, with an approximately 50% reduction in the current response intensity (Figure 2D). This clearly shows that this redox reaction at the FTO/OTMS electrode (72 h) was blocked due to the formation of highly ordered and well-packed self-assembled OTMS layers with low defect density. The moderate blocking effect for the shorter deposition time can be attributed to the formation of a layer with more holes and defects, allowing electron tunneling. The recorded changes confirm that the extent of blocking is different for each modification, which depends primarily on the nature and the method of layer formation, and, thus, on the chemical structure formed on the electrode surface.

3.2. Double-Alkoksysilane Modification

The differences observed for the APTMS- and OTMS-modified electrodes prompted us to synthesize the systems resulting from the action of a silane solution with two alkoxysilanes at the same time. The analysis of the impedance spectra shows that the deposition time is the decisive factor in the final surface properties of the modified FTO/APTMS_OTMS electrode. The Nyquist plot shows two different curves, depending on the duration of the FTO electrode silanization process (Figure 3A). The tendency of the changes is completely different compared to the previously described FTO/APTMS and FTO/OTMS modifications. After 16 h of deposition, the shape of the impedance spectrum corresponds to the characteristics of the FTO/APTMS electrodes, while after 72 h, it is similar to the FTO/OTMS electrodes. These changes are reflected in the Rct values. Initially, the resistance slightly decreases to 55 Ω and then increases sharply after the next modification, up to 22.5 kΩ (Table 3). In the case of the modification of FTO/APTMS_OTMS (72 h), due to the shape of the Nyquist plot, a different model of the equivalent circuit, R(QR), was used, disregarding the Warburg impedance representing the diffusion element. In the case of the FTO/APTMS_OTMS (16 h) electrode, the Q value is close to the electrical capacity of the output electrode. The parameters Rct and Q suggest that the cross-linking of the material takes place in a shorter time than in the longer layer-formation process. Thus, extending the modification time may lead to secondary changes on the electrode surface. Again, the results obtained from the EIS measurements correlate with the CV results (Figure 3B). The difference in the potential of the oxidation and reduction peaks of the system $[Fe(CN)_6]^{3-/4-}$ for the FTO/APTMS_OTMS (72 h) modification is 1.19 V, where in the case of a shorter deposition time, the value of ΔE oscillates around 0.26 V. Together with an increase in the separation of peaks, there is also a dramatic decrease in the intensities of the current response. The intensity of the current for the anode and cathode response for the FTO/APTMS_OTMS electrode (72 h) is three times lower than for the FTO/APTMS_OTMS (16 h) or FTO (Table 3). Hence, the longer deposition of the APTMS and OTMS mixture causes the reorganization of the ordered structure, leading to the formation of a steric hindrance for the access of $[Fe(CN)_6]^{3-/4-}$ ions to the electrode surface.

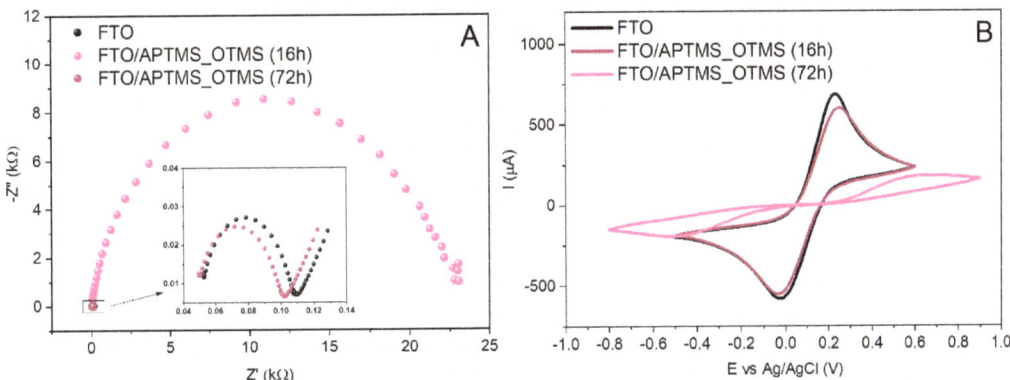

Figure 3. Graph of the Nyquist (**A**) and cyclic voltammetry (**B**) values obtained for the modified FTO/APTMS_OTMS electrodes after different deposition times (16 h, 72 h), recorded in a 0.5 M aqueous solution of KCl (EIS) and Na_2SO_4 (CV), containing a 5 mM redox system $[Fe(CN)_6]^{3-/4-}$.

The specific electrochemical behavior of the FTO/APTMS_OTMS electrodes prompted us to change the conditions of the experiment. The mixed silane layer has re-formed, but this is now as a result of the two-step surface modification of the FTO electrode (Figure 4).

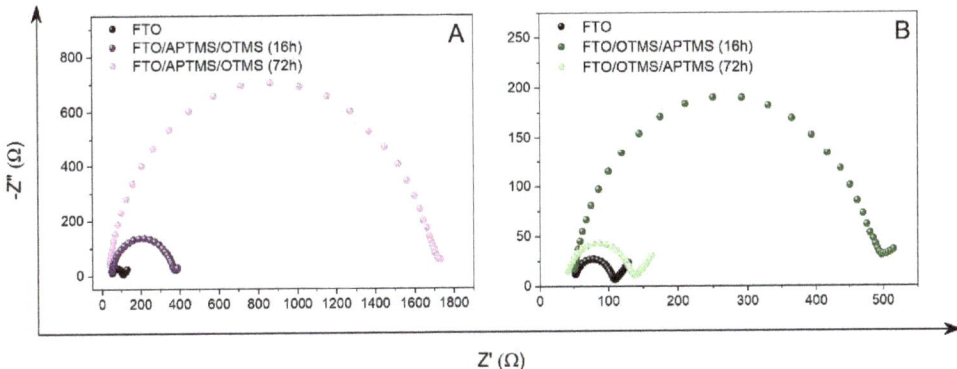

Figure 4. *Cont.*

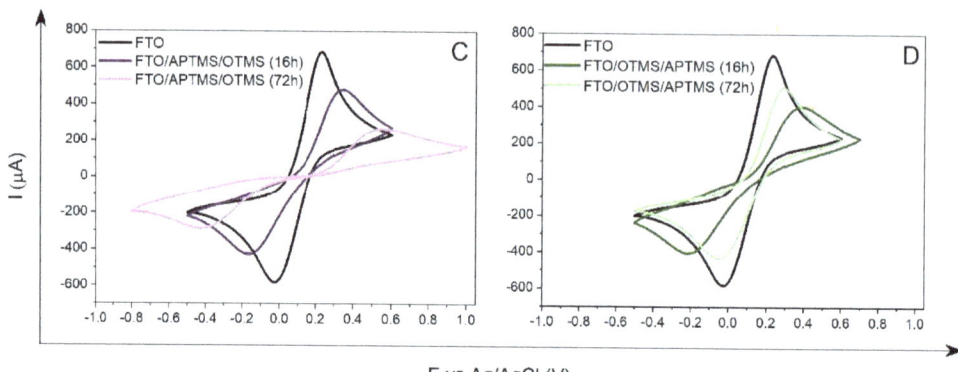

Figure 4. Graphs of the Nyquist (**A,B**) and cyclic voltammetry (**C,D**) values obtained for modified electrodes as a result of two-stage FTO/APTMS/OTMS (**A,C**) and FTO/OTMS/APTMS (**B,D**) silanization, after different time depositions (16 h, 72 h), recorded in a 0.5 M aqueous solution of KCl (EIS) and Na_2SO_4 (CV), containing a 5 mM redox system $[Fe(CN)_6]^{3-/4-}$.

3.3. Two-Step Mixed Alkoksysilane Modification

APTMS and OTMS monosubstrate-modifying solutions were used for the silanization process. We obtained two new electrodes—FTO/APTMS/OTMS and FTO/OTMS/APTMS —using a different order of alkoxysilane deposition. Based on our expectations, the materials modified in this way should show similar conductive properties to the FTO/APTMS_OTMS electrode. It would seem that, regardless of the order of deposited alkoxysilanes, the obtained layers will have an almost identical surface structure and, consequently, the resistance to charge transfer at the electrode/solution interface will be at the same level. The obtained results undoubtedly contradict the expected results. Regardless of the duration of the silanization process, the electrochemical properties of the resulting layers depend essentially on the second deposition step. Despite the fact that a mixed layer of silanes is formed on the surface of the FTO electrode, we observed the dominance of one of the silanes in the impact on the new structural features. In this respect, this effect is comparable to the electrodes obtained as a result of the action of the alkoxysilane mixture (FTO/APTMS_OTMS), where the deposition time was decisive for the predominant proportion of one of the components. On the other hand, for the electrodes obtained at certain stages (FTO/APTMS/OTMS and FTO/OTMS/APTMS), this fact was determined by the substrate participating in the second stage of silanization, as can be concluded based on the electrochemical parameters of the monosubstrate electrodes (FTO/APTMS and FTO/OTMS). Depending on which of the silanes was applied last (APTMS or OTMS), the values of Rct, Q, and ΔE are, respectively, lower or higher than for the one-component modifications (FTO/APTMS or FTO/OTMS) (Table 3). Thus, in the case of the FTO/APTMS/OTMS (16 h) modification, a greater share of the silane layer is attributed to long octyl chains (Figure 4A). The Rct value for this surface is 336.2 Ω, which is only 90.2 Ω lower than the monosubstrate derivative, FTO/OTMS (16 h). This change is influenced by the presence of aminopropyl groups on the surface, which lowers the charge transfer resistances while increasing the capacity of the double layer. APTMS seems to be responsible for organizing the newly formed structure on the surface of the FTO electrode, marking the places where OTMS can be deposited, which determines the lipophilic nature of the surface. For the second modification of FTO/OTMS/APTMS (16 h), the effect of double silanization is different (Figure 4B). Despite the two-stage process, the conductive properties of this electrode are comparable to the FTO/OTMS electrode (16 h). The resistance value is about 440 Ω, which confirms the similarity to the monosubstrate derivative (FTO/OTMS (16 h)). However, the Q value indicates that APTMS contributes, to some degree, to forming this structure, although this is not evident in the Rct value.

It is probable that a well-organized OTMS layer blocks access to the free places on the electrode surface. The APTMS molecules are not able to overcome this steric hindrance within 16 h and settle on the surface, between the OTMS chains. Nevertheless, the analysis of impedance spectra confirms this thesis because a longer silanization time significantly reduces the surface resistance by about 340 Ω, which proves the reorganization of the structure. Depositing the APTMS particles on the electrode surface increases their share in the newly formed structure, thus lowering the charge transfer resistance. The effect of the denser packing of silanes on the FTO/OTMS/APTMS electrode surface (72 h) confirms the lower capacity of the double layer, compared to the Q value, after 16 h of deposition. In contrast to the FTO/APTMS/OTMS electrode (72 h), we observed an increase in the charge transfer resistance for the model system $[Fe(CN)_6]^{3-/4-}$ from 336 to 1664 Ω and a decrease in the double layer capacitance, which confirms the increasing share of OTMS in blocking the surface. Undoubtedly, the impedance results correlate with the CV measurements (Figure 4C,D). The ΔE value represents the above-described electrochemical properties of the obtained electrodes. For the FTO/APTMS/OTMS electrode, the ΔE value almost doubled with increasing deposition time, while for the FTO/OTMS/APTMS modification, we observed a twofold decrease in the ΔE value (Table 3).

Apart from the deposition time of alkoxysilanes, the temperature is undoubtedly another factor influencing the effectiveness of the silanization process. It has been proven that by increasing the reaction temperature, we can simultaneously reduce the deposition time. APTMS and OTMS monosubstrate silanizing solutions were again selected for modification, under changed process conditions. By increasing the temperature to 50 °C and reducing the deposition time to 30 min, two new electrodes were obtained, namely, FTO/APTMS (0.5 h) and FTO/OTMS (0.5 h). The EIS and CV measurements recorded for the new layers were analogous to those obtained after the longer deposition times of APTMS and OTMS (72 h) at room temperature. This proves that the electrochemical properties of the newly formed structures are comparable with those of the FTO/APTMS (72 h) and FTO/OTMS (72 h) electrodes. However, the silanization process in this case is more difficult to control. By simultaneously selecting two variable deposition parameters (time and temperature), from which the resulting structure is conditioned, we do not have the guarantee of obtaining the same defined electrode surface each time. Therefore, we decided to keep the modification at room temperature by changing only the deposition time of the silanes.

The tests showed that the silanized electrode materials were characterized by different stability levels over time. The electrochemical parameters that we took into account were the surface resistance and the reversibility of the redox process. The example of the FTO/APTMS electrode (Figure 5A,B) shows that 6 days after modification, the surface resistance increases significantly, as well as the separation between the oxidation and reduction peaks. In the case of electrodes where OTMS dominates in the structure (Figure 5C,D), this effect could already be noticed on the second day. Due to the fact that the observed changes were significant, we did not set any threshold values, but all tests for the newly formed layers, as well as the subsequent stages of synthesis on the surface, were performed immediately after the deposition process, meaning that the measurement conditions were identical for each electrode.

3.4. Wettability

For further characterization, we investigated the wettability of the electrodes modified with alkoxysilanes used for electrochemical tests. The FTO electrodes that were modified with APTMS, PTMS (aliphatic APTMS), and OTMS were measured by means of the contact angle measurements, as shown in Figure 6.

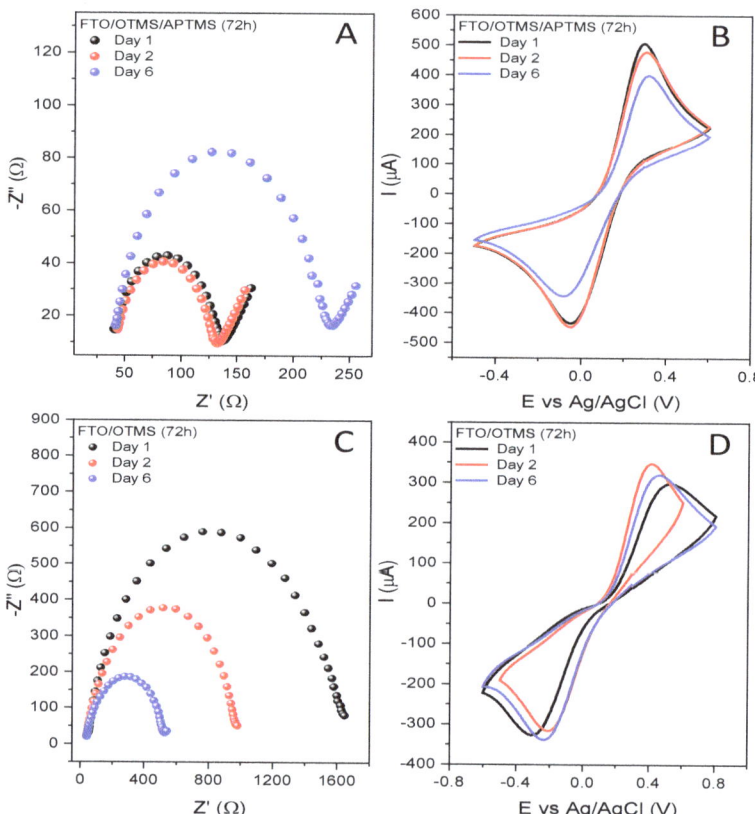

Figure 5. Graphs of Nyquist (**A**,**C**) and cyclic voltammetry (**B**,**D**) obtained for modified electrodes as a result of two-stage FTO/OTMS/APTMS (**A**,**B**) silanization and one-stage FTO/OTMS (**C**,**D**) after different time of storage (1–6 days), recorded in a 0.5 M aqueous solution of KCl (EIS) and Na$_2$SO$_4$ (CV) containing a 5 mM redox system [Fe(CN)$_6$]$^{3-/4-}$.

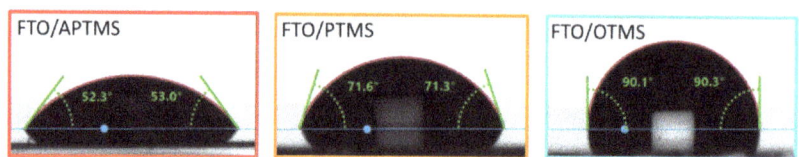

Figure 6. Pictures of the water contact angle (WCA) measurements of modified FTO/APTMS, FTO/PTMS, and FTO/OTMS electrodes.

The contact angle (WCA) value for the unmodified FTO electrode is approximately 56°. The hydroxyl groups formed as a result of activation (oxidation) of the substrate are responsible for the hydrophilic nature of the surface. Based on the FTO/APTMS, FTO/PTMS, and FTO/OTMS modifications, we observed that the contact angle changed depending on the structure of the formed silane layer. For the FTO/APTMS electrode, the angle value decreased to 53°, but for the FTO/PTMS and FTO/OTMS electrodes, the value increased significantly to 71° and 90°, respectively. The FTO/OTMS surface shows a higher WCA value than the FTO/PTMS, due to the presence of long aliphatic chains in the structure. On the surface, highly ordered self-assembling OTMS layers with low defect density are formed, giving the electrode a hydrophobic character. Undoubtedly, the FTO/PTMS electrode also has more hydrophobic properties than the FTO/APTMS. In turn, the presence of an amino group in the side chain of the alkoxysilane (FTO/APTMS) completely changes the surface properties of the electrode. Compared to its alkyl derivative (FTO/PTMS), it has a lower angle value, which proves the good wettability of the FTO/APTMS electrode surface. The obtained values of the contact angles are close to the values reported in the literature, even in relation to the alkoxysilane layers formed on other electrode substrates [24,48,49]. However, it should be remembered that the wettability of the surface strongly depends on the nature of the formed silane layer, and, thus, on the conditions in which it was formed, which may cause discrepancies in the given angle values [50].

4. Conclusions

Based on the EIS and CV measurements, we assessed the operating time of the modified FTO electrodes, from which one can conclude that the electrochemical parameters of the new materials change over time. The FTO/APTMS and FTO/OTMS/APTMS electrodes show the highest reproducibility of results. Regardless of the deposition time, these electrodes are electrochemically stable for about two days, which suggests that the dominant share of APTMS in the newly formed layers extends the life of the electrode. In the case of other electrodes where the structure is dominated by OTMS, it is necessary to immediately perform planned measurements or the subsequent modification stages immediately after silanization, because their electrochemical parameters change over time. In turn, the stability of the FTO/APTMS_OTMS electrode depends upon the timeframe of silane deposition. A greater dispersion of the measurement results was noted for a longer modification time. In contrast, the FTO/APTMS_OTMS electrode surface (16 h) is reactive for two days. Slight differences in the stability of the resulting structures suggest that the silanized FTO electrodes should not be stored. All tests for the newly formed layers, as well as the subsequent stages of synthesis on the surface, should be performed immediately after the deposition process [51].

The discussion of the results found in the course of the conducted research allows for the conclusion that deposition time plays a key role in the formation of self-organizing silane layers. A longer silanization process leads to a reorganization of the structure. Consequently, the newly formed structure causes the layer to have new properties that differ from the intermediate and initial properties. Apart from the interaction time of the silanizing solution with the surface of the FTO electrode, it is also important whether the solution is monocomponent or binary. In the case of two-step modification with alkoxysilanes, the electrochemical properties of the modified surfaces depend on the second deposition step. Comparing the values of Rct and Q, we can state that the best predisposition to create an ordered SAM structure was shown by the APTMS one-component solution. In turn, the FTO/APTMS_OTMS electrode (72 h) (Figure 7) showed the greatest blocking effect for the electrochemical process of the redox system $[Fe(CN)_6]^{3-/4-}$. Taking into account the shorter time of the silanization process, the charge transfer resistance at the electrode/solution interface increased in the order:

Figure 7. Schematic presentation of changes in charge transfer resistance for the investigated electrode materials.

However, after 72 h of deposition, the direction of the changes was exactly the opposite. Moreover, the modification of the FTO electrode surface, using selected self-assembled layers of organosilane molecules, enables the fine-tuning of their hydrophobic/hydrophilic properties by controlling the side chain length and the nature of the molecule.

This research confirms the high efficiency of the silane groups' deposition on FTO glass substrates. By selecting the appropriate alkoxysilanes, one can control the surface properties of the electrode materials and, thus, regulate the electron transfer process at the electrode/solution interface.

Author Contributions: Conceptualization, A.D. and A.W.; methodology, A.D.; investigation, A.D. and A.W.; writing—original draft preparation, A.D. and A.W.; writing—review and editing, A.D., D.Z. and A.W.; visualization, A.D. and A.W.; resources, T.O.; supervision, D.Z., T.O. and A.W. All authors have read and agreed to the published version of the manuscript.

Funding: This research received no external funding.

Institutional Review Board Statement: Not applicable.

Informed Consent Statement: Not applicable.

Data Availability Statement: The data presented in this study are available on request from the corresponding author.

Conflicts of Interest: The authors declare no conflict of interest.

References

1. Swebocki, T.; Niedziałkowski, P.; Cirocka, A.; Szczepańska, E.; Ossowski, T.; Wcisło, A. In Pursuit of Key Features for Constructing Electrochemical Biosensors—Electrochemical and Acid-Base Characteristic of Self-Assembled Monolayers on Gold. *Supramol. Chem.* **2020**, *32*, 256–266. [CrossRef]
2. Prodromidis, M.I. Impedimetric Immunosensors—A Review. *Electrochim. Acta* **2010**, *55*, 4227–4233. [CrossRef]
3. Ho, J.A.; Hsu, W.-L.; Liao, W.-C.; Chiu, J.-K.; Chen, M.-L.; Chang, H.-C.; Li, C.-C. Ultrasensitive Electrochemical Detection of Biotin Using Electrically Addressable Site-Oriented Antibody Immobilization Approach via Aminophenyl Boronic Acid. *Biosens. Bioelectron.* **2010**, *26*, 1021–1027. [CrossRef]
4. Niedzialkowski, P.; Slepski, P.; Wysocka, J.; Chamier-Cieminska, J.; Burczyk, L.; Sobaszek, M.; Wcislo, A.; Ossowski, T.; Bogdanowicz, R.; Ryl, J. Multisine Impedimetric Probing of Biocatalytic Reactions for Label-Free Detection of DEFB1 Gene: How to Verify That Your Dog Is Not Human? *Sens. Actuators B Chem.* **2020**, *323*, 128664. [CrossRef]
5. Niedziałkowski, P.; Bojko, M.; Ryl, J.; Wcisło, A.; Spodzieja, M.; Magiera-Mularz, K.; Guzik, K.; Dubin, G.; Holak, T.A.; Ossowski, T.; et al. Ultrasensitive Electrochemical Determination of the Cancer Biomarker Protein SPD-L1 Based on a BMS-8-Modified Gold Electrode. *Bioelectrochemistry* **2021**, *139*, 107742. [CrossRef] [PubMed]
6. Lin, D.; Tang, T.; Jed Harrison, D.; Lee, W.E.; Jemere, A.B. A Regenerating Ultrasensitive Electrochemical Impedance Immunosensor for the Detection of Adenovirus. *Biosens. Bioelectron.* **2015**, *68*, 129–134. [CrossRef] [PubMed]
7. Duffy, G.F.; Moore, E.J. Electrochemical Immunosensors for Food Analysis: A Review of Recent Developments. *Anal. Lett.* **2017**, *50*, 1–32. [CrossRef]
8. Jayanthi, V.S.P.K.S.A.; Das, A.B.; Saxena, U. Recent Advances in Biosensor Development for the Detection of Cancer Biomarkers. *Biosens. Bioelectron.* **2017**, *91*, 15–23. [CrossRef]
9. Gabriunaite, I.; Valiūnienė, A.; Poderyte, M.; Ramanavicius, A. Silane-Based Self-Assembled Monolayer Deposited on Fluorine Doped Tin Oxide as Model System for Pharmaceutical and Biomedical Analysis. *J. Pharm. Biomed. Anal.* **2020**, *177*, 112832. [CrossRef]
10. Miyata, T.; Hikosaka, T.; Minami, T. High Sensitivity Chlorine Gas Sensors Using Multicomponent Transparent Conducting Oxide Thin Films. *Sens. Actuators B Chem.* **2000**, *69*, 16–21. [CrossRef]
11. Minami, T. Chapter Five—Transparent Conductive Oxides for Transparent Electrode Applications. In *Semiconductors and Semimetals*; Svensson, B.G., Pearton, S.J., Jagadish, C., Eds.; Oxide Semiconductors; Elsevier: Amsterdam, The Netherlands, 2013; Volume 88, pp. 159–200.
12. Shim, Y.-S.; Moon, H.G.; Kim, D.H.; Jang, H.W.; Kang, C.-Y.; Yoon, Y.S.; Yoon, S.-J. Transparent Conducting Oxide Electrodes for Novel Metal Oxide Gas Sensors. *Sens. Actuators B Chem.* **2011**, *160*, 357–363. [CrossRef]
13. Xu, J.; Wang, Y.; Hu, S. Nanocomposites of Graphene and Graphene Oxides: Synthesis, Molecular Functionalization and Application in Electrochemical Sensors and Biosensors. A Review. *Microchim Acta* **2017**, *184*, 1–44. [CrossRef]
14. Kim, C.-L.; Jung, C.-W.; Oh, Y.-J.; Kim, D.-E. A Highly Flexible Transparent Conductive Electrode Based on Nanomaterials. *NPG Asia Mater.* **2017**, *9*, e438. [CrossRef]
15. Takamatsu, S.; Takahata, T.; Muraki, M.; Iwase, E.; Matsumoto, K.; Shimoyama, I. Transparent Conductive-Polymer Strain Sensors for Touch Input Sheets of Flexible Displays. *J. Micromech. Microeng.* **2010**, *20*, 075017. [CrossRef]
16. Lee, K.-T.; Liu, D.-M.; Liang, Y.-Y.; Matsushita, N.; Ikoma, T.; Lu, S.-Y. Porous Fluorine-Doped Tin Oxide as a Promising Substrate for Electrochemical Biosensors—Demonstration in Hydrogen Peroxide Sensing. *J. Mater. Chem. B* **2014**, *2*, 7779–7784. [CrossRef]

17. Banyamin, Z.; Kelly, P.; West, G.; Boardman, J. Electrical and Optical Properties of Fluorine Doped Tin Oxide Thin Films Prepared by Magnetron Sputtering. *Coatings* 2014, *4*, 732–746. [CrossRef]
18. Bierwagen, O. Indium Oxide—A Transparent, Wide-Band Gap Semiconductor for (Opto)Electronic Applications. *Semicond. Sci. Technol.* 2015, *30*, 024001. [CrossRef]
19. Ouerfelli, J.; Djobo, S.O.; Bernède, J.C.; Cattin, L.; Morsli, M.; Berredjem, Y. Organic Light Emitting Diodes Using Fluorine Doped Tin Oxide Thin Films, Deposited by Chemical Spray Pyrolysis, as Anode. *Mater. Chem. Phys.* 2008, *112*, 198–201. [CrossRef]
20. Jasiecki, S.; Czupryniak, J.; Ossowski, T.; Schroeder, G. FTO Coated Glass Electrode Functionalization with Transition Metal Cations Receptors via Electrostatic Self-Assembly. *Int. J. Electrochem. Sci.* 2013, *8*, 12543–12556.
21. Ahuja, T.; Rajesh; Kumar, D.; Tanwar, V.K.; Sharma, V.; Singh, N.; Biradar, A.M. An Amperometric Uric Acid Biosensor Based on Bis[Sulfosuccinimidyl] Suberate Crosslinker/3-Aminopropyltriethoxysilane Surface Modified ITO Glass Electrode. *Thin Solid Film.* 2010, *519*, 1128–1134. [CrossRef]
22. Kim, C.O.; Hong, S.-Y.; Kim, M.; Park, S.-M.; Park, J.W. Modification of Indium–Tin Oxide (ITO) Glass with Aziridine Provides a Surface of High Amine Density. *J. Colloid Interface Sci.* 2004, *277*, 499–504. [CrossRef] [PubMed]
23. Pruna, R.; Palacio, F.; Martínez, M.; Blázquez, O.; Hernández, S.; Garrido, B.; López, M. Organosilane-Functionalization of Nanostructured Indium Tin Oxide Films. *Interface Focus* 2016, *6*, 20160056. [CrossRef] [PubMed]
24. Muthurasu, A.; Ganesh, V. Electrochemical Characterization of Self-Assembled Monolayers (SAMs) of Silanes on Indium Tin Oxide (ITO) Electrodes—Tuning Electron Transfer Behaviour across Electrode–Electrolyte Interface. *J. Colloid Interface Sci.* 2012, *374*, 241–249. [CrossRef] [PubMed]
25. Göbel, G.; Talke, A.; Lisdat, F. FTO—An Electrode Material for the Stable Electrochemical Determination of Dopamine. *Electroanalysis* 2018, *30*, 225–229. [CrossRef]
26. Wang, P.; Li, S.; Kan, J. A Hydrogen Peroxide Biosensor Based on Polyaniline/FTO. *Sens. Actuators B Chem.* 2009, *137*, 662–668. [CrossRef]
27. Naderi Asrami, P.; Mozaffari, S.A.; Saber Tehrani, M.; Aberoomand Azar, P. A Novel Impedimetric Glucose Biosensor Based on Immobilized Glucose Oxidase on a CuO-Chitosan Nanobiocomposite Modified FTO Electrode. *Int. J. Biol. Macromol.* 2018, *118*, 649–660. [CrossRef]
28. Valiūnienė, A.; Kavaliauskaitė, G.; Virbickas, P.; Ramanavičius, A. Prussian Blue Based Impedimetric Urea Biosensor. *J. Electroanal. Chem.* 2021, *895*, 115473. [CrossRef]
29. sadat Vajedi, F.; Dehghani, H. A High-Sensitive Electrochemical DNA Biosensor Based on a Novel ZnAl/Layered Double Hydroxide Modified Cobalt Ferrite-Graphene Oxide Nanocomposite Electrophoretically Deposited onto FTO Substrate for Electroanalytical Studies of Etoposide. *Talanta* 2020, *208*, 120444. [CrossRef]
30. Terracciano, M.; Rea, I.; Politi, J.; De Stefano, L. Optical Characterization of Aminosilane-Modified Silicon Dioxide Surface for Biosensing. *J. Eur. Opt. Soc.-Rapid Publ.* 2013, *8*, 13075. [CrossRef]
31. Acres, R.G.; Ellis, A.V.; Alvino, J.; Lenahan, C.E.; Khodakov, D.A.; Metha, G.F.; Andersson, G.G. Molecular Structure of 3-Aminopropyltriethoxysilane Layers Formed on Silanol-Terminated Silicon Surfaces. *J. Phys. Chem. C* 2012, *116*, 6289–6297. [CrossRef]
32. Taglietti, A.; Arciola, C.R.; D'Agostino, A.; Dacarro, G.; Montanaro, L.; Campoccia, D.; Cucca, L.; Vercellino, M.; Poggi, A.; Pallavicini, P.; et al. Antibiofilm Activity of a Monolayer of Silver Nanoparticles Anchored to an Amino-Silanized Glass Surface. *Biomaterials* 2014, *35*, 1779–1788. [CrossRef] [PubMed]
33. Ashur, I.; Jones, A.K. Immobilization of Azurin with Retention of Its Native Electrochemical Properties at Alkylsilane Self-Assembled Monolayer Modified Indium Tin Oxide. *Electrochim. Acta* 2012, *85*, 169–174. [CrossRef]
34. Sanli, S.; Ghorbani-Zamani, F.; Moulahoum, H.; Gumus, Z.P.; Coskunol, H.; Odaci Demirkol, D.; Timur, S. Application of Biofunctionalized Magnetic Nanoparticles Based-Sensing in Abused Drugs Diagnostics. *Anal. Chem.* 2020, *92*, 1033–1040. [CrossRef] [PubMed]
35. Venkata Jagadeesh, R.; Lakshminarayanan, V. Electrochemical Investigation on Adsorption Kinetics of Long Chain Alkylsilanes and Influence of Solvents on Their Self-Assembly and Electron Transfer Behavior on Indium Tin Oxide. *J. Appl. Electrochem.* 2020, *50*, 1129–1138. [CrossRef]
36. Li, H.; Su, T.A.; Camarasa-Gómez, M.; Hernangómez-Pérez, D.; Henn, S.E.; Pokorný, V.; Caniglia, C.D.; Inkpen, M.S.; Korytár, R.; Steigerwald, M.L.; et al. Silver Makes Better Electrical Contacts to Thiol-Terminated Silanes than Gold. *Angew. Chem. Int. Ed.* 2017, *56*, 14145–14148. [CrossRef]
37. Wang, Y.; Huang, J.-T. Transparent, Conductive and Superhydrophobic Cellulose Films for Flexible Electrode Application. *RSC Adv.* 2021, *11*, 36607–36616. [CrossRef]
38. Moses, P.R.; Wier, L.M.; Lennox, J.C.; Finklea, H.O.; Lenhard, J.R.; Murray, R.W. X-ray Photoelectron Spectroscopy of Alkylaminesilanes Bound to Metal Oxide Electrodes. *Anal. Chem.* 1978, *50*, 576–585. [CrossRef]
39. Zhu, M.; Lerum, M.Z.; Chen, W. How To Prepare Reproducible, Homogeneous, and Hydrolytically Stable Aminosilane-Derived Layers on Silica. *Langmuir* 2012, *28*, 416–423. [CrossRef]
40. Cirocka, A.; Zarzeczańska, D.; Wcisło, A. Good Choice of Electrode Material as the Key to Creating Electrochemical Sensors—Characteristics of Carbon Materials and Transparent Conductive Oxides (TCO). *Materials* 2021, *14*, 4743. [CrossRef]
41. Kern, W. *Others Handbook of Semiconductor Wafer Cleaning Technology*; Noyes Publication: Park Ridge, NJ, USA, 1993; pp. 111–196.

42. Cirocka, A.; Zarzeczańska, D.; Wcisło, A.; Ryl, J.; Bogdanowicz, R.; Finke, B.; Ossowski, T. Tuning of the Electrochemical Properties of Transparent Fluorine-Doped Tin Oxide Electrodes by Microwave Pulsed-Plasma Polymerized Allylamine. *Electrochim. Acta* **2019**, *313*, 432–440. [CrossRef]
43. Kwok, D.Y.; Neumann, A.W. Contact Angle Measurement and Contact Angle Interpretation. *Adv. Colloid Interface Sci.* **1999**, *81*, 167–249. [CrossRef]
44. Kwok, D.Y.; Neumann, A.W. Contact Angle Measurements and Interpretation: Wetting Behavior and Solid Surface Tensions for Poly(Alkyl Methacrylate) Polymers. *J. Adhes. Sci. Technol.* **2000**, *14*, 719–743. [CrossRef]
45. Kwok, D.Y.; Gietzelt, T.; Grundke, K.; Jacobasch, H.-J.; Neumann, A.W. Contact Angle Measurements and Contact Angle Interpretation. 1. Contact Angle Measurements by Axisymmetric Drop Shape Analysis and a Goniometer Sessile Drop Technique. *Langmuir* **1997**, *13*, 2880–2894. [CrossRef]
46. Orazem, M.E.; Pébère, N.; Tribollet, B. Enhanced Graphical Representation of Electrochemical Impedance Data. *J. Electrochem. Soc.* **2006**, *153*, B129. [CrossRef]
47. Lasia, A. *Electrochemical Impedance Spectroscopy and Its Applications*; Springer-Verlag: New York, NY, USA, 2014; ISBN 978-1-4614-8932-0.
48. Song, X.; Zhai, J.; Wang, Y.; Jiang, L. Self-Assembly of Amino-Functionalized Monolayers on Silicon Surfaces and Preparation of Superhydrophobic Surfaces Based on Alkanoic Acid Dual Layers and Surface Roughening. *J. Colloid Interface Sci.* **2006**, *298*, 267–273. [CrossRef] [PubMed]
49. Kulkarni, S.A.; Ogale, S.B.; Vijayamohanan, K.P. Tuning the Hydrophobic Properties of Silica Particles by Surface Silanization Using Mixed Self-Assembled Monolayers. *J. Colloid Interface Sci.* **2008**, *318*, 372–379. [CrossRef] [PubMed]
50. Zeng, X.; Xu, G.; Gao, Y.; An, Y. Surface Wettability of (3-Aminopropyl)Triethoxysilane Self-Assembled Monolayers. *J. Phys. Chem. B* **2011**, *115*, 450–454. [CrossRef]
51. Siqueira Petri, D.F.; Wenz, G.; Schunk, P.; Schimmel, T. An Improved Method for the Assembly of Amino-Terminated Monolayers on SiO_2 and the Vapor Deposition of Gold Layers. *Langmuir* **1999**, *15*, 4520–4523. [CrossRef]

Disclaimer/Publisher's Note: The statements, opinions and data contained in all publications are solely those of the individual author(s) and contributor(s) and not of MDPI and/or the editor(s). MDPI and/or the editor(s) disclaim responsibility for any injury to people or property resulting from any ideas, methods, instructions or products referred to in the content.

Article

Fabrication and Characterization of an Electrochemical Platform for Formaldehyde Oxidation, Based on Glassy Carbon Modified with Multi-Walled Carbon Nanotubes and Electrochemically Generated Palladium Nanoparticles

Andrzej Leniart [1,*], Barbara Burnat [1], Mariola Brycht [1], Maryia-Mazhena Dzemidovich [1,2] and Sławomira Skrzypek [1]

[1] University of Lodz, Faculty of Chemistry, Department of Inorganic and Analytical Chemistry, Tamka 12, 91-403 Lodz, Poland; barbara.burnat@chemia.uni.lodz.pl (B.B.); mariola.brycht@chemia.uni.lodz.pl (M.B.); maryia.mazhena.dzemidovich@edu.uni.lodz.pl (M.-M.D.); slawomira.skrzypek@chemia.uni.lodz.pl (S.S.)
[2] University of Lodz, Doctoral School of Exact and Natural Sciences, Matejki 21/23, 90-231 Lodz, Poland
* Correspondence: andrzej.leniart@chemia.uni.lodz.pl; Tel.: +48-42-635-5783

Citation: Leniart, A.; Burnat, B.; Brycht, M.; Dzemidovich, M.-M.; Skrzypek, S. Fabrication and Characterization of an Electrochemical Platform for Formaldehyde Oxidation, Based on Glassy Carbon Modified with Multi-Walled Carbon Nanotubes and Electrochemically Generated Palladium Nanoparticles. *Materials* **2024**, *17*, 841. https://doi.org/10.3390/ma17040841

Academic Editors: Aivaras Kareiva and Gediminas Niaura

Received: 6 January 2024
Revised: 30 January 2024
Accepted: 7 February 2024
Published: 9 February 2024

Copyright: © 2024 by the authors. Licensee MDPI, Basel, Switzerland. This article is an open access article distributed under the terms and conditions of the Creative Commons Attribution (CC BY) license (https://creativecommons.org/licenses/by/4.0/).

Abstract: This study outlines the fabrication process of an electrochemical platform utilizing glassy carbon electrode (GCE) modified with multi-walled carbon nanotubes (MWCNTs) and palladium nanoparticles (PdNPs). The MWCNTs were applied on the GCE surface using the drop-casting method and PdNPs were produced electrochemically by a potentiostatic method employing various programmed charges from an ammonium tetrachloropalladate(II) solution. The resulting GCEs modified with MWCNTs and PdNPs underwent comprehensive characterization for topographical and morphological attributes, utilizing atomic force microscopy and scanning electron microscopy along with energy-dispersive X-ray spectrometry. Electrochemical assessment of the GCE/MWCNTs/PdNPs involved cyclic voltammetry (CV) and electrochemical impedance spectroscopy conducted in perchloric acid solution. The findings revealed even dispersion of PdNPs, and depending on the electrodeposition parameters, PdNPs were produced within four size ranges, i.e., 10–30 nm, 20–40 nm, 50–60 nm, and 70–90 nm. Additionally, the electrocatalytic activity toward formaldehyde oxidation was assessed through CV. It was observed that an increase in the size of the PdNPs corresponded to enhanced catalytic activity in the formaldehyde oxidation reaction on the GCE/MWCNTs/PdNPs. Furthermore, satisfactory long-term stability over a period of 42 days was noticed for the GCE/MWCNTs/PDNPs(100) material which demonstrated the best electrocatalytic properties in the electrooxidation reaction of formaldehyde.

Keywords: electrocatalysis; atomic force microscopy; scanning electron microscopy; voltammetry; electrochemical impedance

1. Introduction

Depleting reserves of fossil fuels, coupled with their negative environmental impact on one hand and increased demand for bioenergy and biofuels on the other, have prompted the exploration of new electrochemical platforms with extensive applications in fuel cells, catalysts, and electrochemical sensors. A specific emphasis is placed on the electrocatalytic oxidation of small organic compounds, including methanol, ethanol, isopropanol, formaldehyde, or formic acid, on various modified electrodes. This is due to their potential as electron donors in fuel cells, enabling the creation of high-power density systems [1]. Although direct applicability of formaldehyde in fuel cells is limited, its electrochemical oxidation plays a crucial role in understanding methanol oxidation, where formaldehyde serves as an intermediate [2]. Moreover, owing to the toxicity of formaldehyde and associated health risks, including chronic inflammation, vomiting, seizures, fetal development problems, cardiovascular diseases, and cancer [3–7], the detection and monitoring of its presence are of high importance.

Various electrochemical platforms, employing metals such as platinum, rhodium, nickel, and palladium as key components, have been utilized for the study of formaldehyde electrooxidation and detection [1,8–20]. Pd is particularly intriguing due to its distinctive physicochemical properties, especially its electrochemical attributes, ascribed to weak intermetallic bonds within the crystal lattice [21]. In addition, Pd demonstrates high stability and a remarkable tolerance to CO, which is formed as a byproduct during the oxidation of formaldehyde [22]. Unique catalytic properties of Pd contribute to its widespread use in both heterogeneous and homogeneous catalysis [23]. Pd serves as a catalyst in diverse processes, including the hydrogenation of alkene to alkanes and the oxidation of various organic compounds, with applications in both liquid and gaseous phases [24–28]. Additionally, it plays a crucial role in environmental technologies, acting as a catalyst for the combustion of petroleum products in automobile engines, facilitating hydrogenation reactions, and serving as a precursor in metallization processes for non-metallic materials such as fiberglass and ceramics used in electronics [29–31]. The versatility of Pd extends to its use in various types of sensors [32–40]. The production of Pd nanoparticles (PdNPs) employs various methods, encompassing physical techniques like physical vapor deposition and magnetron sputtering, chemical methods such as chemical vapor deposition and chemical reduction of metal ions, as well as electrochemical deposition [38,41,42].

It is important to emphasize that the choice of substrate onto which PdNPs are deposited significantly influences the effectiveness of its electrocatalytic activity. One such substrate is carbon, which plays a crucial role as the foundation for fuel cells, catalysts, and electrodes, highlighting its wide application in the field of energy and chemical processes. From an electrochemical perspective, the primary and preferred electrode material is glassy carbon (GC), renowned for its unique properties, including high electrical and thermal conductivity, hardness, resistance to extreme temperatures and chemicals, and stability at various polarization potentials [43–46]. To enhance existing properties or discover new ones for glassy carbon electrode (GCE), various modifications are applied using different techniques and materials. Carbon nanomaterials, known for their unique properties, are often employed for these modifications. They not only increase the surface area but also provide a substructure/substrate for subsequent materials used in the modification of electrochemical platforms. Carbon nanotubes (CNTs), specifically single- or multi-walled structures (SWCNTs or MWCNTs), are commonly used as carbon nanomaterials for surface modification [47–50] due to their unique features such as large specific surface area, high electrical and thermal conductivity, and substantial high mechanical strength [51]. Despite the superior properties of CNTs, they do not exhibit electrocatalytic activity toward formaldehyde oxidation; therefore, further modifications are required [52].

To date, the modification of CNTs with PdNPs to obtain a CNT/PdNP composite has been achieved through various methods, including chemical reduction of Pd(II) ions with sodium borohydride [13,53], formaldehyde [52,54], and sodium citrate in an ethylene glycol solution [55]. Other techniques involve the thermal decomposition of palladium acetate [39], RF magnetron sputtering [56], and electrochemical reduction of the MWCNT/lignosulfonate–Pd^{2+} composite [57]. However, these methods are time-consuming and involve multiple steps that are challenging to control, consequently affecting the final properties and the cost of the obtained electrode materials. As an alternative, one can propose an approach employing the electrochemical reduction of Pd(II) ions with charge control, offering full and precise control over the production of PdNPs, overcoming the challenges associated with time-consuming and complex procedures.

The objective of this work is to create an electrochemical platform of GCE/MWCNTs/PdNPs that exhibits improved electrocatalytic properties through the synergistic effect of MWCNTs and PdNPs. Significantly, the focus extends beyond the creation of the platform. This research places crucial importance on understanding and refining the straightforward electrochemical procedure of PdNP production, emphasizing controlled and precise nanoparticle synthesis. Additionally, the study aims to comprehensively characterize the morphology, topography, and elemental composition of the surfaces of the fabricated platforms, along with assessing their electrochemical and electrocatalytic properties in the electrooxidation reaction of formaldehyde.

2. Materials and Methods

2.1. Chemical Reagents

All reagents utilized in the experiments were of analytical purity and did not undergo additional purification processes. Solutions were prepared using triple-distilled water. The following solutions were employed for the measurements:

- hydrochloric acid solution (HCl, 35–38%, Avantor Performance, Gliwice, Poland) at a concentration of 0.1 mol L^{-1}, containing ammonium tetrachloropalladate(II) ($(NH_4)_2PdCl_4$, Ventron GMBH, Karlsruhe, Germany) at a concentration of 1.0 mmol L^{-1};
- perchloric acid solution ($HClO_4$, 95%, Avantor Performance) at a concentration of 0.1 mol L^{-1} both without and with the addition of formaldehyde (HCHO, 37%, Avantor Performance) at a concentration of 0.1 mol L^{-1}.

Prior to each measurement, the solutions underwent deoxygenation using argon with a rating of 5.0 (Linde Gaz Poland, Kraków, Poland).

2.2. Preparation of Electrodes

Six types of working electrodes were employed in the study, i.e., a glassy carbon electrode (GCE, diameter of 3 mm, L-Chem, Horka nad Moravou, Czechia), a GCE modified with multi-walled carbon nanotubes (GCE/MWCNTs), and four modified variations of GCE/MWCNTs containing different amounts of PdNPs (GCE/MWCNTs/PdNPs).

The preparation of the GCE surface involved manual polishing on felt using an aqueous Al_2O_3 suspension (ATM GMBH, Blieskastel, Germany) with a grain size of 0.3 µm. Following this, the electrode surface was rinsed with triple-distilled water, underwent a 5 min cleaning cycle in an ultrasonic cleaner, and was finally dried under an argon stream.

For the preparation of a GCE modified with MWCNTs (>95%, diameter 8–9 nm, length 5 µm, Sigma-Aldrich, Poznań, Poland) by the drop-drying method, a suspension of the MWCNTs in dimethylformamide (DMF, 99.8%, Avantor Performance, Gliwice, Poland) at a concentration of 0.5 mg L^{-1} was used. For this purpose, 5 µL of the MWCNT suspension was dropped onto the surface of the GCE and then the coated electrode was dried for 24 h, enabling the formation of a durable MWCNT layer. The GCE/MWCNTs underwent subsequent modification with PdNPs via electrochemical deposition from a 0.1 mol L^{-1} hydrochloric acid solution containing 1.0 mmol L^{-1} $(NH_4)_2PdCl_4$. The PdNP synthesis was carried out using the coulometric method at a constant polarization potential, applying programmed and defined charge values of −5, −20, −50, and −100 mC. The deposition potential of 0.1 V ensured that the charge flowing during electrolysis was solely associated with the reduction reaction of Pd(II) ions to metallic Pd. Following the PdNP deposition process, each of the resulting GCE/MWCNTs/PdNPs underwent rinsing with triple-distilled water and drying using an argon stream. This procedure generated four variations of GCE/MWCNTs/PdNPs, each featuring different PdNPs contents, denoted as GCE/MWCNTs/PdNPs(5), GCE/MWCNTs/PdNPs(20), GCE/MWCNTs/PdNPs(50) and GCE/MWCNTs/PdNPs(100), containing 0.3, 1.1, 2.8, and 5.5 µg of metallic Pd, respectively (the mass of Pd was calculated based on Faraday's laws).

2.3. The Research Methodology

The research strategy involved two primary approaches. The first approach focused on analyzing the topography and surface morphology of the working electrodes, along with elemental analysis of the fabricated GCE/MWCNTs/PdNPs. Characterization of the working electrodes involved atomic force microscopy (AFM) and scanning electron microscopy (SEM) coupled with energy dispersive X-ray spectrometry (EDX). AFM analyses were conducted utilizing a Dimension Icon system (Bruker, Santa Barbara, CA, USA), employing the intermittent contact method (tapping mode). A commercial TESPA V2 silicon probe (Bruker, Santa Barbara, CA, USA) with a nominal spring constant of 42 N m^{-1} and a resonance frequency of 320 kHz served as the AFM probe. SEM measurements were performed using a FEI Nova NanoSEM 450 microscope (FEI, Hillsboro, OR, USA), equipped with an electron gun with thermal field emission (Schottky emitter). SEM measurements were conducted with an accelerating voltage of 10 kV using a through-the-lens detector (TLD). Elemental surface composition was determined from EDX spectra collected with an EDX analyzer (Ametek Inc., Berwyn, PA, USA).

The second approach involved conducting an electrochemical evaluation of the working electrodes. This encompassed their characterization in a $HClO_4$ solution using cyclic voltammetry (CV) and electrochemical impedance spectroscopy (EIS), as well as determining the electrocatalytic properties of GCE/MWCNT/PdNPs in the formaldehyde electrooxidation process. All electrochemical tests were carried out in a 15 mL electrochemical cell using a classic three-electrode configuration, where the working electrode was either GCE, GCE/MWCNTs, or GCE/MWCNTs/PdNPs with varying PdNPs contents, a silver chloride electrode (Ag | AgCl in 3 mol L^{-1} KCl solution, Mineral, Łomianki-Sadowa, Poland) served as the reference electrode, while a platinum wire (99.99%, The Mint of Poland, Warsaw, Poland) was used as an auxiliary electrode. All potential values in this work are referenced to Ag | AgCl. The electrochemical measurements were conducted using a PGSTAT 128 N potentiostat/galvanostat (Metrohm AUTOLAB B.V., Utrecht, The Netherlands), equipped with a frequency response module (FRA2) for electrochemical impedance spectroscopy measurements. Additionally, an M164 type electrode stand (MTM Anko Instruments, Kraków, Poland) was employed. The cyclic voltammograms were recorded in $HClO_4$ solution, in both the presence and absence of formaldehyde, using a scan rate of 100 mV s^{-1}. The EIS spectra were obtained over a frequency range of 10,000–0.01 Hz with an amplitude of 10 mV and 50 measuring points.

3. Results

3.1. Microscopic Characterization of the Working Electrodes

The assessment of topography and surface morphology for each type of working electrode, along with additional elemental analysis for GCE/MWCNTs/PdNPs with various PdNPs content, was conducted at three randomly selected places on the electrode surface. Figure 1 displays representative images depicting surface topography (AFM) and surface morphology (SEM) for the GCE and the GCE/MWCNTs, while Figure 2 presents AFM and SEM images of GCE/MWCNTs modified with different PdNPs contents (GCE/MWCNTs/PdNPs(5), GCE/MWCNTs/PdNPs(20), GCE/MWCNTs/PdNPs(50), and GCE/MWCNTs/PdNPs(100)).

The findings from both AFM and SEM images reveal distinct differences in surface topography and morphology among the examined electrodes. Notably, the unmodified GCE (Figure 1A) displays slight surface irregularities, probably stemming from polishing, evident as small scratches with depths measuring a few nanometers. At higher magnifications, the fine-grained structure of the GCE surface becomes evident. Moreover, the GCE surface appears homogeneous and clean.

In the case of GCE/MWCNTs (Figure 1B), the surface of the GCE appears to be entirely covered with multi-walled carbon nanotubes (MWCNTs) that are bent and intertwined. Additionally, areas containing a higher quantity of carbon nanotubes forming aggregates can be observed. The generated layer of MWCNTs on the surface of the glassy carbon electrode leads to a significant increase in surface area, as indicated by the height scale in the AFM images.

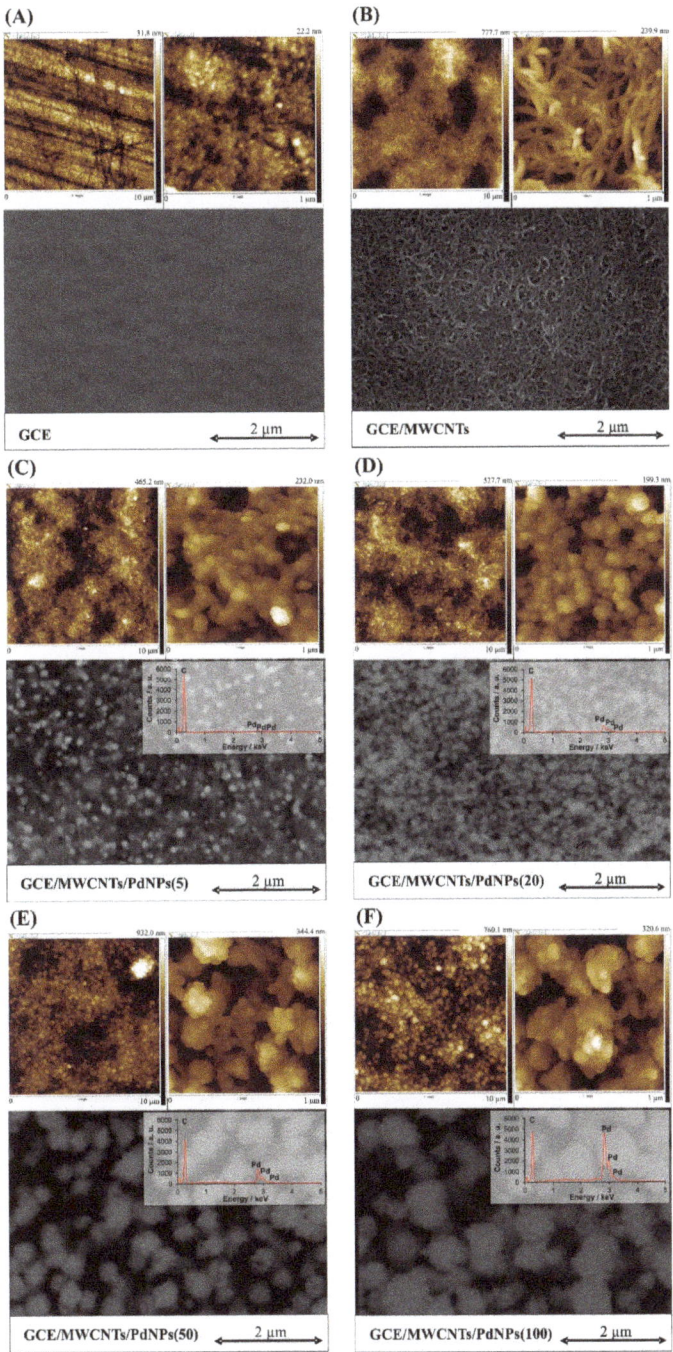

Figure 1. AFM (10 µm² and 1 µm² scanning areas) and SEM images (magnitude 100,000×) of the (**A**) GCE, (**B**) GCE/MWCNTs, (**C**) GCE/MWCNTs/PdNPs(5), (**D**) GCE/MWCNTs/PdNPs(20), (**E**) GCE/MWCNTs/PdNPs(50), and (**F**) GCE/MWCNTs/PdNPs(100) surfaces. Insets display EDX spectra for the electrodes with PdNPs.

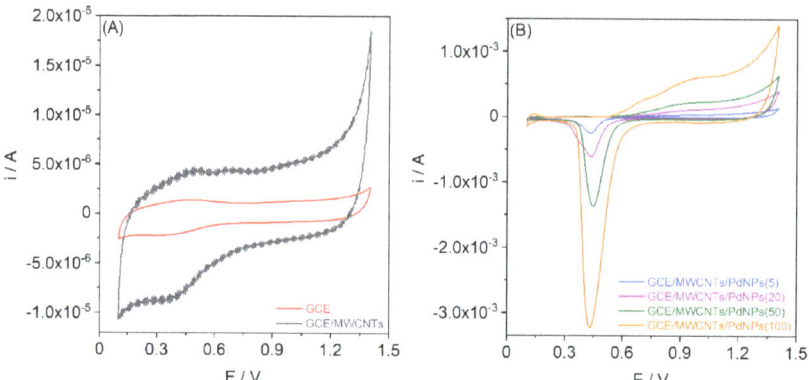

Figure 2. Cyclic voltammograms (scan rate of 100 mV s^{-1}) recorded in 0.1 mol L^{-1} HClO$_4$ solution for (**A**) the GCE and GCE/MWCNTs, as well as (**B**) the GCE/MWCNTs/PdNPs.

SEM and AFM images of the GCE/MWCNTs/PdNPs (Figure 1C–F) clearly illustrate both CNTs and electrodeposited PdNPs on the surface. In the case of GCE/MWCNTs/PdNPs(5) (Figure 1C), where the smallest amount of Pd (0.3 µg) was deposited, visible PdNPs measuring between 10 nm to 30 nm are observed. With a larger amount of deposited Pd (1.1 µg) on the GCE/MWCNTs/PdNPs(20), larger and more densely packed PdNPs are formed (Figure 1D), ranging from 20 nm to 40 nm. Following the deposition of 2.8 µg and 5.5 µg of Pd, subsequent AFM and SEM images of the GCE/MWCNTs/PdNPs(50) (Figure 1E) and GCE/MWCNTs/PdNPs(100) (Figure 1F) demonstrate the formation of even larger PdNPs. In the case of GCE/MWCNTs/PdNPs(50), the PdNPs measure between 50 nm and 60 nm, while for GCE/MWCNTs/PdNPs(100), their size ranges from 70 nm to 90 nm.

Analysis of the results indicates that the nucleation and further growth of PdNPs are prominently observed during the initial stages of electrodeposition, primarily occurring on CNTs until they are fully coated. Subsequent electrodeposition processes primarily contribute to the enlargement of these nanoparticles. The applied procedure of electrochemical generation of PdNPs leads to obtaining nanoparticles that should be considered as larger particles than clusters [58].

To analyze the surfaces of the tested electrodes using AFM images with scanning areas of 1 µm^2, topographic parameters were determined utilizing the AFM image analysis program NanoScope Analysis v. 1.50 (Bruker). These parameters included real surface area (Sr; derived from AFM images), surface area difference (SAD; ratio of the geometric surface area (scanning area) to the surface area obtained from three-dimensional (3D) AFM images, expressed as a percentage), root mean square roughness coefficient (Rq; the mean square deviation between points on the tested surface and those on the optimal plane), average roughness coefficient (Ra; the arithmetic mean values of absolute deviations between points on the tested surface and those on the optimal plane), and the maximum height (Rmax; the vertical distance between the highest and lowest data points on the tested surface area after to the plane fit operation). The values of these parameters are detailed in Table 1.

Upon analyzing the determined topographic parameters outlined in Table 1, it is evident that the application of MWCNTs onto the GCE surface amplified all its topographic parameters. The roughness coefficients, Rq and Ra, underwent a 10-fold increase, while the SAD escalated by almost 40 times. The deposition of a small quantity of PdNPs, specifically 0.3 and 1.1 µg, reduced the topographic parameters compared with the GCE/MWCNTs. This occurred due to the partial filling of holes (gaps) between entangled nanotubes by the PdNPs, resulting in a partial smoothing of the surface. However, the deposition of larger amounts of Pd led to an increase in topographic values that is attributed to the formation of progressively larger PdNPs on the surface of the GCE/MWCNTs (Figure 1E,F). It can be stated that the deposition of Pd under the conditions applied by us leads to the formation PdNPs of various sizes, which can be easily controlled by adjusting the value of the charge during electrodeposition.

Table 1. Determined topographic parameters characterizing the surface of the tested electrodes for scanning areas of 1 µm^2.

Electrodes	Sr (µm^2)	SAD (%)	Rq (nm)	Ra (nm)	Rmax (nm)
GCE	1.02	2.3	2.4	1.9	20.8
GCE/MWCNTs	1.89	88.9	25.9	20.3	204
GCE/MWCNTs/PdNPs(5)	1.51	50.7	29.9	23.1	241
GCE/MWCNTs/PdNPs(20)	1.48	48.2	26.7	20.8	218
GCE/MWCNTs/PdNPs(50)	1.89	88.5	51.0	40.8	304
GCE/MWCNTs/PdNPs(100)	1.81	81.2	49.2	39.2	308

Additionally, elemental analysis was performed for the GCE/MWCNTs/PdNPs to confirm that the presence of PdNPs and to determine the Pd content using a standard-free quantitative analysis. The outcomes of the quantitative elemental composition for each of the GCE/MWCNTs/PdNPs are detailed in Table 2.

Table 2. Quantitative elemental composition for each of the GCE/MWCNTs/PdNPs.

Electrodes	Mass Content (%)	
	C(K)	Pd(L)
GCE/MWCNTs/PdNPs(5)	90.15	9.85
GCE/MWCNTs/PdNPs(20)	72.99	27.01
GCE/MWCNTs/PdNPs(50)	28.29	71.71
GCE/MWCNTs/PdNPs(100)	11.32	88.48

3.2. Electrochemical Characterization of the Working Electrodes

Cyclic voltammograms for each electrode were recorded in a 0.1 mol L^{-1} HClO$_4$ solution, spanning a potential range from +0.1 V to +1.4 V, using a potential scan rate of 100 mV s^{-1} (Figure 2).

As can clearly be seen from Figure 2A, the CV voltammograms for both the GCE and GCE/MWCNTs display similar characteristics, without distinct current peaks. The considerably higher current values for the GCE/MWCNTs stem from the increased capacity at the electrode–solution interface. This augmentation results from the application of the MWCNT layer onto the relatively smooth GCE surface, leading to a significant surface development. The initial increase in the anodic current at a potential around +1.4 V correlates with the oxygen evolution reaction. However, the CV voltammograms for GCE/MWCNTs/PdNPs exhibit a markedly different profile compared with the GCE and GCE/MWCNTs (Figure 2B). This divergence in shape is directly linked to the presence of PdNPs. The increase in the anodic current observed from a potential of approximately +0.6 V corresponds to the oxidation of PdNPs. The electrooxidation process of Pd is intricate and can manifest through diverse pathways [14,15]. Principally, this process involves the passivation of Pd, leading to the formation of Pd(II) oxide (PdO). Concurrently, the cathodic peak detected at a potential ranging from +0.47 to +0.45 V relates to the reduction of PdO previously formed during the scanning in the anodic direction. Notably, an increase in the Pd quantity on the electrodes correlates with amplified peak currents, corresponding to the electrooxidation of Pd and the subsequent electroreduction of PdO.

Subsequently, the EIS measurements were carried out within a frequency range of 10,000–0.01 Hz with a signal amplitude of 10 mV. The results of EIS measurements, depicted as Nyquist plots, are presented in Figure 3. The obtained EIS characteristics reveal a reduction in impedance following each modification of the GCE. The GCE/MWCNTs exhibited an impedance value of 1.7×10^5 Ω cm^2 at the lowest frequency (10 mHz), notably one order of magnitude lower than the impedance value for the unmodified GCE (1.3×10^6 Ω cm^2). However, for GCE/MWCNTs/PdNPs, the electrode resistance decreased with increasing Pd content. The impedance value for the GCE/MWCNTs/PdNPs(5) at 10 mHz was equal to 1.6×10^5 Ω cm^2, decreasing to 2.9×10^4 Ω cm^2 for the GCE/MWCNTs/PdNPs(100). This decrease in electrode impedance value indicates increased electrochemical activity, and the reduction–oxidation processes are anticipated to occur more readily on the electrode.

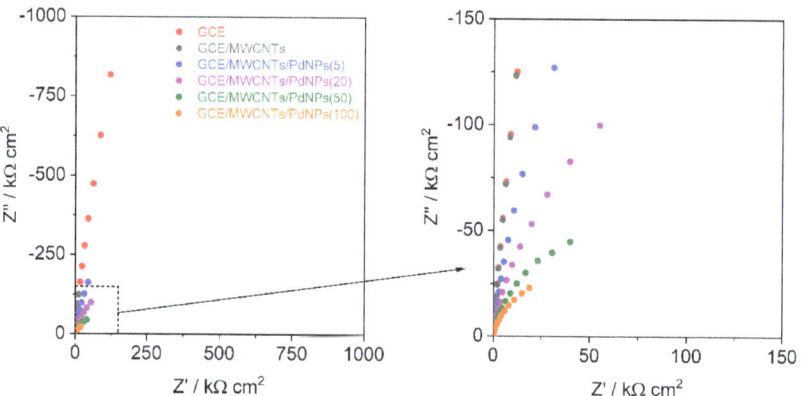

Figure 3. Impedance characteristics (Nyquist plots) recorded in 0.1 mol L^{-1} HClO$_4$ solution for the GCE, GCE/MWCNTs, and GCE/MWCNTs/PdNPs. Frequency range of 10,000–0.01 Hz (amplitude 10 mV, 50 measuring points).

3.3. The Electrocatalytic Activity of Each Electrode in the Process of Formaldehyde Oxidation

The evaluation of the electrocatalytic activity of each of the GCE/MWCNTs/PdNPs with various amounts of deposited Pd was conducted using CV in a 0.1 mol L^{-1} HClO$_4$ solution containing formaldehyde at a concentration of 0.1 mol L^{-1}. The recorded cyclic voltammograms spanning the potential range from +0.1 V to +1.4 V with a potential scan rate of 100 mV s^{-1} are presented in Figure 4A for GCE and GCE/MWCNTs and in Figure 4B for GCE/MWCNTs/PdNPs.

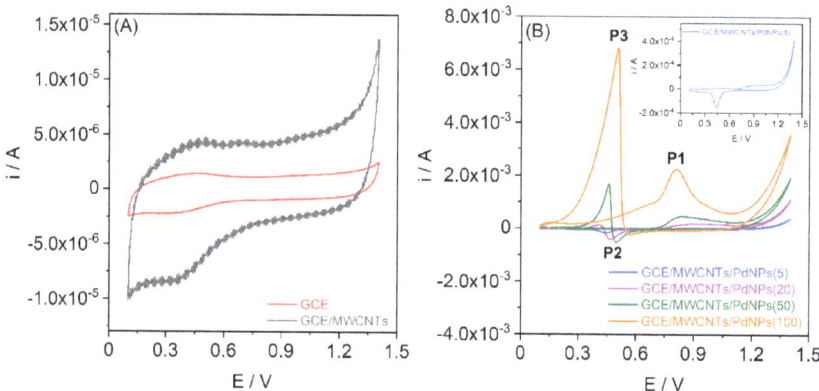

Figure 4. Cyclic voltammograms (scan rate of 100 mV s^{-1}) recorded in 0.1 mol L^{-1} HClO$_4$ solution with 0.1 mol L^{-1} formaldehyde for (**A**) GCE and GCE/MWCNTs, (**B**) GCE/MWCNTs/PdNPs.

As evident from Figure 4A, the shape of the cyclic voltammograms for the GCE and GCE/MWCNTs remain consistent with those in the HClO$_4$ solution without formaldehyde

(Figure 2A). In both cases, no current peaks are visible within the tested potential range, indicating the inactivity of these electrodes toward the electrooxidation of formaldehyde. In the presence of MWCNTs, the electrode surface becomes more developed resulting in an increase capacitive current. However, for the GCE/MWCNTs/PdNPs, there is a distinct difference in the shapes of the cyclic voltammograms (Figure 4B). Cyclic voltammograms for the GCE/MWCNTs/PdNPs(5) closely resemble those for the $HClO_4$ solution (Figure 2B), suggesting that on an electrode with a small amount of deposited PdNPs, formaldehyde electrooxidation occurs to a limited extent. In contrast, the other GCEs modified with MWCNTs/PdNPs exhibit peak P1 in the cyclic voltammograms at a potential ranging from +0.81 to +0.92 V (Table 3), which is associated with the oxidation of both formaldehyde and PdNPs. Further polarization in the anodic direction leads to a decrease in current, attributed to the blocking of the PdNP surface by the adsorption of carbon monoxide (CO), an intermediate product of formaldehyde oxidation, on the electrode surface. Reversing the polarization to the cathodic direction results in the appearance of a cathodic peak P2 at a potential spanning from +0.57 to +0.47 V (Table 3), which is associated with the reduction of PdO. As a result of further cathodic polarization, a second anodic peak P3, commonly referred to as the "inversion" peak, emerges. This peak reaches its maximum at a potential occurring in the range from +0.50 to +0.40 V (Table 3). The appearance of this peak is attributed to the regained electroactivity of the PdNPs, leading to further formaldehyde oxidation due to CO desorption and possible conversion to carbon dioxide (CO_2). This process unblocks the electrode surface previously hindered by adsorbed CO [59,60].

Table 3. Peak potential and peak current values for the GCE/MWCNTs/PdNPs derived from CVs presented in Figure 4B.

Electrodes	P1		P2		P3	
	E (V)	i (A)	E (V)	i (A)	E (V)	i (A)
GCE/MWCNTs/PdNPs(5)	0.97	3.22×10^{-5}	0.44	-1.43×10^{-4}	-	-
GCE/MWCNTs/PdNPs(20)	0.92	1.92×10^{-4}	0.47	-3.53×10^{-4}	0.40	4.95×10^{-4}
GCE/MWCNTs/PdNPs(50)	0.84	4.64×10^{-4}	0.50	-4.82×10^{-4}	0.46	2.16×10^{-3}
GCE/MWCNTs/PdNPs(100)	0.81	2.02×10^{-3}	0.57	-1.63×10^{-4}	0.50	6.97×10^{-3}

Based on the cyclic voltammograms, it can be concluded that the electrooxidation of formaldehyde involving PdNPs is a complex process that may follow various mechanisms [55,61]. The electrooxidation of formaldehyde can be represented by the reaction Equations (1)–(3):

$$HCHO + H_2O \rightleftharpoons CH_2(OH)_2 \longrightarrow CO_2 + 4H^+ + 4e^- \qquad (1)$$

$$HCHO \longrightarrow CO_{(ads)} + 2H^+ + 2e^- \qquad (2)$$

$$CO_{(ads)} + H_2O \longrightarrow CO_2 + 2H^+ + 2e^- \qquad (3)$$

The enhanced mechanism of electrocatalytic performance of formaldehyde oxidation on the GCE modified with MWCNTs and PdNPs is attributed to several key factors. Pd, known for its catalytic activity in numerous organic reactions, plays a crucial role in this enhancement. The deposition of PdNPs with a nanopore cavity onto the electrode material significantly increases its electroactive surface area, providing more active sites for electrocatalysis [54,62]. The nanopores create a unique structure that facilitates a rapid and efficient electron transfer pathway from the exterior to the interior of the electrode. This is especially crucial for formaldehyde oxidation, where the accelerated electron transfer significantly enhances the overall electrocatalytic performance. Furthermore, the incorporation of MWCNTs contributes to the overall improvement. MWCNTs possess excellent electrical conductivity and mechanical strength, enhancing the electrical and structural properties of the modified electrode. This, in turn, promotes efficient charge transfer during electrocatalytic reactions [55,63]. In summary, the combination of PdNPs and MWCNTs synergistically amplifies the electrocatalytic performance of the modified GCE by increasing the electroactive surface area, improving electrical conductivity, and facilitating fast electron transfer through the nanopore cavity structure.

3.4. The Long-Term Stability Study

The long-term stability of the developed electrode material showcasing the best electrocatalytic properties in the electrooxidation reaction of formaldehyde, i.e., GCE/MWCNTs/PdNPs(100), was investigated through cyclic voltammetric experiments in a 0.1 mol L^{-1} formaldehyde solution over a period of 42 days. Measurements were systematically conducted at 14-day intervals. As depicted in Figure 5, the cyclic voltammograms displayed minimal variations in peak currents, with a relative standard deviation value below 10%. These results indicate that the GCE/MWCNTs/PdNPs(100) exhibited satisfactory stability during the tested period.

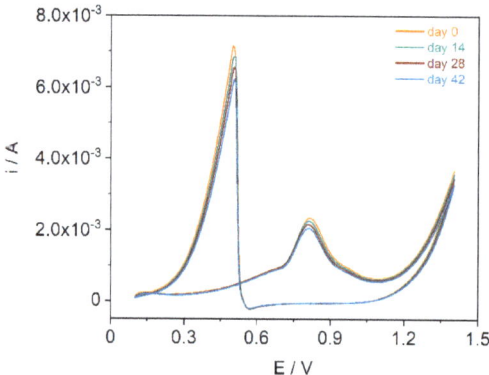

Figure 5. Cyclic voltammograms (scan rate of 100 mV s^{-1}) recorded in 0.1 mol L^{-1} HClO$_4$ solution with 0.1 mol L^{-1} formaldehyde for GCE/MWCNTs/PdNPs(100) over a period of 42 days.

4. Conclusions

This paper presents the complete procedure for preparing modified GCEs using MWCNTs and PdNPs, characterizing their morphology and electrochemical properties and demonstrating their electrocatalytic properties in the formaldehyde electrooxidation process. PdNPs were electrochemically obtained on a GCE covered with MWCNTs using a charge-controlled potentiostatic method, employing four different deposition charges (-5, -20, -50, and -100 mC). This allowed the preparation of four GCE/MWCNTs/PdNPs, each containing varying amounts of Pd (0.3, 1.1, 2.8, and 5.5 µg, respectively). The fabricated GCE/MWCNTs/PdNPs as well as unmodified GCE and GCE/MWCNTs underwent topographical characterization using AFM and SEM, as well as electrochemical assessment using CV and EIS. Based on the presented findings, the following conclusions were drawn:

(i) the amount of Pd deposited on the electrode depends on the charge utilized during electrodeposition: higher charges lead to a greater amount of Pd on the electrode;
(ii) the electrodeposition process produced PdNPs, and their size varied depending on the applied charge (amount of deposited Pd): a higher deposition charge (amount of deposited Pd) resulted in larger PdNPs;
(iii) small deposition charges result in surface smoothing, while very large deposition charges lead to a significant development of the surface;
(iv) the resistance of the GCE/MWCNTs/PdNPs was lower compared with that of the GCE and GCE/MWCNTs, and it decreased with an increase in the amount of Pd on the electrode;

(v) GCE and GCE/MWCNTs exhibited no electrocatalytic activity in the formaldehyde electrooxidation;

(vi) GCE/MWCNTs/PdNPs demonstrated electrocatalytic activity in the formaldehyde electrooxidation. This activity increased with the growth of PdNP size (amount of deposited Pd).

To conclude, the presented method of fabricating GCE/MWCNTs/PdNPs is promising for the development of electrochemical platforms suitable for the electrooxidation of various small organic molecules. The material produced may find applications as an electrochemical sensor, catalyst, or environmental remediation device for the removal of small organic molecules such as formaldehyde.

Author Contributions: Conceptualization, A.L.; methodology, A.L., B.B., M.B., M.-M.D. and S.S.; validation, A.L., B.B. and M.B.; formal analysis, A.L., B.B., M.B. and M.-M.D.; investigation, A.L., B.B., M.B. and M.-M.D.; resources, A.L.; data curation, A.L., B.B. and M.B.; writing—original draft preparation, A.L., B.B. and M.B.; writing—review and editing, A.L., B.B., M.B., M.-M.D. and S.S.; visualization A.L., B.B., M.B. and M.-M.D.; supervision, A.L.; funding acquisition, S.S. All authors have read and agreed to the published version of the manuscript.

Funding: This research received no external funding.

Institutional Review Board Statement: Not applicable.

Informed Consent Statement: Not applicable.

Data Availability Statement: The raw data supporting the conclusions of this article will be made available by the authors on request.

Conflicts of Interest: The authors declare no conflict of interest.

References

1. Hassaninejad-Darzi, S.K. A Novel, Effective and Low Cost Catalyst for Formaldehyde Electrooxidation Based on Nickel Ions Dispersed onto Chitosan-Modified Carbon Paste Electrode for Fuel Cell. *J. Electroceramics* **2014**, *33*, 252–263. [CrossRef]
2. Samjeské, G.; Miki, A.; Osawa, M. Electrocatalytic Oxidation of Formaldehyde on Platinum under Galvanostatic and Potential Sweep Conditions Studied by Time-Resolved Surface-Enhanced Infrared Spectroscopy. *J. Phys. Chem. C* **2007**, *111*, 15074. [CrossRef]
3. Weng, X.; Chon, C.H.; Jiang, H.; Li, D. Rapid Detection of Formaldehyde Concentration in Food on a Polydimethylsiloxane (PDMS) Microfluidic Chip. *Food Chem.* **2009**, *114*, 1079–1082. [CrossRef]
4. Zhang, Y.; Yang, Y.; He, X.; Yang, P.; Zong, T.; Sun, P.; Sun, R.C.; Yu, T.; Jiang, Z. The Cellular Function and Molecular Mechanism of Formaldehyde in Cardiovascular Disease and Heart Development. *J. Cell. Mol. Med.* **2021**, *25*, 5358–5371. [CrossRef]
5. Kang, D.S.; Kim, H.S.; Jung, J.H.; Lee, C.M.; Ahn, Y.S.; Seo, Y.R. Formaldehyde Exposure and Leukemia Risk: A Comprehensive Review and Network-Based Toxicogenomic Approach. *Genes Environ.* **2021**, *43*, 13. [CrossRef]
6. Möhner, M.; Liu, Y.; Marsh, G.M. New Insights into the Mortality Risk from Nasopharyngeal Cancer in the National Cancer Institute Formaldehyde Worker Cohort Study. *J. Occup. Med. Toxicol.* **2019**, *14*, 4. [CrossRef] [PubMed]
7. Kwak, K.; Paek, D.; Park, J.T. Occupational Exposure to Formaldehyde and Risk of Lung Cancer: A Systematic Review and Meta-Analysis. *Am. J. Ind. Med.* **2020**, *63*, 312–327. [CrossRef] [PubMed]
8. Habibi, B.; Delnavaz, N. Electrocatalytic Oxidation of Formic Acid and Formaldehyde on Platinum Nanoparticles Decorated Carbon-Ceramic Substrate. *Int. J. Hydrogen Energy* **2010**, *35*, 8831. [CrossRef]
9. Karczmarska, A.; Adamek, M.; El Houbbadi, S.; Kowalczyk, P.; Laskowska, M. Carbon-Supported Noble-Metal Nanoparticles for Catalytic Applications—A Review. *Crystals* **2022**, *12*, 584. [CrossRef]
10. Senila, M.; Cadar, O.; Senila, L.; Böringer, S.; Seaudeau-Pirouley, K.; Ruiu, A.; Lacroix-Desmazes, P. Performance Parameters of Inductively Coupled Plasma Optical Emission Spectrometry and Graphite Furnace Atomic Absorption Spectrometry Techniques for Pd and Pt Determination in Automotive Catalysts. *Materials* **2020**, *13*, 5136. [CrossRef]
11. Xu, B.; Chen, Y.; Zhou, Y.; Zhang, B.; Liu, G.; Li, Q.; Yang, Y.; Jiang, T. A Review of Recovery of Palladium from the Spent Automobile Catalysts. *Metals* **2022**, *12*, 533. [CrossRef]
12. Priskila Rahael, K.; Penina Tua Rahantoknam, S.; Kahfi Hamid, S.; Kuo, Y.; Lin, C.-C.; Nigussa, K.N. A Study of Properties of Palladium Metal as a Component of Fuel Cells. *Mater. Res. Express* **2019**, *6*, 105540. [CrossRef]
13. Noroozifar, M.; Khorasani-Motlagh, M.; Ekrami-Kakhki, M.S.; Khaleghian-Moghadam, R. Electrochemical Investigation of Pd Nanoparticles and MWCNTs Supported Pd Nanoparticles-Coated Electrodes for Alcohols (C1–C3) Oxidation in Fuel Cells. *J. Appl. Electrochem.* **2014**, *44*, 233. [CrossRef]

14. Eshagh-Nimvari, S.; Hassaninejad-Darzi, S.K. Electrocatalytic Performance of Nickel Hydroxide-Decorated Microporous Nanozeolite Beta-Modified Carbon Paste Electrode for Formaldehyde Oxidation. *Electrocatalysis* **2023**, *14*, 365–380. [CrossRef]
15. Su, L.; Cheng, Y.; Shi, J.; Wang, X.; Xu, P.; Chen, Y.; Zhang, Y.; Zhang, S.; Xinxin, L. Electrochemical Sensor with Bimetallic Pt–Ag Nanoparticle as Catalyst for the Measurement of Dissolved Formaldehyde. *J. Electrochem. Soc.* **2022**, *169*, 047507. [CrossRef]
16. Xu, S.; Jiang, L.; Huang, X.; Ju, W.; Liang, Y.; Tao, Z.; Yang, Y.; Zhu, B.; Wei, G. Efficient Formaldehyde Sensor Based on PtPd Nanoparticles-Loaded Nafion-Modified Electrodes. *Nanotechnology* **2023**, *35*, 025704. [CrossRef] [PubMed]
17. Hosseini, S.R.; Raoof, J.B.; Ghasemi, S.; Gholami, Z. Pd-Cu/Poly (o-Anisidine) Nanocomposite as an Efficient Catalyst for Formaldehyde Oxidation. *Mater. Res. Bull.* **2016**, *80*, 107–119. [CrossRef]
18. Xue, J.; Li, X.; Jin, X.; Wang, X.; Zhen, N.; Song, T.; Lan, T.; Qian, S.; Zhang, H.; Liu, J. Electrocatalytic Oxidation of Formaldehyde Using Electrospinning Porous Zein-Based Polyimide Fibers. *Mater. Lett.* **2022**, *320*, 132318. [CrossRef]
19. Pötzelberger, I.; Mardare, C.C.; Burgstaller, W.; Hassel, A.W. Maximum Electrocatalytic Oxidation Performance for Formaldehyde in a Combinatorial Copper-Palladium Thin Film Library. *Appl. Catal. A-Gen.* **2016**, *525*, 110–118. [CrossRef]
20. Cremers, C.; Jurzinsky, T.; Bach Delpeuch, A.; Niether, C.; Jung, F.; Pinkwart, K.; Tübke, J. Electrocatalyst for Direct Alcohol Fuel Cells. *ECS Trans.* **2015**, *69*, 795–807. [CrossRef]
21. Wieckowski, A. (Ed.) *Interfacial Electrochemistry*, 1st ed.; Routledge: New York, NY, USA, 2017; ISBN 9780203750469.
22. Soleh, A.; Saisahas, K.; Promsuwan, K.; Saichanapan, J.; Thavarungkul, P.; Kanatharana, P.; Meng, L.; Mak, W.C.; Limbut, W. A Wireless Smartphone-Based "Tap-and-Detect" Formaldehyde Sensor with Disposable Nano-Palladium Grafted Laser-Induced Graphene (NanoPd@LIG) Electrodes. *Talanta* **2023**, *254*, 124169. [CrossRef]
23. Tsuji, J. *Palladium Reagents and Catalysts*; Wiley: Hoboken, NJ, USA, 2004; ISBN 9780470850329 | Online: 9780470021200.
24. Baldauf, M.; Kolb, D.M. Formic Acid Oxidation on Ultrathin Pd Films on Au(Hkl) and Pt(Hkl) Electrodes. *J. Phys. Chem.* **1996**, *100*, 11375–11381. [CrossRef]
25. Lu, Q.; Wei, X.Z.; Zhang, Q.; Zhang, X.; Chen, L.; Liu, J.; Chen, Y.; Ma, L. Efficient Selective Hydrogenation of Terminal Alkynes over Pd–Ni Nanoclusters Encapsulated inside S-1 Zeolite. *Microporous Mesoporous Mater.* **2024**, *365*, 112883. [CrossRef]
26. Mansy, H.E.; Khedr, M.H.; Abdelwahab, A. Palladium Functionalized Carbon Xerogel Nanocomposite for Oxygen Reduction Reaction. *Mater. Chem. Phys.* **2024**, *313*, 128797. [CrossRef]
27. Xu, S.; Kim, E.H.; Wei, A.; Negishi, E.I. Pd- and Ni-Catalyzed Cross-Coupling Reactions in the Synthesis of Organic Electronic Materials. *Sci. Technol. Adv. Mater.* **2014**, *15*, 044201. [CrossRef]
28. Xu, W.; Liu, C.; Xiang, D.; Luo, Q.; Shu, Y.; Lin, H.; Hu, Y.; Zhang, Z.; Ouyang, Y. Palladium Catalyst Immobilized on Functionalized Microporous Organic Polymers for C–C Coupling Reactions. *RSC Adv.* **2019**, *9*, 34595–34600. [CrossRef] [PubMed]
29. Lee, S.; Cho, H.; Kim, H.J.; Hong, J.W.; Lee, Y.W. Shape-and Size-Controlled Palladium Nanocrystals and Their Electrocatalytic Properties in the Oxidation of Ethanol. *Materials* **2021**, *14*, 2970. [CrossRef] [PubMed]
30. Khoperia, T.N.; Tabatadze, T.J.; Zedgenidze, T.I. Formation of Microcircuits in Microelectronics by Electroless Deposition. *Electrochim. Acta* **1997**, *42*, 3049–3055. [CrossRef]
31. Antolini, E. Palladium in Fuel Cell Catalysis. *Energy Environ. Sci.* **2009**, *2*, 915–931. [CrossRef]
32. Eswaran, M.; Rahimi, S.; Pandit, S.; Chokkiah, B.; Mijakovic, I. A Flexible Multifunctional Electrode Based on Conducting PANI/Pd Composite for Non-Enzymatic Glucose Sensor and Direct Alcohol Fuel Cell Applications. *Fuel* **2023**, *345*, 128182. [CrossRef]
33. Sikeyi, L.L.; Matthews, T.; Adekunle, A.S.; Maxakato, N.W. Electro-Oxidation of Ethanol and Methanol on Pd/C, Pd/CNFs and Pd–Ru/CNFs Nanocatalysts in Alkaline Direct Alcohol Fuel Cell. *Electroanalysis* **2020**, *32*, 2681–2692. [CrossRef]
34. Nachaki, E.O.; Ndangili, P.M.; Naumih, N.M.; Masika, E. Nickel-Palladium-Based Electrochemical Sensor for Quantitative Detection of Formaldehyde. *ChemistrySelect* **2018**, *3*, 384–392. [CrossRef]
35. He, K.; Jin, Z.; Chu, X.; Bi, W.; Wang, W.; Wang, C.; Liu, S. Fast Response-Recovery Time toward Acetone by a Sensor Prepared with Pd Doped WO3 Nanosheets. *RSC Adv.* **2019**, *9*, 28439–28450. [CrossRef]
36. Yao, G.; Zou, W.; Yu, J.; Zhu, H.; Wu, H.; Huang, Z.; Chen, W.; Li, X.; Liu, H.; Qin, K. Pd/PdO Doped WO3 with Enhanced Selectivity and Sensitivity for Ppb Acetone and Ethanol Detection. *Sens. Actuators B Chem.* **2024**, *401*, 135003. [CrossRef]
37. Lee, B.; Cho, S.; Jeong, B.J.; Lee, S.H.; Kim, D.; Kim, S.H.; Park, J.-H.; Yu, H.K.; Choi, J.-Y. Highly Responsive Hydrogen Sensor Based on Pd Nanoparticle-Decorated Transfer-Free 3D Graphene. *Sens. Actuators B Chem.* **2024**, *401*, 134913. [CrossRef]
38. Joudeh, N.; Saragliadis, A.; Koster, G.; Mikheenko, P.; Linke, D. Synthesis Methods and Applications of Palladium Nanoparticles: A Review. *Front. Nanotechnol.* **2022**, *4*, 1062608. [CrossRef]
39. Haghighi, B.; Hamidi, H.; Bozorgzadeh, S. Sensitive and Selective Determination of Hydrazine Using Glassy Carbon Electrode Modified with Pd Nanoparticles Decorated Multiwalled Carbon Nanotubes. *Anal. Bioanal. Chem.* **2010**, *398*, 1411–1416. [CrossRef]
40. Chen, A.; Ostrom, C. Palladium-Based Nanomaterials: Synthesis and Electrochemical Applications. *Chem. Rev.* **2015**, *115*, 11999–12044. [CrossRef] [PubMed]
41. Scholl, H.; Blaszczyk, T.; Leniart, A.; Polanski, K. Nanotopography and Electrochemical Impedance Spectroscopy of Palladium Deposited on Different Electrode Materials. *J. Solid State Electrochem.* **2004**, *8*, 308–315. [CrossRef]
42. Hubkowska, K.; Pająk, M.; Czerwiński, A. The Effect of the Iridium Alloying and Hydrogen Sorption on the Physicochemical and Electrochemical Properties of Palladium. *Materials* **2023**, *16*, 4556. [CrossRef]
43. Van der Linden, W.E.; Dieker, J.W. Glassy Carbon as Electrode Material in Electro- Analytical Chemistry. *Anal. Chim. Acta* **1980**, *119*, 1–24. [CrossRef]

44. McCreery, R.L. Advanced Carbon Electrode Materials for Molecular Electrochemistry. *Chem. Rev.* **2008**, *108*, 2646–2687. [CrossRef]
45. Zittel, H.E.; Miller, F.J. A Glassy-Carbon Electrode for Voltammetry. *Anal. Chem.* **1965**, *37*, 200–203. [CrossRef]
46. Dekanski, A.; Stevanović, J.; Stevanović, R.; Nikolić, B.Ž.; Jovanović, V.M. Glassy Carbon Electrodes: I. Characterization and Electrochemical Activation. *Carbon* **2001**, *39*, 1195–1205. [CrossRef]
47. Yáñez-Sedeño, P.; Pingarrón, J.M.; Riu, J.; Rius, F.X. Electrochemical Sensing Based on Carbon Nanotubes. *TrAC Trends Anal. Chem.* **2010**, *29*, 939–953. [CrossRef]
48. Pacios Pujadó, M. *Carbon Nanotubes as Platforms for Biosensors with Electrochemical and Electronic Transduction*; Springer: Berlin/Heidelberg, Germany, 2012; ISBN 978-3-642-31420-9.
49. Leniart, A.; Brycht, M.; Burnat, B.; Skrzypek, S. An Application of a Glassy Carbon Electrode and a Glassy Carbon Electrode Modified with Multi-Walled Carbon Nanotubes in Electroanalytical Determination of Oxycarboxin. *Ionics* **2018**, *24*, 2111–2121. [CrossRef]
50. Leniart, A.; Brycht, M.; Burnat, B.; Skrzypek, S. Voltammetric Determination of the Herbicide Propham on Glassy Carbon Electrode Modified with Multi-Walled Carbon Nanotubes. *Sens. Actuators B Chem.* **2016**, *231*, 54–63. [CrossRef]
51. Jorio, A.; Dresselhaus, G.; Dresselhaus, M.S. (Eds.) *Carbon Nanotubes*; Topics in Applied Physics; Springer: Berlin/Heidelberg, Germany, 2008; Volume 111, ISBN 978-3-540-72864-1.
52. Gao, G.Y.; Guo, D.J.; Li, H.L. Electrocatalytic Oxidation of Formaldehyde on Palladium Nanoparticles Supported on Multi-Walled Carbon Nanotubes. *J. Power Sources* **2006**, *162*, 1094–1098. [CrossRef]
53. Ulas, B. Optimization of Electrode Preparation Conditions by Response Surface Methodology for Improved Formic Acid Electrooxidation on Pd/MWCNT/GCE. *Ionics* **2023**, *29*, 4603–4616. [CrossRef]
54. Zhu, Z.Z.; Wang, Z.; Li, H.L. Self-Assembly of Palladium Nanoparticles on Functional Multi-Walled Carbon Nanotubes for Formaldehyde Oxidation. *J. Power Sources* **2009**, *186*, 339. [CrossRef]
55. Liao, H.; Qiu, Z.; Wan, Q.; Wang, Z.; Liu, Y.; Yang, N. Universal Electrode Interface for Electrocatalytic Oxidation of Liquid Fuels. *ACS Appl. Mater. Interfaces* **2014**, *6*, 18055–18062. [CrossRef]
56. Gamboa, A.; Fernandes, E.C. Resistive Hydrogen Sensors Based on Carbon Nanotubes: A Review. *Sens. Actuators A. Phys.* **2024**, *366*, 115013. [CrossRef]
57. Płócienniczak-Bywalska, P.; Rębiś, T.; Leda, A.; Milczarek, G. Lignosulfonate-Assisted In Situ Deposition of Palladium Nanoparticles on Carbon Nanotubes for the Electrocatalytic Sensing of Hydrazine. *Molecules* **2023**, *28*, 7076. [CrossRef] [PubMed]
58. Ishida, T.; Murayama, T.; Taketoshi, A.; Haruta, M. Importance of Size and Contact Structure of Gold Nanoparticles for the Genesis of Unique Catalytic Processes. *Chem. Rev.* **2020**, *120*, 464–525. [CrossRef] [PubMed]
59. Zhang, X.-G.; Arikawa, T.; Murakami, Y.; Yahikozawa, K.; Takasu, Y. Electrocatalytic Oxidation of Formic Acid on Ultrafine Palladium Particles Supported on a Glassy Carbon. *Electrochim. Acta* **1995**, *40*, 1889–1897. [CrossRef]
60. Safavi, A.; Maleki, N.; Farjami, F.; Farjami, E. Electrocatalytic Oxidation of Formaldehyde on Palladium Nanoparticles Electrodeposited on Carbon Ionic Liquid Composite Electrode. *J. Electroanal. Chem.* **2009**, *626*, 75–79. [CrossRef]
61. Olivi, P.; Bulhões, L.O.S.; Léger, J.M.; Hahn, F.; Beden, B.; Lamy, C. The Electrooxidation of Formaldehyde on Pt(100) and Pt(110) Electrodes in Perchloric Acid Solutions. *Electrochim. Acta* **1996**, *41*, 927–932. [CrossRef]
62. Cai, Z.-X.; Liu, C.-C.; Wu, G.-H.; Chen, X.-M.; Chen, X. Palladium Nanoparticles Deposit on Multi-Walled Carbon Nanotubes and Their Catalytic Applications for Electrooxidation of Ethanol and Glucose. *Electrochim. Acta* **2013**, *112*, 756–762. [CrossRef]
63. Balogun, S.A.; Fayemi, O.E.; Hapeshi, E.; Cannilla, C.; Bonura, G.; Balogun, S.A.; Fayemi, O.E. Effects of Electrolytes on the Electrochemical Impedance Properties of NiPcMWCNTs-Modified Glassy Carbon Electrode. *Nanomaterials* **2022**, *12*, 1876. [CrossRef]

Disclaimer/Publisher's Note: The statements, opinions and data contained in all publications are solely those of the individual author(s) and contributor(s) and not of MDPI and/or the editor(s). MDPI and/or the editor(s) disclaim responsibility for any injury to people or property resulting from any ideas, methods, instructions or products referred to in the content.

Article

The Use of an Acylhydrazone-Based Metal-Organic Framework in Solid-Contact Potassium-Selective Electrode for Water Analysis

Paweł Kościelniak [1,*], Marek Dębosz [1], Marcin Wieczorek [1], Jan Migdalski [2], Monika Szufla [3], Dariusz Matoga [3] and Jolanta Kochana [1]

[1] Department of Analytical Chemistry, Faculty of Chemistry, Jagiellonian University, Gronostajowa 2, 30-387 Kraków, Poland; marek.debosz@doctoral.uj.edu.pl (M.D.); marcin.wieczorek@uj.edu.pl (M.W.); jolanta.kochana@uj.edu.pl (J.K.)
[2] Department of Analytical Chemistry and Biochemistry, Faculty of Materials and Ceramics, AGH University of Science and Technology, A. Mickiewicza 30, 30-059 Kraków, Poland; migdal@agh.edu.pl
[3] Department of Inorganic Chemistry, Faculty of Chemistry, Jagiellonian University, Gronostajowa 2, 30-387 Kraków, Poland; monika.szufla@doctoral.uj.edu.pl (M.S.); dariusz.matoga@uj.edu.pl (D.M.)
* Correspondence: pawel.koscielniak@uj.edu.pl

Citation: Kościelniak, P.; Dębosz, M.; Wieczorek, M.; Migdalski, J.; Szufla, M.; Matoga, D.; Kochana, J. The Use of an Acylhydrazone-Based Metal-Organic Framework in Solid-Contact Potassium-Selective Electrode for Water Analysis. *Materials* 2022, *15*, 579. https://doi.org/10.3390/ma15020579

Academic Editors: Sławomira Skrzypek, Mariola Brycht, Barbara Burnat and Anastasios J. Tasiopoulos

Received: 15 November 2021
Accepted: 10 January 2022
Published: 13 January 2022

Publisher's Note: MDPI stays neutral with regard to jurisdictional claims in published maps and institutional affiliations.

Copyright: © 2022 by the authors. Licensee MDPI, Basel, Switzerland. This article is an open access article distributed under the terms and conditions of the Creative Commons Attribution (CC BY) license (https://creativecommons.org/licenses/by/4.0/).

Abstract: A solid-contact ion-selective electrode was developed for detecting potassium in environmental water. Two versions of a stable cadmium acylhydrazone-based metal organic framework, i.e., JUK-13 and JUK-13_H2O, were used for the construction of the mediation layer. The potentiometric and electrochemical characterizations of the proposed electrodes were carried out. The implementation of the JUK-13_H2O interlayer is shown to improve the potentiometric response and stability of measured potential. The electrode exhibits a good Nernstian slope (56.30 mV/decade) in the concentration range from 10^{-5} to 10^{-1} mol L^{-1} with a detection limit of 2.1 µmol L^{-1}. The long-term potential stability shows a small drift of 0.32 mV h^{-1} over 67 h. The electrode displays a good selectivity comparable to ion-selective electrodes with the same membrane. The K-JUK-13_H2O-ISE was successfully applied for the determination of potassium in three certified reference materials of environmental water with great precision (RSD < 3.00%) and accuracy (RE < 3.00%).

Keywords: water analysis; potentiometry; ion-selective electrode; potassium; metal organic frameworks

1. Introduction

Water is one of the most plentiful and essential compounds on the Earth and one of the most critical to life. Moreover, it is a good solvent for many substances, whose levels should be controlled. Water analysis is crucial in areas such as public health or environmental studies [1]. It covers the monitoring of such parameters such as physicochemical, biological and chemical properties. Water monitoring is also a determinant of a country's development and indicates all the actions undertaken in order to reduce water pollution and improve water quality [2]. Investigations of water quality can be carried out using a lot of analytical techniques including classical ones, such as titrimetry or gravimetry, and modern ones, such as atomic absorption spectrometry (AAS), inductively coupled plasma-mass spectrometry (ICP-MS), photometry, or UV-Vis spectrophotometry [3]. The majority of them require sample pretreatment and do not comply with the principles of green chemistry [2].

Potassium occurs widely in the environment, including all natural waters. It is a necessary element for the normal functioning of a human body, as it maintains the normal osmotic pressure in all cells, and ensures proper functioning of the muscles and nerves. Moreover, it is vital for synthesizing proteins and metabolizing carbohydrates. An elevated level of potassium in the blood can lead to serious diseases such as kidney failure, heart attack, or diabetes. The need for controlling the potassium concentration in water and other sources is undoubtedly valuable not only from the diagnostic point of view, but also from the point of view of water quality assurance [4,5].

An analytical technique which is very simple to make, cheap, and fulfills the requirements of green chemistry is potentiometry. It is possible to determine the analytes directly at the sampling site with the use of portable analytical devices [6]. The potentiometric measurements are very often performed via in-line monitoring of water quality in various processes, without the use of special gases (as acetylene in FAAS, argon in ICP-OES or ICP-MS) or other special media or cooling agents for different spectrometers. Due to the use of several different electrodes sensitized for different analytes, it is possible to perform a multicomponent analysis at the same time in a very quick and easy way [7–10]. Moreover, the energy consumption in potentiometric measurements is much lower in comparison with other modern instrumental techniques [11]. The classical electrodes with inner filling solution are characterized by particularly good metrological parameters, fast response, and stability of measured potential. However, the presence of the inner filling solution hinders the miniaturization and the modification of the electrode shape. The development of solid-contact ion-selective electrodes allowed for the elimination of the above problems and kept the analytical parameters, i.e., stability and reproducibility of potential, at a similar level. The stabilization of the potential is offered due to the use of the ion-to-electron transducer layer playing the role of the inner filling solution in this type of electrodes. Numerous materials have been proposed as the solid contact, such as conducting polymers, carbon materials, nanomaterials, intercalation compounds, ionic liquids, or molecular redox couples [12,13]. However, the search for the perfect transducer material is still ongoing. The ideal material should be characterized by a reversible transition from ionic to electronic conduction, high exchange current density, stable chemical composition and possibly high hydrophobicity so that the formation of water between transducer and membrane interface is minimalized [14]. So far, the proposed materials, utilized as solid contact, do not meet all the mentioned requirements. Therefore, new materials are constantly proposed, the application of which would contribute to obtaining the desired metrological parameters of the solid contact (SC) electrodes. Recently, the use of organic metal frameworks as an ion-to-electron transducer was carried out and the presented results were encouraging [15].

Metal organic frameworks (MOFs) are an attractive sub-class of highly ordered and porous materials with two- or three-dimensional structures. MOFs are compounds composed of metal clusters (or ions) and bridging organic linkers as initially coined by Yaghi et al. [16] and further defined as IUPAC recommendations [17]. They possess numerous attractive properties including permanent porosity, abundant structures, high-surface area, good thermal stability, scalability and processability; all these properties contribute to the fact that MOFs found usage in a lot of different fields such as drug delivery, proton conduction, gas capture, separation, chemical and electrochemical sensing, catalysis, energy storage, etc. [18–21]. The possibility of the implementation of a multitude of strategies of MOFs' synthesis can lead to materials with desirable properties for a given application [15]. Unfortunately, the majority of MOFs are electrical insulators which significantly limits the possibility of their use in electrochemistry, especially in potentiometry where they can play the role of a conductive ion-to-electron layer. However, some of the strategies for the improvement of the electric conductivity of MOFs are known. They include procedures such as the incorporation of ionic guest species, modification of the material with the use of redox-active or conductive compounds [22–24]. The use of MOFs in potentiometry is rare and only a few publications on this topic can be found in the literature. Mendecki et al. [15] presented the use of conductive MOFs as ion transducers in the solid contact potassium and nitrate selective electrodes. The proposed sensors exhibited perfect sensing properties, i.e., a near-Nernstian response and wide linear range. The utilized materials inhibited the formation of a water layer that led to the potential drift of only ca. 11 $\mu V\ h^{-1}$. In Mahmoud's work [25], Cu-MOF was utilized as an ionophore (modifier) in a carbon paste electrode for the detection of Al^{3+} ions in polluted water and pharmaceutical samples. The fabricated electrode was highly sensitive for Al^{3+} ions and allowed to obtain results with high precision and accuracy during the determination of the target ion in real samples.

The present work describes the preliminary studies of the implementation of two versions of a stable cadmium acylhydrazone-based metal organic framework, differing with guest molecules, as ion-to-electron transducers in solid contact ion-selective electrodes sensitive towards potassium ions. For the modification of a working glassy carbon electrode, we used a three-dimensional microporous MOF, $\{[Cd_2(oba)_2(tdih)_2]\cdot 7H_2O\cdot 6DMF\}_n$ (JUK-13), built of 4,4′-oxybis(benzenedicarboxylate) (oba^{2-}) and terephthalaldehyde diisonicotinoylhydrazone (tdih) linkers (see Figure S1, Supplementary Material for linker formulas) [26]. Additionally, we used the same framework after exchanging DMF for water molecules, that is $\{[Cd_2(oba)_2(tdih)_2]\cdot 13H_2O\}_n$ denoted as JUK-13_H2O [26]. Notably, the diacylhydrazone linker (tdih) decorates the pores of the MOF with the –CO-NH-N- groups that are potentially capable of chelating metal ions in solution through the formation of five-membered rings with O- and N-donor coordination bonds (Figure S1). We hypothesized that this ability can be advantageous for cation detection in aqueous solutions. Moreover, importantly, in the family of mixed-linker acylhydrazone-carboxylate MOFs known to date, JUK-13 stands out due to its high stability in water, confirmed previously by repeatable water vapor adsorption-desorption cycles and immersions in water, and due to its high N_2 uptake and high pore volume in a two-dimensional channel system, as confirmed by crystal structure and adsorption isotherms [26]. The comparison of XRD patterns between pristine JUK-13 and JUK-13_H2O clearly indicates that JUK-13 undergoes a phase transition upon guest exchange (Figure S2). IR spectra demonstrate that this transition involves a rearrangement of the linker since characteristic bands corresponding to symmetric and asymmetric stretching of carboxylates considerably differ after guest exchange. Thermogravimetric analyses additionally confirm this exchange and demonstrate thermal stability of both materials up to approximately 300 °C (Figure S2). This phase transition, occurring upon immersion of JUK-13 in water, involves rearrangements of carboxylates and an exchange of guest molecules from DMF and water (present in the as-synthesized material) for water molecules only, which can be clearly observed by TGA and IR spectroscopy. The specific surface area of JUK-13 was determined by BET surface analysis to be 1010 $m^2\ g^{-1}$ (Figure S3). In contrast, the analogue of JUK-13 functionalized by sulfonic groups (JUK-13-SO3H) easily degrades in liquid water, whereas it is stable under humid conditions [27]. The initial potentiometric characterizations were carried out for the electrodes modified by JUK-13 and JUK-13_H2O, and the latter was selected for further electrochemical studies. In order to verify the applicability of the proposed electrode, the sensor was utilized to analyze certified reference materials.

2. Materials and Methods

2.1. Reagents and Solutions

The following reagents were utilized for the ion-selective membrane preparation: potassium ionophore I (valinomycin) (Sigma Aldrich, Darmstadt, Germany), bis(2-ethylhexyl) sebacate (DOS) (Sigma Aldrich, Darmstadt, Germany), potassium tetrakis(pentafluorophenyl) borate (KTFAB, 97%) (Alfa Aesar, Ward Hill, MA, USA), high molecular weight poly(vinyl chloride) (PVC) and tetrahydrofuran (THF) (Sigma Aldrich, Darmstadt, Germany). Other chemicals, such as KCl, NaCl, LiCl, 1 mol L^{-1} HCl, ethanol, $CaCl_2$ $2H_2O$, NH_4Cl, and $MgCl_2$ $6H_2O$ were obtained from Merck Milipore. CH_3COOLi $2H_2O$ (Chempur, Piekary Śląskie, Poland) was also utilized. $\{[Cd_2(oba)_2(tdih)_2]\cdot 7H_2O\cdot 6DMF\}_n$ (JUK-13) and its hydrated counterpart, $\{[Cd_2(oba)_2(tdih)_2]\cdot 13H_2O\}_n$ (JUK-13_H2O) was synthesized according to literature procedure with a solvent-based approach [26]. Optical images of JUK-13 crystals in polarized light are shown in Supplementary Material (Figure S4). Notably, JUK-13 can also be obtained by an alternative solvent-free mechanochemical method.

Stock standard solutions of KCl (1 mol L^{-1}), NaCl (1 mol L^{-1}), LiCl (1 mol L^{-1}), $CaCl_2 \cdot 2H_2O$ (1 mol L^{-1}), NH_4Cl (1 mol L^{-1}), and $MgCl_2 \cdot 6H_2O$ (1 mol L^{-1}) were prepared by dissolving adequate amounts of chloride salts in distilled water. Working solutions of potassium chloride, in the range of 10^{-5}–10^{-1} mol L^{-1}, were prepared immediately prior to the use in two solvents: distilled water (for the calibration procedure) and 10^{-1} mol L^{-1} lithium acetate (for the determination of potassium by calibration curve method). Ultrapure water (18.2 MΩ·cm) from an HLP 5 system (Hydrolab, Straszyn, Poland) was utilized throughout the work.

For the determination of selectivity coefficients, the solutions of each chloride salt, in the concentration range of 10^{-3}–10^{-1} mol L^{-1}, were prepared in distilled water. The 10^{-1} mol L^{-1} lithium acetate solution was prepared by dissolving an adequate amount of substance in distilled water and was utilized as an Ionic Strength Adjuster.

Certificate Reference Materials of Environmental Water, EnviroMAT Waste Water PlasmaCal, EnviroMAT Ground Water ES-H, EnviroMAT Drinking Water EP-H-1 (SCP SCIENCE, Baie D'Urfé, QC, Canada), were prepared by 50-fold dilution with lithium acetate solution.

2.2. Instrumentation

The potentiometric measurements were performed using a 16-channel Lawson Labs voltmeter equipped with the EMF Suite software (version 2.0.). Potentials were measured against Ag|AgCl|3 mol L^{-1} KCl|1 mol L^{-1} CH_3COOLi reference electrode (Mineral, Warsaw, Poland). The bare or modified glassy carbon electrode (GCE, φ = 3 mm, BASi, West Lafayette, IN, USA) was utilized as a working electrode.

Electrochemical impedance spectroscopy (EIS) was executed using an Autolab Frequency Response Analyzer System (AUT20.FRA2-Autolab, Eco Chemie, B.V, The Netherlands). The measurement cell was composed of a tested electrode (the working electrode), Ag|AgCl wire (the reference electrode) and Pt wire (the counter electrode). The EIS measurements were carried out in 10^{-1} mol L^{-1} potassium chloride. The impedance spectra were recorded in a frequency range of 100 kHz–10 mHz, using a sinusoidal excitation signal with an amplitude of 10 mV. Before recording the spectra, the open-circuit potential (OCP) was measured.

All measurements were performed at room temperature.

2.3. Electrode Preparation

The surface of glassy carbon electrodes was polished with 0.3 μm Al_2O_3, rinsed and ultrasonicated in distilled water for 5 min. After ultrasonication, the glassy carbon electrodes were rinsed with distilled water, ethanol and again with distilled water. Then, the electrodes were air-dried.

The one component suspension containing 5 mg mL^{-1} of each Metal Organic Framework (JUK-13 and JUK-13_H2O) was prepared by weighing appropriate amounts of each MOF and mixing them with 0.5 mL of ethanol and then homogenizing with the use of an ultrasonic bath for 20 min.

The potassium selective membrane (K^+-ISM) cocktail was composed of 1.06% (w/w) ionophore (Valinomycin), 0.34% (w/w) ion exchanger (KTFAB, 97%), 32.90% (w/w) polymer (PCV), and 65.80% (w/w) plasticizer (DOS). The components were dissolved in 1 mL of THF to produce a solution with 20% dry weight.

A total of 12.5 μL of MOF suspension was drop cast on the glassy carbon surface using an automatic pipette. The obtained MOF layer was allowed to dry in ambient conditions. Then, 40 μL of K^+-ISM cocktail was deposited by drop-casting it onto the MOF layer in two consecutive steps (2 × 20 μL). The prepared solid-contact potassium-selective electrode was left overnight to allow the evaporation of the THF from the membranes. This procedure was utilized to prepare a series of electrodes. Once the THF solution evaporated, each electrode was conditioned in 10^{-2} mol L^{-1} KCl for at least 12 h. A calibration curve was received in the concentration range of 10^{-5}–10^{-1} mol L^{-1}.

2.4. Potentiometric Water Layer Test

The potentiometric water layer test was performed with fully conditioned solid-contact potassium-selective electrodes by recording the potential sequentially in 10^{-1} mol L^{-1} KCl, 10^{-1} mol L^{-1} NaCl and once again in 10^{-1} mol L^{-1} KCl. The solutions were stirred continuously during the measurements.

2.5. Selectivity Coefficients

The selectivity coefficients ($K_{I,J}^{pot}$) of the tested electrodes were determined by the separate solution method (SSM) [28]. The procedure was based on the measurements of the electromotive force (EMF) in the concentration range of 10^{-3}–10^{-1} mol L^{-1} of chlorides of different cations. The values of selectivity coefficients were calculated by using the measured potentials in the solutions of main and interfering ion as parameters in the Nikolsky–Eisenman equation.

3. Results

3.1. Potentiometric Response and Potential Stability of MOF-Containing Potentiometric Potassium Sensor

After conditioning, the studied solid contact electrodes were calibrated in potassium chloride solutions with concentrations ranging from 10^{-5} to 10^{-1} mol L^{-1}. For comparative purposes, the calibration was carried out for a coated wire electrode (CWE) as well, following the same scheme. The obtained calibration curves are shown in Figure 1. All tested electrodes exhibited good sensitivity to K^+ ions in the same linear range. The CWE provided a linear response in the tested concentration range of K^+ with a slope of 53.94 mV/decade. The use of JUK-13 as the ion-to-electron transducer does not have any impact on the response of a potassium selective membrane. This is confirmed by the obtained slope value of the calibration curve for the respective material. Moreover, the use of JUK-13_H2O caused the sensitivity of the electrode to increase by 2 mV/decade in comparison with the sensitivity of the CWE electrode.

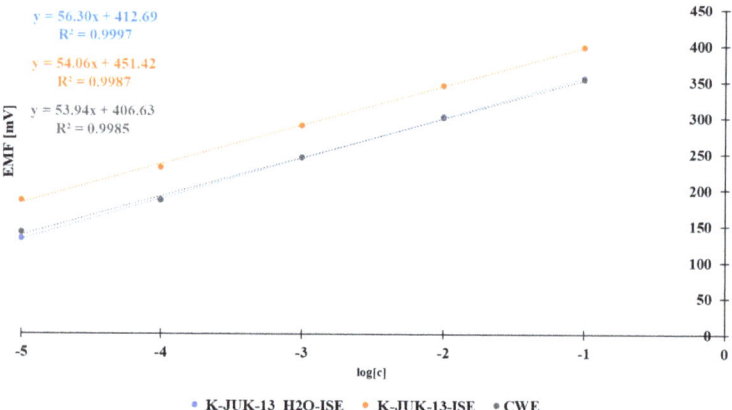

Figure 1. Calibration graphs for the constructed electrodes with MOFs as solid contact and CW electrode in the main ion concentration range 10^{-5}–10^{-1} mol L^{-1}.

In order to verify the suitability of the proposed materials as the ion-to-electron transducers, the measurements of potential stability for the tested electrodes were performed in the following fashion. The potential readings were carried out in 10^{-1} mol L^{-1} potassium chloride solution for 67 h. During the measurements, the solution was stirred with a magnetic stirrer. For comparison, the stability measurements were also carried out for CWE. The obtained changes in the potential of the studied electrodes and a coated wire electrode over time are shown in Figure 2.

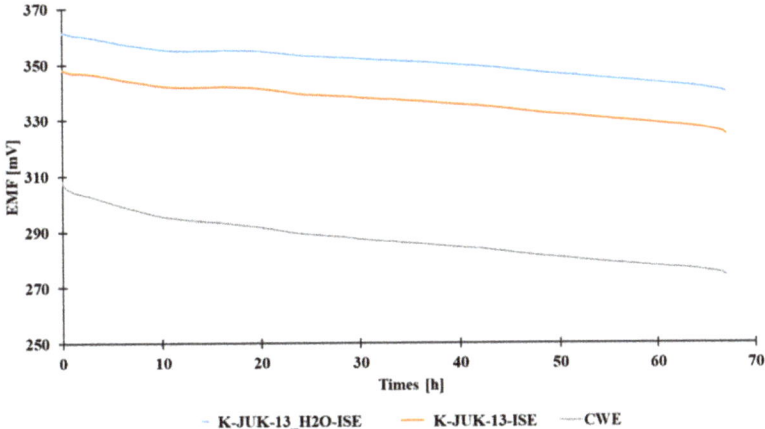

Figure 2. The change in potential of the tested electrodes with MOFs as ion-to-electron transducers and a coated wire electrode over time. The measurements were carried out in 10^{-2} mol L^{-1} KCl.

As shown in Figure 2, the use of the proposed MOFs as ion-to-electron transducers did not eliminate the potential drift. Only a slight reduction in the potential drift of solid contact electrodes in comparison with the CWE electrode is observed, which is confirmed by the data collected in Table 1.

Table 1. Long-term potential stability of the studied solid contact and coated wire electrodes.

Time (h)	Potential (mV)		
	K-JUK-13_H2O-ISE	K-JUK-13-ISE	CWE
0.00	361.58	348.27	308.28
67.00	339.90	324.70	274.21
Drift (mV h^{-1})	0.32	0.35	0.51

The coated wire electrode exhibited the potential drift equal 0.51 mV h^{-1} and in the case of SC electrodes the values were lower by around 0.20 mV h^{-1}. Among the tested versions of metal-organic frameworks, the use of JUK-13_H2O allowed for obtaining the lowest potential drift.

Based on the obtained sensitivities and potential drift values, the JUK-13_H2O was utilized in further studies.

3.2. Limit of Detection

In order to determine the limit of the detection (LOD) of the studied SC and CW electrodes, the calibration measurements in the concentration range 10^{-7} to 10^{-1} mol L^{-1} were carried out. The LOD was calculated as the intersection of two lines as shown in Figure 3. For SC-type and CW-type electrodes, the LOD equals 2.1×10^{-6} mol L^{-1} and 1.8×10^{-6} mol L^{-1} K$^+$, respectively. The utilized MOF has no impact on the LOD value.

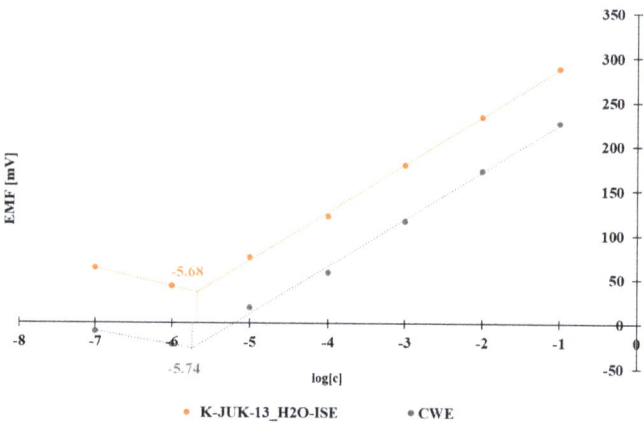

Figure 3. Determination of the limit of detection for K-MOF-ISE and CWE.

3.3. Selectivity Coefficients

The potentiometric selectivity coefficients of the tested electrodes, with and without the MOF, using the chloride salts of different cations, were determined by the separate solution method. The obtained values are shown in Table 2. In order to evaluate the obtained selectivity coefficients, the data obtained for a potassium selective electrode with multi-walled carbon nanotubes modified with octadecylamine are presented in Table 2. Each of the tested electrode groups utilized the ion-selective membrane with the same composition.

Table 2. Comparison of the potentiometric selectivity coefficients for proposed K-JUK-13_H2O-ISE, coated wire electrode (CWE) and K-OD-MWCNTs-ISE (with the same ion-selective membrane) ($n = 3$, $\alpha = 0.05$).

Interferent	$K_{K,J}^{pot}$		
	K-JUK-13_H2O-ISE	CWE	K-OD-MWCNTs-ISE
Li^+	-3.90 ± 0.09	-4.19 ± 0.43	-3.69 ± 0.66
Ca^{2+}	-3.84 ± 0.46	-3.71 ± 0.41	-4.18 ± 0.30
Mg^{2+}	-4.99 ± 0.20	-4.97 ± 0.22	-3.99 ± 0.76
NH_4^+	-2.21 ± 0.42	-2.61 ± 1.48	-1.64 ± 0.71
Na^+	-4.23 ± 0.13	-4.43 ± 0.01	-3.89 ± 0.27
H^+	-4.00 ± 0.13	-4.02 ± 0.01	-4.73 ± 1.91

The results proved that the CWE and SC electrodes exhibit good selectivity towards their primary ion. The determined selectivity coefficients were similar to the values reported in the literature [29,30] and they are close to the coefficients obtained for an SC-type electrode with OD-MWCNTs [7]. The selectivity coefficients should, by definition, be dependent on the composition of the utilized ion-selective membrane and not on the type of the utilized ion-to-electron transducer. The obtained results confirmed this rule. Among the interfering ions, the most notable interfering cations for the tested electrode are ammonium ions, while sodium ions show the weakest influence on the electrode response.

3.4. Water Layer Test

The formation of an undesirable water layer between the polymeric ion-selective membrane and ion-to-electron transducer can lead to potential instability of solid contact electrode and cause mechanical failure. The composition of this layer varies upon sample changes leading to many unfavorable processes, for instance, sensitivity to the CO_2, slow equilibrium process or longer time of response.

The water layer test was carried out according to the procedure suggested by Fibbioli et al. [31]. Before the test, the electrodes, i.e., K-JUK-13_H2O-ISE (SC-ISE) and CWE, were conditioned in 10^{-1} mol L^{-1} of primary ion solution. First, the potential of the electrodes was recorded in 10^{-1} mol L^{-1} KCl for 1 h, then in a solution of the interfering ion (10^{-1} mol L^{-1} NaCl). After 3 h, the interfering ion solution was replaced by a primary ion solution. The obtained results are shown in Figure 4.

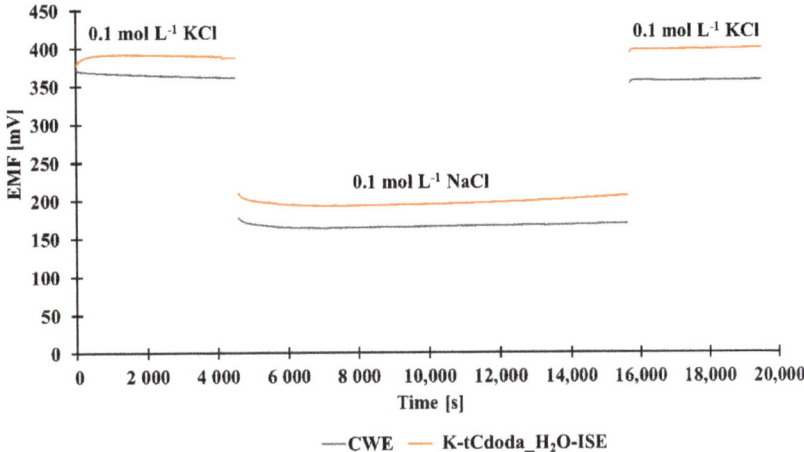

Figure 4. Water layer test of K-JUK-13_H2O-ISE and CWE.

The obtained results suggest that there is a positive drift of the potential when the solution of the primary ion is substituted with the solution of the interfering ion for both types of electrodes. It is caused by the leaching of primary ions from an ion-selective membrane contributing to the growth of the potassium ion concentration in the near membrane solution layer and to the increase in the electrodes' potential. After placing the electrodes back to the primary ion solution, the potential of the tested electrodes returned to the values close to the initial ones.

The data indicate that the use of glassy carbon electrodes with the MOF as a solid contact leads to a positive result of the water layer test similarly as in the case of the coated wire electrode.

The proposed electrodes can work for one month or longer after the first calibration, but their response is poorer and the linear range is narrower each day; this is caused by the leaching of the components from the membrane. For analytical studies, the electrodes were maximally utilized for one week after the first calibration, when the difference in sensitivity did not exceed 3%.

3.5. Electrochemical Impedance Spectroscopy

EIS measurements aimed to study the function of the MOF as a transducer layer and to verify whether the tested material possesses good electrical and ionic conductivity or/and sufficiently high redox capacitance providing the stability of measured potential during the flow of a little electric charge through the system. The spectra were measured in 10^{-1} mol L^{-1} potassium chloride solution. The impedance spectra of the studied electrode with the MOF as the transducer layer are shown in Figure 5.

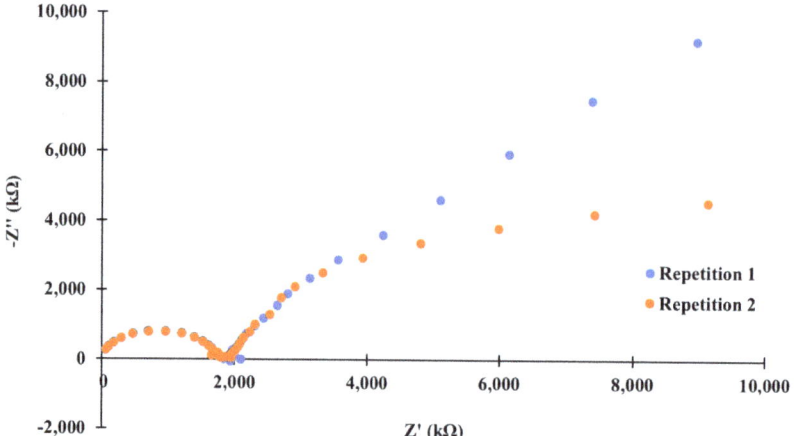

Figure 5. Impedance spectra of the studied solid-contact potassium-selective electrode with JUK-13_H2O as an ion-to-electron transducer.

The obtained spectra possess a semi-circle in the high-frequency range corresponding to the membrane resistance connected parallelly with electrical capacitance. However, in the case of the low-frequency range, there is a branch indicating the presence of a charge transfer resistance between ion-to-electron transducer, in this case, the MOF or electrode substrate-glassy carbon, and an ionically conducting membrane. Since the selected metal organic framework is insoluble in typical solvents, it forms a colloidal suspension, which leads to the fact that the shape of the recorded spectra can be affected by the intergranular resistance. The yielded spectra correspond in their shape to the typical spectra recorded for coated wire electrodes. This indicates that blocking of the charge flow between the mediation layer and the electrode substrate or an irreversible redox reaction is taking place. This was also confirmed by the fact that the measured open circuit potential (OCP) changed its value between the first and the second repetition from 0.26 V to 0.38 V. As a result of the redox reaction, the change in the ratio of the reduced to the oxidized form in the ion-to-electron transducer occurred, contributing to the change in the measured potential. On the basis of the obtained results, it can be concluded that the proposed material exhibits poor ionic and electrical conductivity and possesses low redox capacitance.

3.6. Analytical Application

The proposed solid-contact potassium-selective electrode was verified by determining the content of potassium in three certified reference materials (CRMs) of environmental waters: wastewater, drinking water and ground water. The calibration curve method was utilized as a calibration method. Each CRM sample was diluted 50-fold and analyzed six times. Along with the values of relative errors (RE (%)), confidence intervals ($n = 6$, $\alpha = 0.05$) were calculated in order to compare the results with certificate values. As an ionic strength adjuster, lithium acetate solution (0.1 mol L^{-1}) was utilized. The results are shown in Table 3.

Table 3. Results of potassium determination in certified reference materials using the calibration curve method (C_o and C_x—certified for CRM and found concentrations respectively; RE—relative error; RSD—relative standard deviation) ($n = 6$, $\alpha = 0.05$).

Sample	C_0 (mmol L^{-1})	C_x (mmol L^{-1})	RE (%)	RSD (%)
Waste Water EU-H-3	1.02	1.05 ± 0.02	2.9	2.4
Drinking Water EP-H-1	0.33	0.33 ± 0.01	1.9	1.7
Ground Water ES-H	0.074	0.073 ± 0.002	−1.4	2.8

The obtained results are satisfying in terms of both accuracy and precision. The values of relative error (RE (%)) and relative standard deviation (RSD (%)) do not exceed 3.00%, which points to a very good precision and accuracy of the results. The values of the considered parameters are favorable from an analytical point of view due to the use of an ionic strength adjuster which has a significant impact on the repeatability of recorded signals.

4. Conclusions

This work demonstrates the first use of acylhydrazone-based MOFs as modifiers of working electrodes for potentiometric sensing in aqueous solutions. The properties of the proposed MOF enable its use for potentiometric detection. The sensor exhibited good performance characteristics, including a near-Nernstian response, wide linear range (from 10^{-5} to 10^{-1} mol L^{-1}), good selectivity, and good potential stability in comparison with the CW-type electrode. Moreover, the proposed sensor was successfully applied for K$^+$ determination in certified reference materials of different water samples with very good accuracy and precision. Naturally, future efforts should focus on rare electrically conductive MOFs as the coexistence of both porosity and high conductivity are desirable for more efficient ion-to-electron transducers. The literature reports on the use of this class of materials in potentiometric measurements indicate that this topic is worth pursuing.

Supplementary Materials: The following supporting information can be downloaded at: https://www.mdpi.com/article/10.3390/ma15020579/s1, Supplementary materials contain additional measurement data (Figures S1–S4) and descriptions for the MOF's: JUK-13_H2O and JUK-13_H2O. Figure S1: (Top) Scheme of synthetic procedure for {[Cd2(oba)2(tdih)2]·guests}n (JUK-13); (Bottom) Microporous structure of JUK-13; Figure S2: Comparison of JUK-13 and JUK-13_H2O MOFs: a) PXRD patterns indicate phase transition upon guest exchange in water. b) IR spectra demonstrate that the phase transition involves rearrangements of carboxylates. c) Thermogravimetric analyses confirm the presence of various guest molecules in JUK-13 and JUK-13_H2O and the stability of both materials up to ca. 300 OC upon pore evacuation; Figure S3: Physisorption isotherm of N2 (77 K) for JUK-13; Figure S4: Optical images of JUK-13 crystals in a polarized light: dark field (top) and bright field (bottom).

Author Contributions: Conceptualization, P.K., M.D., M.W.; methodology, M.D., J.M.; validation, P.K., J.M., D.M., J.K.; formal analysis, M.D., M.S.; investigation, M.D., J.M., M.S.; resources, D.M.; data curation, M.D., M.S.; writing—original draft preparation, P.K, M.D., M.W., D.M.; writing—review and editing—P.K., M.D., M.W., M.S., D.M.; visualization, P.K., M.D.; supervision, P.K., J.M., J.K.; project administration, P.K., M.W.; funding acquisition, P.K., M.W., M.D., D.M. All authors have read and agreed to the published version of the manuscript.

Funding: P.K. and M.W. thank for financial support received from the National Science Centre, Poland (Opus, 2017–2020, grant no. 2016/23/B/ST4/00789). M.D. has been supported by the EU Project POWR.03.02.00-00-I004/16. D.M. gratefully acknowledges the National Science Centre (NCN), Poland) for the financial support (Grant no. 2019/35/B/ST5/01067).

Institutional Review Board Statement: Not applicable.

Informed Consent Statement: Not applicable.

Data Availability Statement: Data available in a publicly accessible repository. The data presented in this study are openly available in Jagiellonian University Repository at DOI: 10.26106/cjky-4381.

Conflicts of Interest: The authors declare no conflict of interest.

References

1. Zakir, H.M. Water: The most precious resource of our life. *Glob. J. Adv. Res.* **2015**, *2*, 1436–1445.
2. Mesquita, R.B.R.; Rangel, A.O.S.S. A review on sequential injection methods for water analysis. *Anal. Chim. Acta* **2009**, *19*, 7–22. [CrossRef]
3. Soylak, M.; Aydin, F.A.; Saracoglu, S.; Elci, L.; Dogan, M. Chemical analysis of drinking water samples from Yozgat, Turkey. *Pol. J. Environ. Stud.* **2002**, *11*, 151–156.
4. Pohl, H.R.; Wheeler, J.S.; Murray, H.E. Sodium and potassium in health and disease. *Met. Ions Life Sci.* **2013**, *13*, 29–47.
5. Wieczorek, M.; Madej, M.; Starzec, K.; Knihnicki, P.; Telk, A.; Kochana, J.; Kościelniak, P. Flow manifold for chemical H-point standard addition method implemented to electrochemical analysis based on the capacitance measurements. *Talanta* **2018**, *186*, 183–191. [CrossRef]
6. Keith, L.H.; Gron, L.U.; Young, J.L. Green analytical methodologies. *Chem. Rev.* **2007**, *107*, 2695–2708. [CrossRef] [PubMed]
7. Dębosz, M.; Kozma, J.; Porada, R.; Wieczorek, M.; Paluch, J.; Gyurcsányi, J.L.; Migdalski, J.; Kościelniak, P. 3D-printed manifold integrating solid contact ion-selective electrodes for multiplexed ion concentration measurements in urine. *Talanta* **2021**, *232*, 122491. [CrossRef]
8. Dębosz, M.; Wieczorek, M.; Paluch, J.; Migdalski, J.; Baś, B.; Kościelniak, P. 3D-printed flow manifold based on potentiometric measurements with solid-state ion-selective electrodes and dedicated to multicomponent water analysis. *Talanta* **2020**, *217*, 121092. [CrossRef] [PubMed]
9. Urbanowicz, M.; Pijanowska, D.G.; Jasiński, A.; Bocheńska, M. Multianalyte Calibration Methods for Potentiometric Integrated Sensors System for Determination of Ions Concentration in a Body Fluids. In Proceedings of the 2018 XV International Scientific Conference on Optoelectronic and Electronic Sensors (COE), Warsaw, Poland, 17–20 June 2018; pp. 1–4.
10. Jasiński, A.; Urbanowicz, M.; Guziński, M.; Bocheńska, M. Potentiometric solid-contact multisensor system for simultaneous measurement of several ions. *Electroanalysis* **2015**, *27*, 745–751. [CrossRef]
11. Abd El-Rahman, M.K.; Zaazaa, H.E.; Eldin, N.B.; Moustafa, A.A. Just-dip-it (potentiometric ion-selective electrode): An innovative way of greening analytical chemistry. *ACS Sustain. Chem. Eng.* **2016**, *4*, 3122–3132. [CrossRef]
12. Lyu, Y.; Gan, S.; Bao, Y.; Zhong, L.; Xu, J.; Wang, W.; Liu, Z.; Ma, Y.; Yan, G.; Niu, L. Solid-contact ion-selective electrodes: Response mechanisms, transducer materials and wearable sensors. *Membranes* **2020**, *10*, 128. [CrossRef]
13. Shao, Y.; Ying, Y.; Ping, J. Recent advances in solid-contact ion-selective electrodes: Functional materials, transduction mechanisms, and development trends. *Chem. Soc. Rev.* **2020**, *49*, 4405–4465. [CrossRef]
14. Lindner, E.; Gyurcsányi, R.E. Quality control criteria for solid-contact, solvent polymeric membrane ion-selective electrodes. *J. Solid State Electrochem.* **2008**, *13*, 51–68. [CrossRef]
15. Mendecki, L.; Mirica, K.A. Conductive metal-organic frameworks as ion-to-electron transducers in potentiometric sensors. *ACS Appl. Mater. Interfaces* **2018**, *10*, 19248–19257. [CrossRef] [PubMed]
16. Yaghi, O.M.; Li, G.M.; Li, H.I. Selective binding and removal of guests in a microporous metal−organic framework. *Nature* **1995**, *378*, 703–706. [CrossRef]
17. Batten, S.R.; Champness, N.R.; Chen, X.M.; Garcia-Martinez, J.; Kitagawa, S.; Öhrström, L.; O'Keeffe, M.; Paik Suh, M.; Reedijk, J. Terminology of metal–organic frameworks and coordination polymers (IUPAC Recommendations 2013). *Pure Appl. Chem.* **2013**, *85*, 1715–1724. [CrossRef]
18. Baumann, A.E.; Burns, D.A.; Liu, B.; Thoi, V.S. Metal-organic framework functionalization and design strategies for advanced electrochemical energy storage devices. *Commun. Chem.* **2019**, *86*, 1–14. [CrossRef]
19. Zhao, F.; Sun, T.; Geng, F.; Chen, P.; Gao, Y. Metal-organic frameworks-based electrochemical sensors and biosensors. *Int. J. Electrochem. Sci.* **2019**, *14*, 5287–5304. [CrossRef]
20. Rasheed, T.; Rizwan, K.; Bilal, M.; Iqbal, H.M.N. Metal-organic framework-based engineered materials—Fundamentals and applications. *Molecules* **2020**, *25*, 1598. [CrossRef]
21. Xu, B.; Zhang, H.; Mei, H.; Sun, D. Recent progress in metal-organic framework-based supercapacitor electrode materials. *Coord. Chem. Rev.* **2020**, *420*, 213438. [CrossRef]
22. Madej, M.; Matoga, D.; Skaźnik, K.; Porada, R.; Baś, B.; Kochana, J. A voltammetric sensor based on mixed proton-electron conducting composite including metal-organic framework JUK-2 for determination of citalopram. *Microchim. Acta* **2021**, *188*, 184. [CrossRef] [PubMed]
23. Cassani, M.C.; Castagnoli, R.; Gambassi, F.; Nanni, D.; Ragazzini, I.; Masciocchi, N.; Boanini, E.; Ballarin, B. A Cu(II)-MOF based on a propargyl carbamate-functionalized isophthalate ligand as nitrite electrochemical sensor. *Sensors* **2021**, *21*, 4922. [CrossRef] [PubMed]
24. Chen, S.S.; Han, P.-C.; Kuok, W.-K.; Lu, J.-Y.; Gu, Y.; Ahamad, T.; Alshehri, S.M.; Ayalew, H.; Yu, H.-H.; Wu, K.C.-W. Synthesis of MOF525/PEDOT composites as microelectrodes for electrochemical sensing of dopamine. *Polymers* **2020**, *12*, 1976. [CrossRef]

25. Mahmoud, N.F.; Fouad, O.A.; Ali, E.A.; Mohamed, G.G. Potentiometric determination of the Al(III) ion in polluted water and pharmaceutical samples by a novel mesoporous copper metal–organic framework-modified carbon paste electrode. *Ind. Eng. Chem.* **2021**, *60*, 2374–2387. [CrossRef]
26. Roztocki, K.; Szufla, M.; Hodorowicz, M.; Senkovska, I.; Kaskel, S.; Matoga, D. Introducing a longer versus shorter acylhydrazone linker to a metal–organic framework: Parallel mechanochemical approach, Nonisoreticular Structures, and Diverse Properties. *Cryst. Growth Des.* **2019**, *19*, 7160–7169. [CrossRef]
27. Szufla, M.; Roztocki, K.; Krawczuk, A.; Matoga, D. One-step introduction of terminal sulfonic groups into a proton-conducting metal–organic framework by concerted deprotonation–metalation–hydrolysis reaction. *Dalton Trans.* **2020**, *49*, 9953–9956. [CrossRef]
28. Buck, P.R.; Lindner, E. IUPAC Recommendation for Nomenclature of Ion-Selective Electrodes. *Appl. Chem.* **1994**, *66*, 2527–2536. [CrossRef]
29. Bühlmann, P.; Pretsch, E.; Bakker, E. Carrier-based ion-selective electrodes and bulk optodes. 2. Ionophores for potentiometric and optical sensors. *Chem. Rev.* **1998**, *98*, 1593–1687. [CrossRef]
30. Band, D.M.; Kratochvil, J.; Wilson, P.A.P.; Treasure, T. Relationship between activity and concentration measurements of plasma potassium. *Analyst* **1978**, *10*, 246–251. [CrossRef]
31. Fibbioli, M.; Morf, W.E.; Badertscher, M.; De Rooij, N.F.; Pretsch, E. Potential drifts of solid-contacted ion-selective electrodes due to zero-current ion fluxes through the sensor membrane. *Electroanalysis* **2000**, *12*, 1286–1292. [CrossRef]

MDPI AG
Grosspeteranlage 5
4052 Basel
Switzerland
Tel.: +41 61 683 77 34

Materials Editorial Office
E-mail: materials@mdpi.com
www.mdpi.com/journal/materials

Disclaimer/Publisher's Note: The title and front matter of this reprint are at the discretion of the . The publisher is not responsible for their content or any associated concerns. The statements, opinions and data contained in all individual articles are solely those of the individual Editors and contributors and not of MDPI. MDPI disclaims responsibility for any injury to people or property resulting from any ideas, methods, instructions or products referred to in the content.

www.ingramcontent.com/pod-product-compliance
Lightning Source LLC
LaVergne TN
LVHW070644100526
838202LV00013B/879